STUDIES IN MEDIEVAL HISTORY AND CULTURE

Edited by

Francis G. Gentry

Professor of German

Pennsylvania State University

A ROUTLEDGE SERIES

Studies in Medieval History and Culture

Francis G. Gentry, *General Editor*

ROOTED IN THE EARTH,
ROOTED IN THE SKY

Hildegard of Bingen and Premodern Medicine

Victoria Sweet

Routledge
New York & London

Published in 2006 by
Routledge
Taylor & Francis Group
270 Madison Avenue
New York, NY 10016

Published in Great Britain by
Routledge
Taylor & Francis Group
2 Park Square
Milton Park, Abingdon
Oxon OX14 4RN

Transferred to Digital Printing 2010

International Standard Book Number-10: 0-415-97634-0 (Hardcover)
International Standard Book Number-13: 978-0-415-97634-3 (Hardcover)

Library of Congress Cataloging-in-Publication Data

Catalog record is available from the Library of Congress

ISBN10: 0-415-97634-0 (hbk)
ISBN10: 0-415-99333-4 (pbk)

ISBN13: 978-0-415-97634-3 (hbk)
ISBN13: 978-0-415-99333-3 (pbk)

Publisher's Note
The publisher has gone to great lengths to ensure the quality of this reprint but points out that some imperfections in the original may be apparent.

Taylor & Francis Group
is the Academic Division of Informa plc.

Visit the Taylor & Francis Web site at
http://www.taylorandfrancis.com

and the Routledge Web site at
http://www.routledge-ny.com

We are just like plants only, as their roots are in the earth so
ours are in the sky.

Plato, *Timaeus* (90A)

Contents

Appendix
The Only Surviving Witness

List of Figures

List of Abbreviations

"Aus Kindheit"—Staab, Franz. "Aus Kindheit und Lehrzeit Hildegards. Mit einer Übersetzung der 'Vita' ihrer Lehrerin Jutta von Sponheim." In *Hildegard von Bingen. Prophetin durch die Zeiten. Zum 900. Geburtstag*, edited by Edeltraud Forster, 58–86. Freiburg: Herder, 1997.

CC—Moulinier, Laurence, ed. *Hildegardis Bingensis Cause et Cure*. Vol. 1. Berlin: Rarissima mediaevalia, 2003.

CCCM—*Corpus Christianorum, Continuatio Medievalis*.

Cetedoc—Tombeur, Paul. "The CETEDOC Library of Christian Latin Texts: CDROM." Turnhout: Brepols, 2002.

Derolez—Derolez, Albert, ed. *Guiberti Gemblacensis epistolae quae in codice B. R. Brux. 5527–5534 inveniuntur*. Vol. 66 and 66A, *CCCM*. Turnhout: Brepols, 1988.

LDO—Derolez, Albert, and Peter Dronke, eds. *Liber divinorum operum*. Vol. 92, *CCCM*. Turnhout: Brepols, 1996.

Lewis and Short—Lewis, Charleton T., and Charles Short. *A Latin Dictionary*. First ed. Oxford: Clarendon Press, 1879.

LI—Portmann, Marie-Louise, and Alois Odermatt. *Wörterbuch der unbekannten Sprache (Lingua ignota) in der Reihenfolge der Manuskripte*. Basle: Basler Hildegard-Gesellschaft, 1986.

Lulofs—Lulofs, H. J. Drossaart, and E. L. J. Poortman, eds. *Nicolaus Damascenus, De Plantis: Five Translations*. Amsterdam: North Holland Publishing Company, 1989.

LVM—Hildegard, of Bingen. *Liber vita meritorum*. Edited by Angela Carlevaris. Vol. 90, *CCCM*. Turnhout: Brepols, 1995.

OLD—Glare, P. G. W. *Oxford Latin Dictionary*. Combined edition, first published 1982 ed. Oxford: Clarendon Press, 1984.

PL—Migne, Jacques-Paul, ed. *Patrologiae Cursus Completus [Series Latina]*. Vols. 221, Paris: Garnier, 1844–1905.

PL197—Migne, Jacques-Paul, ed. *S. Hildegardis Abbatissae Opera Omnia*. Vol. 197, *Patrologiae Cursus Completus*. [*Series Latina*]. Paris: Garnier, 1855.

Physica—Hildegard, of Bingen, *Liber simplicis medicinae* [=*Physica*]. In *S. Hildegardis Abbatissae Opera Omnia*, PL197, cols. 1117–1352.

"Reform und Reformgruppen"—Staab, Franz. "Reform und Reformgruppen im Erzbistum Mainz. Vom 'Libellus de Willigis consuetudinibus' zur 'Vita domnae Juttae inclusae.'" In *Reformidee und Reformpolitik im Spätsalisch-Frühstaufischen Reich*, edited by Stefan Weinfurter, 119–187. Mainz: Selbstverlag der Gesellschaft für mittelrheinische Kirchengeschichte, 1992.

Scivias—Hildegard, of Bingen, *Scivias*. In *S. Hildegardis Abbatissae Opera Omnia*, PL197, cols. 383–738.

Van Acker—Van Acker, Lieven, ed. *Hildegardis Bingensis Epistolarium*. Vols. 91, 91A, *CCCM*. Turnhout: Brepols, 1991, 1993; Van Acker, Lieven, and M. Klaes-Hachmoller, eds. *Hildegardis Bingensis epistolarium pars tertia CCLI-CCCXC*. Vol. 91B, *CCCM*. Turnhout: Brepols, 2001.

Vita (1993)—Klaes, Monika, ed. *Vita Sanctae Hildegardis*. Vol. 126, *CCCM*. Turnhout: Brepols, 1993.

Vita (1998)—Klaes, Monika, ed. *Vita Sanctae Hildegardis, Leben der heiligen Hildegard von Bingen; Canonizatio Sanctae Hildegardis, Kanonisation der Heiligen Hildegard*. Freiburg: Herder, 1998.

Preface

Two experiments turned me from a life of medicine to a life of medicine and history, but not the kind of experiments expected from a physician. They were not scientific experiments but "historical experiments" and the first was my abortifacient garden.

A dozen years ago I replanted my garden, and partly in response to the abortion controversy with the boycott of RU-486, and partly in response to my own curiosity, I decided to create a garden of plants reported cross-culturally in the folk literature (Chinese, Ayurvedic, premodern European and Native American) as useful for "bringing down the period." These included rue, thyme, wormwood, and lavender. The wormwood did very well; it flourished and grew from a small plant to a large and thriving bush, with powdery, silvery, and very bitter leaves. The thyme next to it died, however, and so, eventually, did the rue and the lavender, and then the adjoining bulbs, the rosemary, and the young olive tree downwind. Indeed, everything, I gradually realized, anywhere near the wormwood died, after a while, especially if it was newly planted and young. So I cut the wormwood down. Yet, even now, nothing grows very well anywhere near its former place of residence. It seems to have caused a prolonged sterility in the earth.

And this made me wonder. How much was wormwood's use in abortifacient remedies the result not only of observation, but also of applying observations of plant life to humans? If wormwood dispatched young plants, would it be useful as an abortifacient? Indeed, how much was folk medical culture in general the result of applying observations of plant life, and death, to the body? The available literature seemed more impressed with the religious and superstitious origins of folk knowledge: its rituals, its trances, its symbolic meanings. But given the rural and agriculturally based cultures from which most folk knowledge originated, it seemed likely that practical observation must also have contributed a great deal to the ancient formularies.

That experiment had to do with practical, applicable knowledge—provocative but not completely surprising. It was my experiment with brewing that really intrigued me. As I began to read the literature, first of Chinese and Ayurvedic medicine, then of European folk medicine (looking for the elusive complement to modern Western medicine), I came upon the surprising fact that beer had not always been made with hops. Sometimes it had been made with dandelion (for bitterness), sometimes with cinnamon, cloves, pepper. I wondered what such beers would taste like; and I wondered what kind of effects such beers might have. Perhaps it was the hops in beer that made beer so sedating? Did beer made without hops have different effects? Was beer perhaps one of the first way of making medically effective potions? So my first beer would be a non-hopped beer, made with licorice, ginger, and pepper—a stimulating beer.

I followed the instructions, and soon I was stirring a huge pot on the stove, very sweet, a sweet soup. Then, as per instructions, I let the mixture cool, added the yeast, poured the two gallons of liquid into the five-gallon white plastic container, put the container on my dining room table, made sure that the lid was on (but not "tightly" as per instructions), and left for the weekend.

As soon as I opened the front door at the end of the weekend I noticed the smell—a pleasant, yeasty smell like a bakery. Then I saw the white plastic container. The top was almost off—pushed off, I saw when I went over, by soft, white, yeasty, and sweet-smelling foam spilling over the top. The foam was filled with tiny little holes, and it was crackling and moving on its own, even as I watched it, spilling over the sides. The most surprising thing of all was the heat. The soup of proto-beer was hot; it was breathing; and it was moving. The four classic, defining principles of life: moisture, heat, breath, movement.

That the practical knowledge of premodern medicine, of whatever version, might be based in experience, in the carrying over of observations from plants to bodies, was not all that surprising. But that theoretical concepts discovered in observations of daily life might also be carried over and applied to the body, was much more surprising and intriguing. Although, when you think about it, perhaps it shouldn't have been.

Hence this study.

Acknowledgments

I could never have completed this study without the help of many people.

The support I have had from the academic communities of Medieval Studies and the History of Medicine, Science, and Technology has been crucial, beginning with Professor George Brown's initial enthusiasm for my project. Professors Guenter Risse, the late Jack Pressman, Warwick Anderson, and Joan Cadden were all extremely generous with their time; they saved me from many (but not all!) infelicitous detours. In addition, the help, as well as the enthusiasm, of Walt Schalick, Faye Marie Getz, Linda Voigts, Michael McVaugh, Luke Demaitre, Monica Green, Karen Reeds, and Beth Haiken have been invaluable. Numerous librarians and libraries have been open to me; I would especially like to mention Stanford University Libraries, and its librarians, John Rawlings and Glen Worthey. Also Michelle Brown of the British Library was very helpful in providing manuscripts and advice at the British Library Manuscript Reading Room.

I am grateful for the financial support of the Angie and Joseph Francis Fellowship, the President's Research Fellowship from the University of California, the Burroughs-Wellcome Fund, and the American Association for the History of Medicine. I would particularly like to thank the committees for the Shryock Medal, the Estes Award, and the Stannard Memorial Award for their confidence in my work.

The patience and flexibility of my medical colleagues, including (but not limited to), Drs. Mary Anne Johnson, Terry Hill, Tim Skovrinski, Paul Isaakson, Anne Fricker, Monica Banchero, Hosea Thomas, Theresa Berta, Grace Dammann, and Craig Wilson, as well as of our secretary, Johnnie Brooks, allowed me to integrate this study into my ongoing life as physician, and vice versa. The Department of Anthropology, History and Social Medicine at the University of California, San Francisco, supported my endeavor during a difficult period, with special thanks to Professors Dorothy Porter,

Brian Dolan, and Adele Clark. Also I would like to thank Lisa Camp de Avalos, my illustrator.

Then there are friends and family. Jenny Fichmann, Typex, Choci, and Company Cat have always been supportive, while providing distractions when they were most needed. My parents and my sisters, Dorothy, Jennifer, and Patricia, as well as nieces and nephews, have been enthusiastic and admiring. David Kleber, Susan McGreivey, Wendy Brown, and Rebecca Moore have been my loyal fans and I appreciate them more than I can say.

Finally I would especially like to thank Routledge Press, Professor Frank Gentry, and Max Novick, for giving me the opportunity to share the results of my work.

Introduction

Sometime in the 1150s Hildegard of Bingen wrote down her instructions for making a potion that would be generally good for health:

> Take ginger, licorice, zedoary, sugar, and fine flour and make a powder.[1] Add the juice of springroot—about as much as a copyist takes up when he dips his pen—and make little cakes. Dry them in March or April, because in these months the sun is temperate. The heat of the ginger and the cold of the zedoary will unite the humors, and the heat and moistness of the sugar will retain and humidify them, and the coldness of the springroot will temper them. For ginger, zedoary, and sugar retain good humors in the body, and springroot expels the bad humors. Prepare this medicine in March and April because it is at this time that the sun and the wind are of good temper (CC, IV, 236:1–26).[2]

What kind of medicine is this and what kind of body is it for? What is the relationship between its ingredients, and temperature and humidity? What are these good and bad humors in need of strengthening or expulsion? Why should it be made only in March or April? And what do the sun and the wind have to do with it?

All of these concepts—the *qualities* of hot and cold, wet and dry, the bodily *humors*, the *elements* of sun and wind (and earth and water)—come out of premodern medicine, that is, out of Western medicine as it was before the triumph of modern medicine in the nineteenth century. In this study, I will look at how they fit together to create a coherent, stable, and usable medical system, so very different from the one in use today.

We know how the analogous systems of China and India "worked" because they are still practiced, and it is possible to go to China or India and experience them in their original languages from active practitioners.

Premodern European medicine, by contrast, is very difficult to get at, since it disappeared in the nineteenth century and there has been little impetus since then to understand it from the inside, so to speak. Current medical textbooks spend only a few sentences on premodern medicine, mainly noting the sterility of its theories and uselessness of its techniques.

Without a living premodern past to study, historians, sociologists, and anthropologists have assumed that modern Western medicine exemplifies some specific, particular "Western mentality" that is intrinsically mechanistic and reductive, and so must have grafted onto it holistic concepts from other cultures. What this study will show, to the contrary, is that the *premodern* medical West was *not* very different from the East. When understood the way a practitioner understood it, premodern Western medicine was holistic, not reductive, ecologic, not mechanistic. To be sure, this Western holistic model did disappear in the nineteenth century, going underground to surface only in old wives' tales, proverbs, and homegrown alternative systems like homeopathy. If, however, we can bring the West's indigenous holism back to the surface, it can provide a native complement to the mechanistic and computational models of the body that now prevail.

AN OBSESSION WITH THE NUMBER FOUR

Up until the end of the 1870s or so, the ancient medical model of the West was still the main way by which the body was comprehended. Almost everyone—patient and doctor, philosopher and physicist—understood, to some degree, the basic concepts of this system: the elements, qualities, and humors. That is to say, they satisfactorily explained the body (Rosenberg 1979). With the triumph of germ and then cellular theory around 1880 however, the ancient system was surprisingly quickly replaced, and by the 1920s was no longer comprehensible as an organic explanatory model. Scholars therefore began to study its development, its content, its rise, and its fall. Their conclusions were summarized in 1964 by Erich Schöner in an influential monograph that has provided the basis for most writing on the subject (e.g., Nutton 2004).

According to Schöner (and here I summarize the basic points of his 193 page survey), the theory of humoral medicine was based on a special Western fascination with the number four, traceable back to the religious numerology of Pythagoras. It was this mystic fascination with "four" that accounted for the fact that there were exactly four elements—earth, air, water, and fire; four qualities—hot, cold, wet, and dry; and four humors—blood, bile, phlegm, and melancholia. It also explained why there were exactly four seasons—spring, summer, autumn, winter; four ages—childhood, youth, adulthood, old age; four organs, four temperaments, four colors, and four tastes.

Using the scanty remaining evidence, Schöner argued for a gradual elaboration of the "fours" starting with the pre-Socratic philosophers. Initially, he explained, they had proposed only one primal element for the basic stuff of the universe: *either* air, water, earth, *or* fire; then two—fire *and* water—but, eventually, all four. Many other concepts, each parsed as a set of four, accrued to these four elements over time, including the four colors, the four organs, the four seasons, ages, times of day, temperaments, qualities, humors, and tastes. In Schöner's view, this elaboration of linked concepts was not empiric but rational and logical. To be sure, he did note that our impression of a single, unified theory, at least at the beginning and for many centuries, was not accurate. For instance, even in the *Corpus Hippocratum*, the "fours" were only one of many theories of health and disease: another theory posited *two*, not four, medical humors and Plato held that there were three. The later Pneumatics even proposed that the *qualities* of hot and cold, wet and dry, and not the humors, were the fundamental concepts in medicine. Nevertheless, the final system was tetradic.

The entire system was eventually pulled together by the ancient master of medicine, Galen (Schöner, 86–93). It was Galen's system that became the authoritative medical system for the Middle Ages. Ptolemy added the planets and zodiacal constellations; and a later author finally added the four directions and the four winds (99).[3] In this view, the essentially symbolic, magical character of the humoral system was proved by its application in the Christian Middle Ages to such theological ideas as the four evangelists and the four rivers of paradise, which began to appear at the corners of the classic diagrams in the late antique period.[4]

Until well into the nineteenth century, then, this schema provided the basis for most Western medical thought.[5] Although the system was not fixed—tetrads were added by new thinkers and diagrams could be various, and even based on the number seven instead of four—its enduring appeal was its intellectual attraction. Once it was accepted, so it is said, its concepts "inevitably" provided the explanation for even new discoveries (Sears, 12).

In an oft-reproduced diagram, Schöner portrayed chronologically this way of understanding the linkages and connections amongst the concepts of the premodern system (fig. 0.1). That is, he placed the concepts that had been developed first at its center; concepts that were developed later he arranged centripetally. He also color-coded the concepts according to the time at which they were presumably developed, and he organized them from earliest at the center, to latest at the periphery. His diagram, then, literally

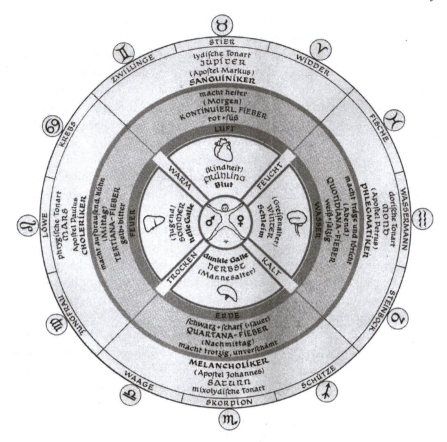

Figure 0.1. Schöner's Diagram of the Tetradic Connections. From Erich Schöner, "Das Viererschema in der antiken Humoralpathologie," *Sudhoffs Archiv* 4 (1964), following p. 114. By permission of Franz Steiner Verlag.

highlighted scholar's time—the chronology of conceptual development—rather than observer's or practitioner's time.

Now, given the kind of rethinking that has been applied to other, non-Western, medical systems, Schöner's formulation has gone remarkably unchallenged. Scholars have continued to use it to explain the premodern medical cosmology as "an elaborate attempt to represent the macrocosm and microcosm, uniting the elements, the humors, the calendar, the signs of the zodiac and so on," whose concepts should be traced to Pythagorean numerology (Parsons, 442). The West used it to understand its world for millenia because, in sum, it had an "obsession with the number four" (Peter Jones 1998, 20).[6]

But why would such an unscientific, unempirical system have lasted so long in a Europe not known for its stasis? In "Humoralism," Vivian Nutton gave a number of plausible reasons: its concepts and correlations were neat and inclusive and had explanatory force; and it was adaptable to many "sciences" such as physiognomy and astrology. It was vague enough to explain almost anything, simple enough that patients could understand it, but complicated enough to provide physicians with "an apparent certitude of an effective system of practice." Last, it had the "authority of antiquity" (288, 290).[7]

And yet, if the humoral model survived for so long out of a medieval respect for its ancient authority, why then were there so many (subtle) variations on it?[8] Would it not have been passed on exactly in the form it was received? If it survived because of its beauty, explanatory strength, and professionalizing power, then why did it ever fall out of favor?

Did practical observation have nothing to do with the system? After all, fire *is* hot and dry like summer, and water *is* cold and wet like winter. Moreover, the entire system bears uncanny resemblances to the great medical systems of the two other premodern civilizations of China and India. They, too, have elements, qualities, and humors, although there are enough differences (numbers of elements, qualities, humors) to preclude a direct inheritance. What is there about a theory of elements, qualities, and humors that would account for its independent development in the two other great cultures of the premodern world?

Most puzzling, however, is how the system as currently understood could have been *used* by a real medical practitioner responsible for real patients. The short version—the four humors, elements, qualities, etc.—has no practical applicable meaning in a real case. The long version, exemplified by Rebecca Flemming's forty-page exposition on ancient medicine, where "proportionality is played out over a constant material substream across which the qualities ebb and flow, bringing transsubstantiation in their wake" (96), is even less applicable. Can it be that practitioners, for millennia, took the trouble of studying, learning, applying, and transmitting a medical theory that had little grounding in an experience of body?

All of this is a way of saying that, despite decades of work, we still do not yet understand *how* the elements, qualities, and humors, meant. What is clear is that they were *not* about substreams or transsubstantiations; they were about something as real, and as practical, as when to pick a plant and how to make a pill.

ANTHROPOLOGY HELPS OUT

What has been missing in our view has been the voice of a *particular* practitioner from a particular place and a particular time—what anthropologists call an informant. Such an informant is not easy to come by. Before 1100, Latin texts, although they were often practical—herbals, short theoretical tracts, and recipes—were all anonymous, even when nominally attached to an apocryphal writer. Early Greek texts, while in theory far more ancient than the Latin, and therefore, perhaps, more authentic, come to us today only in fragments or as late manuscript versions. It is impossible to know whether such late versions *do* reflect early practice. In any case, none of them can be connected to an historical practitioner. There is Galen, of course, the definitive spokesman of medical antiquity. He *was* a practitioner, and much about his life is known, but he was also a rhetorician. It is difficult to ascertain how much the twenty volumes of his writing actually reflected his practice; moreover, they are often self-contradictory.[9]

After 1150, though, and until the arrival of a true vernacular literature, and perhaps not even then (Getz 1990), medical writers, even when they were practitioners, participated in a scholastic discourse.[10] It is never altogether certain whether their writings reflected their actual practice or, instead, broadcast their knowledge of academic texts.[11]

This is where Hildegard comes in. She is a compelling subject in her own right as we shall see in Chapter One, and to which anyone who knows her music can attest. Among her many accomplishments was the completion of an unusual, indeed a unique, medical text c. 1150—*Causes and Cures*. It answers our requirements reasonably well. This text *was* written by a practitioner, not an academic, and as a practical manual, not as a piece of rhetoric, by a particular person about whom much is known. As one of the earliest medical texts by a European practitioner—if not the earliest—it offers our best chance of understanding how the premodern system of elements, qualities, and humors provided such a successful medical model for so long a time.

The approach of this book, then, is quasi-anthropological: it approaches *Causes and Cures* as if we were student practitioners of Hildegard's in 1150 in the Rhineland. By that I mean that we will mainly limit ourselves to the texts that would have been available to such a student—popular medieval encyclopedias, short summaries of medical theory, herbals, and practical infirmary texts. As for Galen, Aristotle, and the Hippocratic Corpus, we will use only those texts by these authors that were available around 1150, and in their twelfth-century Latin translation. Because Hildegard's nonmedical work came out of a very different, far more rhetorical stance than did *Causes and Cures*, we will use it sparingly. But the remarkable 1100-word glossary of her invented, secret language, the *Lingua ignota*, will prove very useful for providing a visual inventory of her medical world and it will be referred to in every chapter. Our interpretative background will be the rich new scholarship on medieval medicine, along with insights from Shigehisa Kuriyama and Francis Zimmermann, on the analogous Chinese and Ayurvedic medical systems.

In Chapters One and Two, we will discover that Hildegard was probably the infirmarian or, as she would have said, the *pigmentarius* of the monastery of Disibodenberg; and that she, most likely, wrote *Causes and Cures* as a medical manual for her student infirmarian. In the next three chapters we will examine *Causes and Cures* in detail, as a manual for understanding premodern medicine from the viewpoint of a particular practitioner. It falls neatly into three distinct parts—Book One, a chapter that explains the basics of medieval science; Book Two, mainly based on the humors and devoted to understanding human physiology; and Books Three, Four, and Five, which function together like a traditional infirmary text—and the next three chapters follow this implicit outline. Thus Chapter Three explores the elements using one specific element, *air*, as a probe; Chapter Four, the humors, and Chapter Five, the medico-botanic concept of *viriditas*, as developed in *Causes and Cures'* use of medicinal plants. Each chapter begins with a summary of the relevant piece of *Causes and Cures*, followed by an in-depth analysis of the material on, respectively, elements, humors, and *viriditas* in *Causes and Cures*, in Hildegard's other work, and in a variety of other, mainly contemporary, sources.

This "anthropological" approach complements the historical, which, in the past decades, has uncovered a remarkable amount of information about premodern medicine—about the variety of its healers, about their medical training, and about their "healthcare delivery system." What Hildegard has to teach is very different: it is how the various concepts of premodern medicine were understood and used by a particular practitioner, in a particular place, at a particular time. As it happens, she had no formal medical training but learned and interpreted the ancient concepts in her own way, using her own experience and from her own perspective. It was a way that made practical—as well as philosophical—sense, as useful for making medicines as it was for understanding the cosmos. And it was an experience of daily, earthly life, from a perspective that is now quite foreign to us—although not completely so, as we shall see.

All translations are mine, and, when reference is to a manuscript, the transcription is mine, unless otherwise noted.

Chapter One
A Wonder-Working Woman

Who was Hildegard of Bingen? And how did it happen that she wrote a practical medical text which, unlike so many of its fellows, survived the centuries with its authorship still intact?

Hildegard was born in 1098 at the time of the First Crusade, and she died in 1179, the year of the Third Lateran Council.[1] She lived, therefore, during the twelfth-century Renaissance, an expansive moment in Europe signaled by the mini-climactic optimum, a growing population, expanding boundaries, technical innovations, and intellectual creativity. During her lifetime, the nation-states of England and France came into being; the early universities and medical schools took form; written law was formulated for the first time since Rome; and inventions, discoveries, and imports from the East appeared, including windmills and watermills, distillation, algebra, and the astrolabe. In France and England, the consolidation of effective royal power meant that there was less violence and more order; resources were freed for the development of science, art, and philosophy.

Hildegard was, however, born in the rural Rhineland, and there a power struggle between Church and State, emperor and bishop, led to a century of shifting alliances, sporadic violence, and uncertainty. Despite the efflorescence of her Europe, Hildegard had to negotiate her life in a small world marked by the storming of towns and even monasteries, the excommunication of emperors, the imprisonment of archbishops, and, most sobering, the exile, and even burning of writers, thinkers, and theologians.

The politics of the Comté Palatinate, where she spent her life, were particularly complex and unstable. Bounded by the river Nahe, the Rhine, the Pfalz, and the city of Mainz (fig. 1.1), the Palatinate was home to the Salian dynasty, which supplied not only religious but also secular leaders. Hence its bishops often sided with the feudal aristocracy, who were their relatives,

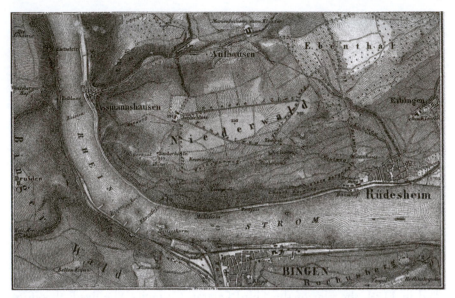

Figure 1.1. A Map of the Rhineland. From Karl Baedeker, *The Rhine* (Leipzig: Baedeker, 1884).

instead of with the Pope. Its politics when emperor and Pope were at odds could be tricky—confusing, inconsistent, and unpredictable.

She was born in the village of Boeckelheim on the Nahe (according to Trithemius) or, perhaps, in Bermersheim (from a study of land donations), the tenth child of pious, well-connected parents, Hildebert and Mecthilde.[2] Two of her older brothers also entered the church: Hugo became cantor and *magister* of the cathedral school in nearby Mainz, and Rorich became a canon in the monastery of Tally, also in the Rhineland. We know only the names of three of her sisters—Irmengard, Odila, and Jutta—but a fourth, Clementia, turns up as a nun at Hildegard's monastery of Rupertsberg, testament, it would seem, to the very impressive charisma of a youngest sister. Two nephews rose high in the Church: Wezelin became the provost of St. Andrew's in Cologne, and Arnold, the Archbishop of Trier.[3] Giselbert, her great nephew, was provost of St. Andreas, and even visited Hildegard at Rupertsberg (Ferrante 1998, "Correspondent," 222, fn. 49).

The family was close. When Hildegard left her first monastery (against its wishes), relatives went with her on the two-day journey to Rupertsberg, across from Bingen on the Rhine. Her brothers, Drutwin and Hugo, gave land and houses to her new monastery[4] and elder brother Hugo even asked Hildegard's advice after a fight with brother Rorich (Van Acker 1993, 465).

Late in life, when Hildegard desperately needed a (male) secretary for help with her last great work, brother Hugo arrived. And when her monastery was under interdict for her shielding of the body of an excommunicate (who was probably a relative), she asked nephews Arnold and Wezelin to mediate with the Pope; and they did.

At the age of eight, she was sent off to live with a distant cousin, fourteen-year old Jutta, daughter of Count Stephen and Lady Sophia of Sponheim.[5] Jutta was probably the seminal encounter in Hildegard's life, and an interesting and provocative woman in her own right. At twelve she had to be stopped from leaving Sponheim as a pilgrim. Instead, arrangements were made for her to be admitted to a monastery, and in the meantime she went to live with the widow (and probably teacher), Uda of Göllheim.[6] It was around the same time that Hildegard arrived, along with one or two other girls, and so, perhaps, all four girls attended some sort of school run by Uda.[7] A few years later, in 1112, all four entered the recently established Benedictine monastery of Disibodenberg.

In her later *Life of St. Disibod*, Hildegard gave Disibodenberg a satisfy-ingly long history but, in fact, it had been an active monastery for only a few years (Van Engen 1998, 34).[8] Founded by Rutthard, the Archbishop of Mainz, in 1098, it was abandoned a few years later when Rutthard was exiled for treason. He started it up again in 1108, this time with more loyal monks from his own monastery in Mainz, and for the rest of the century Disiboden-berg was protected by the Archbishops of Mainz. This relationship meant that, although Disibodenberg was very new, it could rely on the experience, the library, and the traditions of a far more ancient institution.

Also, of prime importance for Hildegard's future, Disibodenberg was re-established as a modern "double" monastery—that is, with women as well as men—and this meant that she had access to the more privileged, more protected, and richer environment of male monastics. Although in theory the nuns were physically separated from the monks, there is some evidence that this was not the case in practice. It is likely, for instance, that the monk-infirmarian trained Hildegard in medicine; another monk, Volmar, would be her secretary until his death, decades later.

When Jutta, Hildegard, and the other girls arrived at Disibodenberg, therefore, it was to a new monastic complex still being built.[9] It would take decades for the all of the monastic structures—church, chapel, hos-pice, dormitories, libraries, infirmary, and scriptorium—to be completed.[10] Throughout her years there, therefore, Hildegard lived with the disruption of construction—of floors being laid, columns carved, and glass smelted—by foreign workers from as far away as Byzantium.[11] Building would be a frequent metaphor in Hildegard's visions (Caviness 1998, 33).

Figure 1.2. The View of the River, Nahe, from Disibodenberg. It flows all the way to the Rhine, at Bingen. Taken by the author, 1998.

Today the walk to Disibodenberg goes up a wide dirt path that climbs steeply up the north side of the plateau. At the top the view is spectacular. To the west and north the hills are covered by vineyards, and there are purple mountains in the far distant background. On the south, steep side of the plateau, two rivers, the Glan and the Nahe, meet and flow together all the way to the Rhine (fig. 1.2).[12] Although very rural, it was (and is) well-populated, with many small villages and towns—in the twelfth century, linked together by a network of roads and rivers.[13]

Soon after their arrival, Jutta (in 1112) and Hildegard (in 1115 or, possibly, in 1112), became Benedictine nuns,[14] and gave their property to Disibodenberg, as was the custom.[15] It appears that a smaller monastery, school, or anchorhold was built for the women, inside of but separate from the male monastic complex and that Jutta became the mistress (*magistra*) or head of the women. In Hildegard's later commentary on the Benedictine Rule, it seems that their discipline was not particularly strict[16]—a well-balanced program of prayer, study, and manual labor, which, in the case of aristocratic nuns, would have meant copying manuscripts, painting, weaving and, probably, gardening.[17]

After her admission to Disibodenberg, though, we hear nothing more about Hildegard for the next two decades. Most likely she became the infirmarian for the women's side of Disibodenberg. Jutta, however, acquired a reputation for her fierce asceticism and her healing powers.[18] Because of her, Disibodenberg became something of a pilgrimage site, especially for the ill, and many women joined what gradually turned into a small women's monastery in the midst of the men's.

In 1136, however, Jutta died of a winter fever, as did the abbot, and the new abbot, Cuno, chose Hildegard to replace her as head of the women. Two years later, in 1138, the monks provided a new and larger house for the now almost twenty nuns, indicating, perhaps, Hildegard's success as *magistra*.

Then in 1140, Hildegard took a decisive step: she confided to her friend, Volmar, also a monk at Disibodenberg, that she wanted permission to begin writing a book of theology, which would reveal and explain the remarkable visions she had been having since she had been a child. This was a dicey request and obtaining permission took several years and quite a bit of politics.[19] But, eventually, permission was granted and, with Volmar's help and that of her friend, Richardis, Hildegard completed her first book, *Scivias*, in 1147 (See fig. 1.3 following page 124). This event marked the beginning of the second part of Hildegard's life. Writing a book of theology was problematic for anyone at the time. Peter Abelard, the famous theologian (and lover of Heloise) had been imprisoned and his book burnt for less; his student, the revolutionary, Arnold of Brescia, would be burned at the stake a few years later. And Hildegard was a woman. But, it turns out, not an unpolitical one.

After *Scivias* was completed, she sent a letter to Bernard of Clairvaux, arguably the most powerful person in Europe at the time, and Abelard and Arnold's implacable enemy. She humbly (even obsequiously) introduced herself, and asked for his opinion of her book.[20] At the same time Abbot Cuno went to Mainz to discuss *Scivias* with the Archbishop; the Pope, too, was informed.[21] He happened to be nearby at the time, assembling the cohorts for a Second Crusade; and he sent Albero, the Archbishop of Verdun, out to Disibodenberg to investigate. Albero met Hildegard, and was, apparently, impressed, since he returned to the Pope with a copy of *Scivias*, and Pope Eugenius himself is said to have read portions of it to the gathering of crusaders.

Next, Hildegard was vouchsafed a vision: she was to leave Disibodenberg and build a new women's monastery at a place she had never heard of—Rupertsberg—eighteen miles or so down the Nahe, across from Bingen on the Rhine.[22]

Despite this vision, however, Cuno absolutely refused to give permission for her to leave.[23] Indeed, a papal document, previously unknown, suddenly appeared; it listed, in detail, all of Disibodenberg's lands (among which were Jutta's). It also stipulated that no changes could be made in those possessions unless all of Disibodenberg's monks unanimously consented (Beyer,

Doc. 552). Hildegard took to her bed and had still another vision: she would never write another word nor have another vision as long as she stayed at Disibodenberg. Meanwhile, the powerful Marchioness von Stade, whose daughter and granddaughter were nuns under Hildegard, and who, eventually, would become a nun herself at Rupertsberg, went to Mainz for a talk with the Archbishop. He listened carefully, donated a lucrative watermill, the Mühlenwert by the Binger Loch, to the project (Silvas, 238) and arranged for Hildegard to buy Rupertsberg outright.

Hildegard made a miraculous recovery. A short while later, she left for Bingen with most (though not all) of Disibodenberg's nuns and with Volmar as prior.[24]

Even today it feels like a far distance from Disibodenberg to Rupertsberg. Eighteen miles as the crow flies, perhaps, but the Nahe is not navigable, so even by car the trip is surprisingly long. Hildegard, the nuns, and their retinue probably rode horses on the country roads, lanes, and paths, and it would have taken them a day or two to reach the Rhine (fig. 1.4).

Bingen is very different from the isolated, north-facing Disibodenberg. It spreads out from the mountain to the banks of the Rhine, and faces east and south towards Mainz and the rest of Europe. The move has been well-described as a move from periphery to center; Bingen, Rupertsberg, and Rudesheim across the Rhine were on a major thoroughfare. Mainz with its cathedral, its scholars,

Figure 1.4. Looking up the Nahe, back toward Disibodenberg. Taken by the author, 1998. The old Roman bridge is still there. Hildegard built her monastery on the right bank. All that remains is a single stone arch, now incorporated into a realtor's office in Bingen.

and its wealth was only twenty miles upriver, and the Rhine was lined with villages, although it became unnavigable just at Bingen, at the Binger Loch. Rupertsberg, which Hildegard planned to settle, lies just across the Nahe from Bingen, over the Roman bridge. When the nuns arrived it was a barely discernible hill, overgrown with brush[25] and home to only a single family.[26]

Hildegard immediately went about making this wilderness into a full-fledged monastic complex (fig. 1.5). She negotiated the purchase of Rupertsberg's land, consolidated its many scattered holdings,[27] and began the building of the church, the nuns' dwellings, the infirmary, and the gardens. She also put together the less-material infrastructure for her convent. She designed special clothing for her nuns, created an 1100-word glossary of a private language for the monastery, and composed original and unusual chants. She formulated her own, gentler version of the Benedictine Rule,[28] completed the medical and encyclopedic texts necessary for the monastic library and infirmary, and set up a scriptorium for the copying and illumination of manuscripts.

For the dedication of the church, completed quite quickly in 1151, she wrote the first musical drama in Europe, the *Ordo virtutem* (Dronke 1981, 100–106).[29] It starred Volmar, the only male in the monastery, as the only male figure in the piece—the devil—who to this day sounds like her monastic nemesis, Abbot Cuno of Disibodenberg.[30] And she began her second major book of

Figure 1.5. Hildegard's Monastery at Rupertsberg. Plate A92, from Daniel Meissner, *Sciagraphia cosmica*, Nuremberg, 1678, volume 1. Typ 620.78.565, Department of Printing and Graphic Arts, Houghton Library, Harvard College Library, Cambridge, Mass., with permission.

theology, the *Liber vitae meritorum*, which, like *Scivias*, was made up of detailed descriptions of a series of visions, followed by her exegesis.

As might be imagined, her reputation grew and spread, and she began to receive letters, visits, and invitations to speak from other abbesses, from abbots, bishops, and cardinals, from Germany, France, England, and even Byzantium. She was consulted mainly for spiritual but sometimes for medical reasons, not only by religious but also by lay people, including most of the royalty in Europe—Henry II, Philip of France, Irene of Byzantium, and the Holy Roman Emperor, Frederick Barbarossa. Her relationship with Frederick is still somewhat mysterious. Somehow Hildegard, although she disapproved of his politics vis à vis the Church, was able to establish a friendly but straightforward relationship with him.[31] Thus, even during a retaliatory visit to the Rhineland,[32] he was persuaded to exempt Rupertsberg from the royal tax.[33]

Although the 1150s were productive and, on the whole, successful for her, this is not to say that Hildegard did not sustain some losses. It is clear that she ruffled feathers and had, if not enemies, then some who were clearly not her friends. Although she did settle the issue of the nuns' dowered lands with the monks of Disibodenberg, there remained discomfort and irritation between the monasteries.[34] On top of her particularly bad relationship with Cuno, some of her own nuns were unhappy with her leadership and left. Even her main protector, the Archbishop of Mainz, was not always entirely taken with her; in a letter to her he compared her to Balaam's ass who was also miraculously able to speak, on occasion.[35]

Perhaps the most upsetting loss was her friend, Richardis. Richardis had been a nun with Hildegard for at least ten years and along with Volmar, had helped with *Scivias* and moved with Hildegard to Rupertsberg. But Richardis' brother Hartwig, Archbishop of Bremen, found her a prestigious appointment as abbess of a monastery in Bassum. He also arranged for Richardis' niece, Adelheid, likewise a nun at Rupertsberg, to become abbess of Gandersheim. Against Hildegard's wishes, Richardis and Adelheid both accepted their new posts and left Rupertsberg and Hildegard was, it has to be said, crushed. She blamed Cuno, and wrote angry, pleading letters to everyone:

"My soul is truly sad . . . because this terrible man [Cuno] took our daughter from our convent!"[36]

"Listen to me," she wrote Hartwig. "Do not ignore my words like your mother, your sister, and the Count!"[37]

"Oh, oh, as mother, how could you, my daughter, have left me like an orphan? I loved the nobility of your ways and your understanding and chastity and your soul and your whole life, so that many even said to me, what are you doing?"[38]

She wrote to Richardis, to Hartwig, to Richardis' mother, to the Archbishop of Mainz, even to the Pope to compel Richardis' return. But Richardis never did come back and died, as Hildegard had predicted, within the year.[39]

In the next decade (1160–1170), things seem to have gone better. She was able to complete the *Liber vitae meritorum*, and she continued her extensive correspondence (which eventually amounted to several hundred letters). She founded a second, sister monastery to Rupertsberg across the Rhine at Eibingen, settling it with thirty nuns, in addition to the fifty at Rupertsberg. She traveled to it by boat twice weekly for administrative, and presumably, spiritual administration.[40] (It survives as a Benedictine nunnery to this day, with a hiatus of only a few centuries.) She composed additional chants and received many visitors, and she began to preach, not only to her own nuns at Rupertsberg and Eibingen, but also to monks and nuns at other monasteries. Indeed, she was invited to speak even in the cathedral towns of Cologne, Mainz, and many more (Kienzle, 299–324).[41] And she began her third and most ambitious work, the remarkable *Liber divinorum operum*.

But by the early 1170s, time had begun to drain away some of her most important resources. In particular, Volmar, her secretary and the prior of Rupertsberg for more than a quarter of a century, died in 1173 (Silvas, 122 and Van Engen 2000, 376),[42] and it was very difficult to replace him. Disibodenberg, although obliged by contract to send a replacement, dragged its heels for more than a year.[43] It took pressure from Pope Alexander III before Disibodenberg sent the monk Godfrey, and he died soon afterwards, in 1176. Hildegard's brother Hugo then came out from Mainz to help with the *Liber divinorum operum*, but he, too, died within the year. Finally Guibert, an admirer and monk from Flanders, left his monastery (against the orders of his abbot) and went out to Rupertsberg; he took up Godfrey's unfinished *Life of Hildegard* (*Vita Hildegardis*) and the *Liber divinorum operum*.

Clearly then, despite Hildegard's future reputation, attitudes about her while she was alive were not unambivalent. The final public event of her life confirms this impression. Now, monasteries often let lay people be buried in their cemeteries, and, in 1178, Hildegard had allowed someone, probably a relative, to be buried in Rupertsberg's graveyard. Unfortunately, this person had been excommunicated. He should not have been buried in consecrated ground, but Hildegard declared that she had heard his repentance before he died. Even so, Christian, the Archbishop of Mainz, ordered her to have the body disinterred.

She refused (Van Engen 2000, 391). Indeed, she went down to the cemetery, located the tomb and with her *baculus*—the staff that signified her independent authority as abbess—made the sign of the cross and had all traces of the burial removed.[44] The Archbishop retaliated by imposing a humiliating punishment—interdict: all church functions at Rupertsberg were suspended. No services could be performed, no chants sung; the candles on the altar were blown out and the routines of monastic life came to a complete halt.

Hildegard tried everything to have the interdict removed—short of obeying the Archbishop and having the disinterment proceed. But it was not until six months later, at the Third Lateran Council (March 5–19, 1179) that the Pope was persuaded to lift it (without the disinterment) and services at Rupertsberg could resume. A few months after, on September 17, Hildegard died in her sleep at the age of eighty-one.

FINDING THE LIFE WITHIN THE *LIFE*

Even before Hildegard's death, however, her monk-secretaries began to write a *Life of Hildegard* (the *Vita Hildegardis*)—with her help. Incorporating personal material from Hildegard herself, Godfrey completed only Book One before his death. Theodorich, a literary monk who hadn't known Hildegard personally, was then commissioned by her friends to complete Godfrey's work. In Books Two and Three of the *Life* he repeated (in a more literary style) Godfrey's information, and then completed the *Vita Hildegardis*. Guibert wrote still a third *Life*, but based on the same material. These are the texts that provide the main source for biographies of Hildegard to this day—including this one.[45]

But Godfrey, Theodorich, Guibert, and even Hildegard herself were not trying to write biography or autobiography in our sense of the word. Rather they were attempting to establish the features of Hildegard's life that would qualify her for sainthood. As Jessica Weinstein pointed out, the purpose of the *Vita Hildegardis* was not biography but hagiography, to demonstrate that Hildegard had in fact been a saint (2000, 90). Indeed, within a generation her congregation of nuns at Rupertsberg had written to the Pope, asking for an official investigation into Hildegard's sainthood. In response a Papal Inquisition was dispatched to Rupertsberg in 1232 to interview the nuns, examine Hildegard's work, and determine whether she had been, verifiably, a saint.

This claim to sainthood is, as a matter of fact, the main reason we have such an extraordinary detailed account of one medieval woman's life, including a sworn list of her authentic writings, the writings themselves, and even a consecutive narrative of the location of all (or almost all) of her bones. It is for this reason, too, that her medical text still has its attribution intact—Hildegard's nuns treasured and preserved it along with everything else.

But it is also the reason it has been so difficult to square the Hildegard of the *Vita*, and, for that matter, of her biographers, with the Hildegard that comes through in the documents, letters, and, especially, in her medicine. Her *Life* emphasized three themes in particular: Hildegard was uneducated and received her knowledge through her visions; she was isolated from the world in a cell; and she was chronically ill. These three themes have been repeated by her biographers to this day. For instance, Anne King-Lenzmeir states that "Hildegard is astonishing . . . a sickly, enclosed, contemplative, Benedictine woman"(21); Änne Bäumer, that she was uneducated, always sick, and immured for forty years. Regine Pernoud writes that she had only learned the psalms, was in fragile health, and that her daily life was that of a Benedictine nun, marked by the canonical hours (19).

Recent research and close study of her writing, however, suggest that the life of Hildegard of Bingen was tailored to fit the proper life of a woman saint. For instance, something as apparently straightforward as the day of her death, which Theodorich tells us was on Sunday (the proper death day of a saint) was, rather, on a Monday.[46] What, then, *were* the criteria of a woman saint? And did Hildegard of Bingen fit them as well as did the Hildegard of the *Vita*?[47]

Requirements of a Female Saint

Scholars of hagiography observed long ago that a saint's *Life* is not meant to be an objective, factual record of a life. Rather, it was supposed to inspire imitation, veneration, and exemplify church doctrine (Robertson 1995, 17).[48] Of course, some relation to actual biography is essential, but there is always, by definition, a polemical purpose for many of its details (Donovan, 8). Consequently the *Life* (as opposed to the life) of a saint had only a few possibilities. He or she could be a martyr, dying for the faith. Or he or she could be a "confessor"—that is, a hermit, a cenobite, or a monk, who died in bed, perhaps, but only after many heroic years of contemplating God and battling the devil (Robertson, 12).

Within these patterns, there are standard elements, as Donald Weinstein and Rudolph Bell demonstrated in their statistical tables of saints' *Lives*. For instance, most saints are from noble families and are either a first or last child (46). As children, they are often marked by some special quality, such as beauty or innocence. A popular *topos* is the old, wise child who prays regularly, attends church happily, and studies hard (29). Usually saints become monastics, although for men and women their reasons differ (48). Thus a male saint enters the monastery as a career choice, as the good son who takes a parentally approved path and receives a good education and ecclesiastical office (48). By contrast, the female saint, who also usually enters a monastery, does so against her parents' wishes, often in order to preserve her virginity from an unwanted marriage. For both, some particular event precipitates their saintly adventure—often the death of a parent (53) or a loved one (55).

To be seen as a saint, however, there are three other requirements—doctrinal purity, heroic virtue, and miraculous intercession after death (141). Since most saints are made not by rulings of papal commissions but by popular acclaim, miracles (143), especially miraculous healing, prophecy, and clairvoyance (147), are crucial. "Heroic virtue" is shown by spreading the word of God, and "reformers of morals, preachers of the apocalypse, or apostles to heretics and unbelievers—often became objects of cult in their own lifetimes (159)."

The Hildegard of the *Vita Hildegardis* fulfills these saintly specifications to a remarkable degree. She came from a noble lineage, was the last child of her parents, and as a child was marked by clairvoyance. She became a monastic, and upon the death of her beloved teacher, began her saintly career. She performed wonders of miraculous healing, prophecy, and clairvoyance, and, what is more, preached moral reform.

But Hildegard was also a woman. Were there particular hagiographic tropes for the woman saint by which we must evaluate her *Life*?

There were, indeed.[49] First, it was much rarer for a woman to become a saint. In the sample of Weinstein and Bell, women made up only 17.5% of saints and, in the twelfth century, even less.[50] Moreover, in general women had many fewer possible avenues of success (227).[51] Instead of proving their sainthood evangelizing or in demonstrations of worldly power, "supernatural power in its various forms was almost twice as prominent [in the lives of females] as in the lives of males. . . ." (228). Thus, although holy women often lived active public lives as spiritual counselors, peacemakers, and healers (229), visions "pervaded the lives of female saints" (228). Illness played an especially prominent, penitential role (156 and 235). So did solitude and enclosure, with women typically "shutting themselves up in a barren cell next to a church" (155).

Also for women, there were two additional saintly career paths, along with martyr-saint and confessor-saint. There was the virgin-saint, whose story, Jocelyn Wogan-Brown argues, is the most important way, in the West, for representing women (3). Virginity was its signal characteristic, established and enforced by the idea (if not always the practice) of being, often literally, walled up in a cell, so-called "strict enclosure." This was important not only for mak-ing a clear separation of the saintly woman's life from everyone else (24) but also for guaranteeing her virginity. For a woman to be a saint, she must, like a wife, be "safely enclosed in her tower, bower, chamber, cave, or cell" (24).

There was one other possibility for a woman saint, in addition to the martyr, confessor, and virgin roles. It was the ancient trope of the powerful, sainted abbess (Wogan-Brown, 189). In fact, this had initially been the com-mon model—"the great monastic lady, withdrawn from worldly power and worldly comfort, but not from the world's misery. . . . She entertained . . . , cared for the sick and poor . . . , protected fugitives . . . , [performed] mira-cles . . . , and usually secured the continued support of noble relatives in her project" (McNamara 8–9). These women, by definition, often were significant landholders, ruled communities, and administered complex estates (Wogan-Brown, 191). But, although "the saintly abbesses . . . remained astonishingly free to invent their own way of life through the application and interpretation of rules written by men" (McNamara, 11), they "had to temper their own talents for innovation and leadership with qualities of humility and reverence that recommended them to the male hierarchy. . . ." (McNamara, 13).

How does the Hildegard of the *Vita Hildegardis* come out against these female *topoi*? Pretty well. In Hildegard's *Life*, several aspects are emphasized: her intellectual inferiority as shown by her lack of schooling, (this to prove "Hildegard's sublimity" according to Weinstein in "Textualizing," 93); her humility and self-censorship, necessary because "the most significant events in Hildegard's life were those that placed her insubordinate and under sus-picion of heresy" (90); her seclusion, as proved by her strict enclosure; and, last, her chronic illness, because "the *Vita* authorizes Hildegard's sick body as select, miraculous and punished" (98).

In short, the *Vita Hildegardis* shows a Hildegard who fulfilled the requirements of a woman saint: she was untrained, unexposed, and invalid. But it also, perhaps unintentionally, reveals a Hildegard who was intellec-tually complex, politically astute, energetic, and long-lived. While it may not be fiction—perhaps Hildegard was a saint precisely *because* her life con-formed to the requirements[52]—nevertheless the *Life* presents biographers, and all those who seek to understand this woman, with a situation of cogni-tive dissonance. What evidence *is* there that she was uneducated, isolated, and chronically ill?

Untrained

It was Hildegard herself who underlined her ignorance of literature—"I hardly had any letters, since an uneducated woman had taught me."[53] Her lack of education was repeated by her contemporaries. Thus Adalbert, the prior of Disibodenberg, recalled (in the translation by Baird) that "your instruction was only that appropriate to a woman; a simple Psalter was your only schoolbook" (172). The theme was taken up by her hagiographers,[54] by the *Acta*,[55] by Trithemius,[56] and by her present-day biographers.

After decades of research, however, scholars have begun to question this notion. Thus Margret Berger notes that "she had the training and ability to use Latin as her literary language" (7); Ines Koring, that she knew the Church Fathers and had done scientific reading (6); Diers, that she was far more educated than she presented (13); Eliza Glaze, that she had read, absorbed, and interpreted Latin medical writings (1998, 125). Constant Mews goes even further, arguing that "she deliberately under-played the extent of her reading of secular literature" (1998, 99). As Barbara Newman dryly suggests, Hildegard perhaps, "exaggerated her educational deficiencies" (1999, fn. 18).

However, because Hildegard almost never mentions sources, what she *did* have access to and *may* have studied has been difficult to pin down. In the 1930s Hans Liebeschütz pointed out hundreds of parallels between Latin authors and Hildegard's writings, concluding that she must have known or, at least, been influenced by many texts, including the *Shepherd* of Hermas, *The Conflict of Virtues and Vices* of Ambrosius Autpertus, the Hermetic treatise *Asclepius*, the *Quaestiones naturales* of Seneca, and the *Mathesis* of Firmicus Maternus.[57]

Indeed, when Peter Dronke took two hundred of her most idiosyncratic usages and compared them with Latin authors he found striking similarities with word usage in St. Benedict's *Rule*, Augustine's *On the Psalms*, Ambrosius Autpertus' *Conflict of Virtues and Vices*, Gregory the Great's *Moralia*, *Gospel Homilies*, and *Ezekiel Commentary*; various works of Bede, Plato's *Timaeus*, the *Book of Nemroth*, the *Cosmographia*, the *Aethicus*, and the *Pantegni*.[58] In the apparatus of the *Liber vitae meritorum*, Angela Carlevaris provides hundreds of analogous citations and argues that Hildegard's sources must have included the Church Fathers, Gregory, Augustine, Eriugena and, probably, Honorius Augustodunensis, William of Conches, Gratian, and Ivo (304–310). Mews adds St. Anselm and William of Hirsau ("Context," 97–99). In her edition of *Causae et Curae*, Laurence Moulinier gives an extensive annotation of more than 100 authors.

Resemblance, of course, does not prove direct textual access, and Charles Burnett rightly allows that there could be common oral sources as well (1998, "Hildegard of Bingen and the Science of the Stars," 120). But the searchable databases of the 221 volume *Patrologia Latina*, the *Cetedoc Library of Christian Latin Texts*, and the *Acta Sanctorum* provide quite convincing evidence that Hildegard indeed knew many authors. As we shall see, it is easy to find whole phrases in Hildegard that are word-for-word identical with numerous Latin authors, especially with Gregory the Great and Ambrosius Autpertus. In the chapters to come on elements, humors, and *viriditas*, such identical phrasing will lead to the inescapable conclusion that Hildegard was widely read.

And the fact is that Hildegard *did* have access to many books. Disibodenberg had a substantial library and was active in book production, which means that it must have borrowed and loaned many books. She herself corresponded with the monks of many monasteries with fine libraries, including St. Jacob's in Mainz, Eberbach, St. Eucharius in Trier, and Gandersheim. She even, presumably, alluded to her reading (though not in the *Vita*), when she wrote that she understood the writings of the "prophets, of the gospels, and the precepts of other saints and of certain philosophers."[59]

A possible resolution of this particular problem has been suggested. Perhaps when Hildegard called herself "uneducated" (*indocta*), she did not mean that she could not read, or had not read anything but the Psalms, but that she did not have the formal training or schooling that theologians—her brother Hugo for instance—had been given (Mews 1998). This may be true, but she was also taking advantage of the trope of female sanctity, where it was *precisely* her lack of learning that authorized the woman saint. Not textual authority but "supernatural experience was a seal of authenticity reassuring her that God was with her, a badge of authority to present to a skeptical male world" (Weinstein 2000, 232).

There was good reason, indeed, for a future woman saint to affirm that all her knowledge came to her not from man but from God alone. To the extent that this was believed, it gave her work unassailable authority and gave herself protection against a potentially fatal literary critique. Hildegard "was careful not to place herself in the same situation as Abelard" who was imprisoned and whose books were burned (Mews 1998, 100).

It was, in fact, *because* Hildegard's contemporaries believed that she was uneducated and, therefore, a pure mouthpiece—a feather on the breath of God, as she wrote—that they permitted her to write and even to preach. As Robert of Val-Roi pointed out: "The apostle does not permit a woman to teach in Church, but this woman is freed from this condition by the assumption of the Spirit, instructed by the *magisterium* of the Spirit" (quoted in Mews 1998, 109). Her literary ignorance as qualifying her for sainthood was more than once pointed out in the *Acta Inquisitionis*, designed to persuade the Pope of Hildegard's sanctity: "And she who, besides the Psalter had never learned her letters, composed many books, through revelation of the Holy Spirit."[60]

Hildegard's protestation of ignorance, then, served numerous purposes. It protected her from accusations of unorthodoxy; it gave her an unassailable authority; and it confirmed her eligibility for female sainthood. However, it is indisputable, simply from the numerous direct, though unacknowledged, quotations found in her work, that she had read widely; and if we are to understand the literary and conceptual context of her medical work, then her claim to inspired non-human knowledge as her only source must be finally given up.

Unexposed

For a potential saint, isolation from the world was another crucial requirement, in part because it evoked the lives of the Desert Fathers, the original monastics. Indeed, from the beginning of monasticism, enclosure was required of monks and nuns, although its meaning varied, as the Latin, *inclusus*—"shut up, shut in, confined, enclosed, or imprisoned" (OLD, s.v. "includo")—implies. For monks, it meant a certain sobriety in the world; thus they were warned not to frequent taverns or embrace women. For women, enclosure served a second and more important function—it guaranteed their purity for the heavenly Spouse, just as a similar enclosure did for the married woman (Wogan-Brown, 26). Indeed, for nuns enclosure was crucial; more than 25% of the regulations in the late fifth century monastic Rule of Caesarius had to do with ensuring that its nuns were adequately enclosed. Although St. Benedict, author of the Benedictine Rule, had little to say about how strictly monastics were to be confined within his monastery, Carolingian legislation of the same period required *strict, full* enclosure, defined by the *Regula monachorum* as confinement to the monastery grounds (Schulenberg, 51–86).

In the early centuries of monasticism, however, there was always a discrepancy between praxis and written norm (McNamara, 63), with some freedom of movement allowed, or simply assumed by abbesses and nuns (65). For instance, despite their enclosure Anglo-Saxon nuns went to Germany, proselytized, and established monasteries (66). Walburga, abbess of the double monastery of Hedenheim, was known to have left the monastery to perform healings (66). That is to note that "however firmly convents may be enclosed in the theories and representations of the churchmen responsible for them, monastic life included a range of options of less than strict segregation, and was frequently permeable as between lay and professed women" (Wogan-Brown, 48). Indeed, the constant warnings by councils that nuns should stay within their convents may hint that precisely the opposite was the case (Mews 1998, 108).

What enclosure for women actually entailed grew much stricter over the centuries. By the thirteenth century it did not simply mean confinement to the monastery's grounds; it meant confinement to a cell, and nuns, at least theoretically, were virtually imprisoned. In theory, women were, literally, walled up: servants passed food into the cell and removed excretions; exercise was impossible and even light was scarce. Indeed, the cell was often compared to a tomb and taking the vows to dying. In theory, the life of an enclosed nun was completely controlled, including her visitors, her reading material, and virtually all of her contact with the outside world.

However, theory was not fact. Indeed, enclosure only became "strict" in the above sense of the word in 1298 with the *Periculoso* of Boniface VIII (Pernoud, 134), and some of what has been inferred about Hildegard's life at Disibodenberg has been based on a retrospective reading. What do we know about Hildegard's *actual* "enclosure"?

Her first bio-hagiographer, Godfrey, tells us that on November 1, 1112, Hildegard was enclosed, using the term *recluditur* without further elaboration.[61] Another bio-hagiographer, the monk Guibert, however, described a very dramatic event in which Hildegard, Jutta, and the other women were literally walled-in to a newly-built stone house near enough to the church so that they could hear the liturgy.[62]

Until the *Vita Juttae* surfaced in 1992 Hildegard's biographers all had adopted the "strict enclosure" interpretation. Thus Trithemius wrote that the women were walled-up[63] and Gronau described the house as having "only two windows for light and air through which came all news" (37). Van Engen detailed their dwelling as "a stone structure, with access sealed off except for a window through which they could receive food and at certain hours converse with others," although he does point out that because of their visitors, the nuns were not isolated in reality (1998, 34). Diers graciously added several rooms and a little garden (8).

Now the kind of enclosure we imagine for Hildegard during these years has an important bearing on the interpretation of her medical work. For instance, if she were imprisoned in a cell during these years (1112–1136), then it is hard to see how she could have trained as the infirmarian for the women, and, perhaps, of the men of Disibodenberg, as I shall argue in Chapter Two. Moreover, if she were strictly enclosed and everything she received—clothes and food, books, and visitors—was proctored, then her contacts with the natural world must have been limited, at least after the age of fifteen, to books and conversation. It would be difficult to explain the vivid sense she conveys of the natural world. It would be especially difficult to explain her medical writings, which are based on medical experience as well as on texts.[64] But because Hildegard's life and writings reveal a woman who *did* understand the secular world—its politics, practical medicine, and natural life—strict enclosure, like the trope of intellectual ignorance, presents another moment of cognitive dissonance.[65]

There is some evidence that Hildegard's enclosure was not strict in the later sense of the word. For instance, in her (hierarchical) 1100-word glossary, the *Lingua ignota*, she differentiated between an enclosed person (*inclusa*) and a nun (*monialis*), where *monialis* (reimonz, #217) comes before, and is different from, *inclusa* (phalischer, #218) (Portmann, 1–4). Also, she never refers to herself as *inclusa* but always as *oblata*, that is, "given as a nun."[66]

Indeed, when she decided to leave Disibodenberg, the Abbot wrote of his surprise that she would want to move from "a lushness of fields and vineyards, and from the beauty of the buildings" to a barren place (Silvas, 226)—a remark difficult to interpret if the women were locked in a stone cell. When the prior of Disibodenberg recalled (with sadness) Hildegard's life at Disibodenberg, he referred to the monks as "we who had known you almost from the cradle, and with whom you lived for many years."[67] This implies an intimacy with the monks as well as freedom of the outdoors not possible in a stone cell.

Too, illuminations in *Scivias* often show men and women together. Thus, in the contemporary portrait of herself painted under her direction (fig. 1.3), Hildegard does not appear to be strictly enclosed; and in Vision 1, part 1 of *Scivias*, it is the nuns who are in the open air, while the monks are in the cells.

From a practical point of view, keeping what eventually amounted to twenty women—and noble, wealthy ones at that—in the strict enclosure described by Hildegard's hagiographers would have required enormous resources—as many as keeping twenty men in prison. Indeed, Mews concluded that the enclosure could not have been strict (1998, fns. 9, 10, 22); and Flanagan writes that "it is hard to see how strict enclosure and isolation from the male side of the community would have been possible."[68] All this evidence suggests that life for Hildegard on Disibodenberg did not take place within the strict confines of a stone cell.

The solution seems to be that originally the nuns' dwelling *was* a stone enclosure, but within a short time, as we have learned from the *Vita Juttae*, it became a school. Thus in the *Vita Juttae*, Jutta is referred to as both "enclosed" (*inclusa*, 174), and as a "nun" (*monacha*, 174); the other nuns are called her "students" (*discipulae*, 174), and the convent a "school" (*schola*,

174) over which Hildegard later "ruled" (*regendam*, 174) (Staab 1992). The ten women under Jutta "passing the time in that place with devotion, paid attention to her admonitions and advice,"[69]and later were taught by Hildegard as teacher (*magistra*).[70] As it became a school and acquired students and other nuns it became an actual monastery within the larger male monastery.[71] Indeed, the nuns moved to bigger quarters on Disibodenberg at least twice, in 1114 and in 1139.[72]

In short, strict enclosure, while hagiographically necessary to establish Hildegard's purity and asceticism, was not entirely accurate. For most of her life, Hildegard did not live as an immured recluse but as a traditional Benedictine nun, confined to her monastery, to be sure, but only up to a point. For interpreting Hildegard's medicine, this means that training and practice as an infirmarian would have been possible.

Invalid

In her *Life*, Hildegard is framed not only as untutored and imprisoned but also as sickly.[73] This theme, too, has been echoed by her biographers ever since. Thus, she was a "sickly child" (Feldmann, 33) and "endured severe and frequent illness from early childhood through advanced age" (Newman 1998, "Three-Part Invention," 198); she was weak and frequently ill (Haverkamp 2000, 15.)

Again, however, her *Life* presents us with a paradox. Although sickly and frail, Hildegard lived to be eighty-one, was wonderfully productive, and, at the age of seventy-three, went on a lengthy preaching tour. The very last image we have of her is hardly one of chronic illness—she makes her way down to the cemetery and stamps her staff on a tomb in defiance of the Pope. And, once again, we are faced with the fact that illness was something of a requirement for a saint, since chronic suffering provided a virtual martyrdom when actual martyrdom was unavailable. For women saints in particular, chronic illness made them that much more enclosed and their achievements, therefore, that much more miraculous.

Indeed, that is what her bio-hagiograhers have always inferred about the effect of illness on Hildegard's life. For example, Trithemius wrote that her work was all the more miraculous given that her frequent illnesses (and moist brain) made her almost completely indisposed for any sort of writing.[74] For modern scholars, Hildegard's poor health has served two other functions: it has provided a reason (other than the practice of medicine), for her interest in things medical;[75] and, under the hypothesis that her illness could be understood as migraine, a physiological explanation for her visions.

What evidence is there to support the hagiographic trope that Hildegard was weak, frail, and constantly suffering from illness?[76] The main evidence comes from Hildegard herself, in the autobiographical pieces of the *Vitae* and in her letters. Thus Hildegard describes herself as a weak and feeble child: between the ages of eight and fifteen, she was so often sick that she was thin and weak.[77] After fifteen, (that is, after she entered the monastery), she mentions no other episodes of illness until she was nearly forty-three when, after her first "pressing" vision, she had some kind of wasting illness (Silvas, 141). There was a *magna pressura dolorum* (a great pressure of pains, or sufferings) and this was followed by a surge of vitality.[78]

She was free of illness, it seems, for nearly seven years, when she experienced a second episode (c. 1147–1150) after Abbott Cuno refused to let her leave Disibodenberg. Although this illness has been understood as having been a long, debilitating sickness that accompanied the move to the Rupertsberg (Newman 1998, 199), Hildegard's own description is telling. After Cuno refused to let her leave and even accused her of vanity, Hildegard writes that:

> One time from weakness of the eyes I could see no light, and I was deprived by such weight of the body that I was not strong enough to get up and I lay filled with terrible pains [*doloribus maximis*]. . . . They all said that I was deceived by my own vanity; and when I heard this, my heart was dismayed, and my flesh and veins dried up. For many days I stayed in bed, since a great voice had forbidden me from ever, in that place, bringing forth or writing another *visio*.[79]

But as soon as Cuno gave permission to leave, she instantly recovered her strength, in what her bio-hagiographers described as a miracle.

For another decade no new episodes of illness were noted; Hildegard's third illness occurred c. 1158–1160. It is described quite differently.[80] It was a very serious illness (Hildegard was expected to die), and it began with a fever and lasted a month. It culminated in a spectacular vision, although she didn't recover completely for nearly three years:

> Another time, God struck me and filled my whole body with airy feathers, so that my veins and blood, flesh and liver, marrow and bones dried up. . . . In this struggle I stayed thirty days, so that from the heat of that airy fire my belly was feverish. . . . Then I saw a vision of the angels who had fought with St. Michael; and after that I gradually began to grow stronger . . . although it took almost three years until I had completely recovered.[81]

> Actually my strength only came back bit by bit and day by day . . . .
> And if that great pressure of pains had not been from God, then I
> would not have been able to live. Although even as I suffered them, I
> preached, sang, and wrote what the Holy Spirit wanted. . . . In fact,
> in these languors I suffered for three years . . . and finally my spirit
> was completely restored, and my body in its veins and its marrow was
> renewed.[82]

For a modern diagnosis we might think of a long, serious, infectious illness—malaria, tuberculosis, or an intraabdominal abscess, for example.

Yet another decade passed without any accounts of particular illnesses,
until 1169–1170, when Hildegard again became ill, with a feeling of heavi-
ness, pain, and burning that lasted more than six months:[83]

> After this I became gravely ill and could not leave my bed. And this
> illness came from the south wind, and from day to day I suffered
> from pains and burnings, so that my spirit almost left my body. And
> this lasted more than six months and that wind perforated my body
> totally, so that I was always in a lot of pain, . . . and this lasted
> almost a year. . . .[84]

This is certainly a different, third type of illness entirely, most likely a different infectious disease of some kind.

Putting together all of Hildegard's own descriptions of her illnesses, it
would appear that she simply had had a series—and a rather short series,
considering her longevity—of separate and distinct illnesses. Nevertheless,
for decades, physician-historians (and others) have struggled to find a unifying diagnosis, and migraine has often been suggested. Is there enough evidence to make this diagnosis? If so, does it contribute in any way to our
understanding of Hildegard and her medicine?

Migraine was initially proposed as an explanation for Hildegard's illness
by the physician and medical historian, Charles Singer, who suggested that
it would account for her longevity and apparent good health in spite of her
complaints of suffering. As a modern syndrome, migraine is a long-term but
episodic illness that leaves no chronic traces. Often beginning in middle age,
a typical attack (of classic migraine) begins with an aura, usually of flashing
lights or images, that evolves into a severe hemicranial headache often associated with nausea and vomiting. Although the aura lasts only for a few minutes, an episode of migraine can last for hours or even days; when it resolves,
some sufferers describe a sense of well-being. There can be cycles of months

when episodes tend to cluster, and they can be triggered by certain foods, certain weathers, menstrual cycles, and stress.

In a tour de force, Singer analyzed several of Hildegard's visions (both text and image) and showed that they displayed characteristics reminiscent of migrainous auras—wavy lines, points of light or fortification spectra, (that is, images of buildings).[85] Migraine was then popularized as her diagnosis by the neurologist Oliver Sacks in his 1970 *Migraine*. Sacks, indeed, went further than Singer and asserted that *all* of Hildegard's visions demonstrated the pathognomic phenomena of migrainous auras. Since then migraine, as an explanatory diagnosis for Hildegard's visions and illnesses, has found its way into most biographies of Hildegard, as well as into the folk understanding of migraines, where, in the words of the Internet, Hildegard is "the unofficial patron saint of the ailment."[86] Indeed, it is now argued that Hildegard must have helped in the design of her illuminations *because* they show the "visual forms that can be associated with the auras experienced by people suffering from migraine" (Caviness 1998, "Hildegard as Designer," 31).

Certainly migraine provides an intriguing solution to the paradox of Hildegard's vitality, productivity, illness, and visions, but it has its problems. For instance, although headaches *were* part of Hildegard's diagnostic purview (they occur in *Causes and Cures* as *dolorem capitis*),[87] she does not describe them for herself, a striking omission, given the pain of migrainous headache. Singer, therefore, suggested that she might have had "atypical migraine"— migrainous auras without headache—or even "hystero-epilepsy" (Singer 1955, 21). Sacks, who devoted an entire chapter in *Migraine* to the multifold presentations of migrainous auras without headaches, did assume that Hildegard, had "amigrainous migraines."

If so, then the migrainous character of Hildegard's illnesses can be demonstrated only by the migrainous character of her visions. Are her visions, then, pathognomic of migraine? Do they *always* show the classic phenomena of lights, points of light, wavy lines, and fortification spectra? Do they conform to the time course and tone of a migrainous aura?

If we look at the illuminations of her visions done at Rupertsberg, probably under her supervision, two-thirds do contain stars, lights, wavy lines or buildings. This is more than the usual in medieval illuminations, but not that much more, since such images were very frequent. If we look instead at her own descriptions of her visions, the evidence is differently ambiguous. The famous description she wrote in reply to Guibert's inquiry about her visions is that:

> The light which I see is not located but yet is more brilliant than the
> sun, nor can I examine its height, length, or breadth, and I name it

the "cloud of living light." And as sun, moon, and stars are reflected in water, so the writings, sayings, virtues, and works of men shine in it before me. . . . But sometimes I behold within this light another light which I name "the Living Light itself." . . . And when I look upon it every sadness and pain vanishes from my memory so that I am again as a simple maid and not as an old woman . . . [88]

This description has been thought to describe the migrainous aura. Other pieces of Hildegard's descriptions have been used to support other elements of a migraine diagnosis. For instance, in support of migraine is the idea that Hildegard felt particularly well after an attack, as she described above, and also:

In this *visio* I felt my veins and marrow completely restored with power, which on account of numerous illnesses I had lacked since a young adult. . . . [89]

And it has been suggested that Hildegard did, in fact, have the headaches of migraine, because she used *pressura* in conjunction with her visions. For instance:

Later in that *visio*, I was forced by great pressure of pains [*magna pressura dolorum*] to show what I saw and heard.[90]

Thus there does seem to be evidence in support of a diagnosis of migraine.

However, if we examine carefully what Hildegard described, the experiences of "light" do *not* seem to conform to the *time course* and *tone* of a typical aura. Thus she described two different kinds of light, not just one. The first was a light which she *always* saw, like the light coming from the sun hidden behind a cloud—the "cloud of living light" (*Acta*, 17D-18A). Since she saw it at all times, it cannot have been an aura. The second was a very bright light, which she seldom saw, and which was more like a flash of the living light (18C). Perhaps this was an aura, although it is curious that it happened so infrequently, if we are to accept that she was often ill from migraines. Moreover, although it has been said that *all* of Hildegard's visions have points of light or lights and Sacks has written that "our literal interpretation would be that she experienced a shower of phosphenes in transit across the visual field. . . ." (Sacks, 114), the word that has been translated as "light" is often *visio* and not *lucis* or *lumen*. *Visio* was not light; it was a visual or mental image (OLD, s.v. "visio").

Moreover, *pressura dolorum* may not have been "painful pressure" as in the powerful headaches of migraine, but a phrase Hildegard used as it was used commonly, to describe any painful event, but especially Christ's sufferings on the Cross.[91]

Surely most puzzling for a diagnosis of migraine is Hildegard's assertion that while in the throes of her experience she was able to understand "the writings of the prophets, the evangelists, and other saints, and certain philosophers without any teaching."[92] Although perhaps she was describing a timeless experience in which she felt or intuited the essence of philosophy, it sounds more like she achieved an actual understanding of the words and ideas of many authors. It is difficult to square this account of knowledge and understanding with a migrainous phenomenon of some kind. Perhaps her migraines, if she had them, were quite separate from her visionary experiences.

To sum up: The longer illnesses that Hildegard described do not fit the sudden, recurrent, and repetitive nature of migraines. Rather, she seems to have had a number of prolonged, distinct illnesses. The question of whether she also had migraine cannot be answered. On one hand, the kind and quality of one kind of "light" that she experienced does seem to fit what many migraine sufferers experience during an attack. On the other hand, Hildegard does not always describe this kind of light with her visions; there were no migrainous headaches, and the experience lasted considerably longer than the few minutes of a migrainous aura. In short, although migraine remains an intriguing and believable hypothesis, there is just not enough evidence to support or reject it. In any case, it appears that the hagiographic picture of Hildegard as invalid was no more true to life than was her lack of education or her strict imprisonment in a stone cell.

Hildegard *was* a wonder-working woman, but not quite in the way her admirers and boosters thought. Instead of succumbing to the archetype of the woman saint—empty-headed, protected, sickly—she used it to live a life that evidently suited her. The framing of her life as saintly did permit her to produce a wonderful body of work: three books of theology with remarkable illuminations; two monasteries; more than three hundred letters; seventy hymns; a musical drama; two original Saint's Lives (not counting her own); a glossary for an unknown language; a nine-book natural science encyclopedia; and a practical medical manual. Moreover, she was not burned at the stake or imprisoned. In fact, all of her work was carefully preserved—*and* she became an unofficial, popular saint, eventually, as well.

It was the very fact of Hildegard's sainthood that ensured the survival of her texts, in her monastery, by her nuns. But, ironically, it was also her saintliness—as an untrained, unexposed, and invalid woman—that ensured skepticism towards her authorship of these very texts. It took decades, indeed, and three "feminist waves" before her theology was once again accepted as having been written by her rather than by her male scribes and secretaries. The illuminations of her theological texts also led to much ink being spilt as to which male monastic scriptorium had *actually* designed and painted them. Only recently have they also been placed back at the scriptorium of Rupertsberg, under Hildegard's direction and, perhaps, by her design (Caviness, "Designer"). Even her natural science text, *Physica*, has finally made it back into the canon of her accepted work. Only the question of her authorship of the medical text, *Causes and Cures*, still raises eyebrows. And, as we shall see in the next chapter, it should.

Because the Hildegard who wrote *Causes and Cures*—down to earth, practical, and medically experienced—cannot possibly be the Hildegard of the *Life*, untrained, unexposed, invalid. Accepting *Causes and Cures* as Hildegard's own work will require that we finally do change our idea of Hildegard, from a miracle into a wonder.

Chapter Two

Gardener of the Body

In the autumn of 1859 a surprising manuscript surfaced in the Royal Library of Copenhagen—Ny kgl. saml. 90b. It was a pleasant manuscript of 182 folios, neatly written on clean parchment, and probably mid-thirteenth century, with generous rubrication and initials characteristic of Benedictine scriptoria. Carl Jessen, the noted historian and medical botanist, had discovered it by accident, and published the first notice of it three years later, in 1862 (Jessen 1862, 104–105). Jessen sadly admitted to having misplaced a page of his notes (104), but was still able to describe the manuscript.

It had few abbreviations and was divided into five books.[1] Book One was a review of natural science and cosmology, organized, like the well-known encyclopedias of Honorius Augustodunensis (*De imagine mundi*) and William of Conches (*Philosophia*), according to the four elements—fire, air, water, earth. So the text began with fiery things—with material on the sun, moon, stars, constellations, and zodiac; then on airy things (mainly the winds); then on waters (salty and sweet, marshy and stagnant, water in pools, rivers, and streams), to conclude with the things of earth (plants, animals, and types of soil).

Book Two was more disorderly but it was similar to other anonymous medical manuals;[2] it included explanations for the physiology of organ systems (brain, lung, kidney, heart) and the effect of the environment on the body (in the form of the traditional non-naturals—food and drink, rest and exercise, emotions and climate). It also had passages on women's fertility, conception, and labor, and practical instructions for bleeding, cupping, scarification, and moxibustion.

Books Three and Four, like most infirmary texts, gave recipes for common symptoms and diseases from head to toe, although neither showed the kind of actual use that medical manuscripts often did show—marginal recipes, notes, and notations. Book Five gave techniques for prognosis using the

appearance of the patient (eyes, face, and voice), along with the classic techniques of pulse-taking, and examinations of the urine, blood, and stool. This
material was, at least on the face of it, quite traditional—and could be found
in widely dispersed tracts such as *On Urines*, *On Pulses*, or *On Phlebotomy*.[3] It
ended with a lunary that predicted personality based on the position of the
moon at conception.[4]

What was surprising about the manuscript was not so much its contents. Jessen was familiar with such "medical miscellanies," which often
turned out to be well-known texts of Galen or Hippocrates bound into a
single manuscript, or sometimes anonymous collections of recipes that were
preceded and followed by passages from Isidore or Bede on cosmology, the
body, and prognostics. What was surprising about the text was its title—*Hildegardis bingensis causae et curae*—Hildegard of Bingen's *Causes and Cures*.[5]

Now it was known that Hildegard had written "something on the
body and how to treat it," because she had mentioned such texts in the 1158
Introduction to her *Liber vitae meritorum*. One natural science text attributed to her, the *Physica*, had survived in several manuscripts and been published in 1533. It had been republished in the *Patrologia Latina*'s volume 197
of Hildegard's collected work in 1855, although it had never received much
scholarly attention.

Ny kgl. saml. 90b, although it did have many similarities to *Physica*,
was definitely not the same text. *Physica* could easily be understood within
the often merely literary encyclopedic tradition of the herbal, the bestiary,
and the lapidary. The new text, *Causes and Cures*, with its specific instructions on formulating recipes, its idiosyncratic and yet traditional techniques
for examining the patient, checking the urine, and bleeding—was clearly a
practical text.[6] If Hildegard had indeed written the text of Ny kgl. saml. 90b
as its title suggested, then she must have practiced medicine.[7]

Jessen himself was circumspect. He simply noted that it appeared to
be the text that Matthew of Westminster in 1298 had called the *Liber compositae medicinae*. Although "one cannot say for sure that it is Hildegard's,
what I can say is that the writing and the manuscript match what we know
of Hildegard. Indeed, her writing is so characteristic that it is impossible to
mistake or imitate. The energy of her visions, her oddities of grammar, the
independence of her thought make deception hardly practical" (105–106).

No matter what your point of view about Hildegard, such a text would
need to be explained. In fact, it led to more than a century of controversy—
about whether Hildegard did write *Causes and Cures* (which never surfaced
again in any other manuscript), and, if she did, how it should be interpreted.
As part of her mystical vision, that is, as theology? As an inspired medicine

useful even for the present? Or simply as a practical text of medieval medicine? The controversy lasted for decades and is only now beginning to be settled.

In a 1917 article on Hildegard's scientific writings, Charles Singer, an admirer of Hildegard's, as well as a careful and intuitive reader of her work, did not consider *Causes and Cures* to be authentic. Rather, it was

> . . . an ill-written document of the thirteenth century . . . [with] none of the characteristics of the acknowledged work of Hildegard. . . . Nothing could be more unlike the ecstatic but well-ordered and systematic work of the prophetess of Bingen than the prosy disorder of the *Causes and Cures*. Linguistically, also, it differs entirely from the typical writings of Hildegard, for it is full of Germanisms, which never interrupt the eloquence of her authentic works. Again, Hildegard's tendency to theoretical speculation, as for instance on the nature of the elements . . . finds no place in the scrappy paragraphs of this apocryphal compilation (Singer 1917, 12–13).[8]

Hermann Fischer, the German botanist and general booster of Hildegard, in his 1927 biography agreed with Singer, although he did accept Hildegard's authorship of *Physica* (which, therefore, merited her a place as *Die erste deutsche Naturforscherin und Ärztin*). To explain *Causes and Cures'* style, which was clearly Hildegardian, and its dating, which had to be earlier than 1233, he suggested that someone other than Hildegard, an admirer, perhaps, had composed a work "faithful to her style" (40), sometime between 1180 and 1233.

But when Marianna Schrader and Adelgundis Führkötter examined the entire corpus of Hildegard's work in 1956, they came to still a different conclusion. Both *Physica* and *Causes and Cures* were authentic, they thought, but no longer in their original form. They suggested that there had originally been a single work by Hildegard that they named the *Liber subtilitatum*. Unlike her visionary works, they wrote, the *Liber subtilitatum* was the outcome of ongoing observation, not religious revelation; it had never been in a finished form but was worked on and added to throughout Hildegard's life (4–6 and 54–59). Only after her death was the text divided into *Physica* and *Causes and Cures*.

Perhaps because of these ongoing questions as to its authenticity, *Causes and Cures* received little attention until the second half of the twentieth century, when interest, but of a peculiar sort, took off.[9] Certainly the *sine qua non* was its translation into German with a gloss by Heinrich Schipperges in 1957. Schipperges, an historian of medieval medicine, did believe that the text (excluding its last section) was authentic (Schipperges 1992[1957], 45), and he placed Hildegard into monastic medicine, which he read as being practical and utilitarian. Although he incorporated this folk and practice model into his interpretation, still he emphasized her spiritual, not physical, approach to the body and highlighted her Benedictine debt.[10] Peter Dronke, who brought the attention of the English-speaking scholarly world to Hildegard's medical writings, followed Schipperges in emphasizing the theological rather than the medical importance of her medicine (1984, 144–201).

At the same time, a second, altogether different approach appeared after World War Two, pioneered by Gottfried Hertzka MD, who based a successful alternative medical practice on *Causes and Cures* and *Physica*.[11] From his work a novel and complete alternative practice has sprung, of passionate Hildegardian practitioners supplied with cookbooks, gardening manuals, and a line of Hildegardian medicines. They believe that "Hildegard produced all of her works, as she has said, through her heavenly or spiritual vision. She did not rely on medical experience or upon traditional learning. . . ." (Strehlow 1988, x). In this interpretation, Hildegard's medicine should be understood as far in advance of her time, "based upon the principle that all internal processes in people can be traced back to their biochemical origin" (Strehlow 1988, xxvii). Perhaps 90 percent of the published literature on Hildegard's medicine relies on and comes out of this movement.[12]

Feminism provided still a third approach, emphasizing the "particular attention paid to women" by Hildegard (Flanagan 1987, 105) and the presumably female differences between her method and traditional ones.[13] Thus in her biography of Hildegard, Barbara Newman focused on gynecology, obstetrics, and Hildegard's attitudes on gender and sexuality (1997[1987], 121–155); so did Joan Cadden (1984) and Marcia Chamberlein (1998).

With the wider, anthropological approach of cutural relativism in the 1990s, along with the greater availability of English and French translations of her texts, scholars next took a new approach to Hildegard's medicine. They tracked down her sources (Glaze 1998; Moulinier 1997, Moulinier 1995, 205–243; Moulinier 2003, lxiii-ci); her relationship to Arabic medicine and the revolutionary translations coming out of the East (Jacquart 1998, "Hildegarde et la physiologie de son temps," 121–134); her connections to women's medical practice (Green 1999); and her medical (Müller 1979), botanical (Müller 1997, *Krankheitsbilder*) and culinary descendants (Weiss-Amer, 87–96). Slowly Hildegard's medicine has come to seem something that can even be used, now and then, as background for mainstream medieval medical practice (Fery-Hue).

Nevertheless, compared to the energy, research, and creativity that has gone into most facets of Hildegard's work, research into her medicine—into its background, its significance for interpreting the rest of her work, and its utility for the history of medicine—has been meager. Partly this because of the company her medicine now keeps; partly, perhaps, to a reluctance of medieval scholars to approach medical subjects; and partly to its contents, which treat abortion, sterility, nocturnal ejaculation—subjects at odds with the notion of a mystical, saintly Hildegard.[14] But mostly this has been due to the ongoing questions about the authenticity of her medical texts, especially of *Causes and Cures*. What do we know about the connection between Hildegard and the *Hildegardis bingensis causae et curae*? And can we arrive at a solution to the question of whether Hildegard did, indeed, write *Causes and Cures*?

In the prologue to her *Liber vitae meritorum*, finished in 1163, Hildegard noted that she had written something "on the subtleties of the different natures of created beings,"[15] and all scholars agree that this was some kind of medical or natural-scientific text. There are two other contemporary witnesses to Hildegard's medical writing. In a letter Volmar mentions her "demonstration of the natures of various created beings,"[16] and in the *Liber divinorum operum*, Hildegard speaks of herself as an uneducated woman through whom wisdom had revealed "certain natural powers of various things."[17] Many scholars have assumed that these three allusions are to a single text by Hildegard.

In contrast, in the 1181 introduction to Book Two of her *Vita*, the monk and author Theodorich noted what has been interpreted as *two* medical texts, namely, "some thing(s) on the nature of people and the elements and the different creatures, and how people can be helped from them."[18] Because the Latin is ambiguous ("que" can be plural or singular), it is not clear whether Theodorich meant that Hildegard had composed one, two, or even three medical/natural-scientific texts. Laurence Moulinier, for instance, has interpreted the phrase as signifying a very early but post-mortem "separation of theoretical from practical," which "is the germ of the separation into the *Liber simplicis medicinae*, centered on the object, and the *Liber compositae medicinae*, centered on man" (1995, 31).

By 1222 there were certainly *two* texts on medicine attributed to Hildegard because Gebeno, the prior of Eberbach, mentioned two—a "book of simple medicine containing eight books on the creation of things and another book on composed medicine about the causes, signs, and cures of various illnesses."[19] In 1233 the nuns at Rupertsberg swore to the papal legates that "all of the texts possessed by the monastery, including . . . the *Liber simplicis medicinae* and the *Liber compositae medicinae* . . . had been written by Hildegard."[20]

But was this text, called by the nuns the *Liber compositae medicinae*, and by Gebeno the *Liber medicine composite de egritudinum causis, signis atque curis* identical to the text in the manuscript discovered by Jessen in 1859? It seems likely. In 1484 Trithemius visited Rupertsberg and read both the *Liber compositae medicinae* and the (separate) *Liber simplicis medicinae* (which must have been the very manuscripts kept at Rupertsberg since Hildegard's death). While he identified the *Liber simplicis medicinae* only as a "liber de herbis" (throwing some doubt on the authenticity of the nine-book *Physica* we have today), he helpfully identified the *Liber compositae medicinae* by its first line—"Deus ante creationem mundi"—the same first line as the manuscript of *Causes and Cures* that surfaced in Copenhagen in 1859.[21]

Given these facts, it seems most likely that Hildegard wrote two medical texts, a *Liber compositae medicinae*, known today as *Causes and Cures* and a *Liber simplicis medicinae*, known today as *Physica*; and yet two scholars of these texts, Laurence Moulinier and Irmgard Müller, have rejected *Causes and Cures* as not authentic. Why?

Moulinier gives little weight to perhaps the strongest point in favor of the authenticity of the text—that Hildegard's nuns swore that *Causes and Cures* was, in fact, written by Hildegard.[22] Instead she focuses on the fact that contemporary witnesses (Hildegard in 1158 and 1172 and Volmar in 1170) appear to refer to one, not two, texts, depending on the interpretation. Elaborating on Schrader and Führkötter's hypothesis of a single initial text by Hildegard (Schrader 1955, 55–56), she contends that the division into two books occurred after Hildegard's death in 1179 but before the Inquisitors' visit in 1233. Both *Causes and Cures* and *Physica*, she argues, whose titles and contents only partly correspond to something Hildegard would have recognized for herself, include interpolations and additions, but it is *Physica*, not *Causes and Cures*, that is the proper descendant of the authentic single text (1995, 23). The numerous parallels between the recipes in Books Three and Four of *Causes and Cures* and sections of the *Physica*, which have long been assumed to mean that *Physica* borrowed from *Causes and Cures*, are more likely to reflect the opposite (1995, 69).

Why so? Because *Causes and Cures* is not the kind of text Moulinier expects from Hildegard. It is disorganized—it does not follow a head-to-toe formula (102); it has several duplicate recipes (102); and the practical Books Three and Four just pop up between the theory of Books One, Two, and Five (103–104 and 2001, 135). Indeed, Books Three and Four seem to be cut off from those that precede them in many ways—by the number of German words they employ, but especially by the eruption of practice into a text which is principally theory.

Instead, Moulinier follows Hermann Fischer and proposes that an admirer wrote *Causes and Cures* sometime between 1180–1220, perhaps related to her canonization (1995, 44). She reiterates these arguments in her (2001) article and in her (2003) introduction (lxii) to *Causes and Cures*. Its first two Books, she hypothesizes, are definitely Hildegardian, and probably continue a text left by her. Books Three and Four were put together from the *Physica* by taking pieces of chapters and reorganizing and recompiling them. Books Five and Six, which show an "oriental astrological determinism," are she believes, "aliens" (2001, 143).

This hypothesis, by a well-known scholar who has edited *Causes and Cures*, now permeates discussions of it and deserves to be examined critically.[23] Some of Moulinier's arguments seem to be more a matter of taste and judgment, or, perhaps, expectation. For instance, while it is true that Books One and Two are mainly theory, and Books Three and Four are practical recipes, it is not exactly true that they "erupt" into the "theory" of Books One, Two, and Five. It was common for "medical miscellanies" to be constructed out of an introduction of theory followed by practical recipes. Moreover, within Book Two is a lengthy, practical passage on bleeding, and another on cupping and scarification. As for its disorderliness, Eliza Glaze, by contrast, found that "the first and most noticeable characteristic of Hildegard's medical writing is the clear sense of organization she imposes on the material" (1998, 131). As for the repetition of recipes in Books Three and Four, Dronke noted that repetition was a characteristic feature of Hildegard's work.[24]

The force of Moulinier's conclusion mainly comes from her assertion that Books Three and Four were cobbled together by someone out of the authentic *Physica*. She bases this argument on the conclusions of Irmgard Müller, who compared Books Three and Four of Ny kgl. saml. 90b (mid-thirteenth-century) to the (fourteenth-century) manuscript of *Physica* (Müller 2000 and Müller 1995).[25] Müller found that there were more than 100 recipes in *Causes and Cures* that overlapped with passages from *Physica*. In particular, single passages in *Causes and Cures* corresponded to *several* passages found in *different* chapters of *Physica*. Therefore, Müller concluded, Books Three and Four must have been a montage of *Physica*, and *Causes and Cures* was not an original work of Hildegard's but a later compilation.[26]

Certainly the strongest evidence against *Causes and Cures*' authenticity is Müller's assertion that Books Three and Four derived from *Physica*. Suppose we look at three of the examples she gives, chosen randomly (eyes closed, pen stabbing).

In the first example, Müller asserts that a passage from *Physica* (in the *Patrologia Latina*) is also found in *Causes and Cures*.[27] (The differences between the two passage are in bold):

1a. *Causes and Cures* (210:9–12)

If someone becomes demented from thinking too much, then let him take **hun** and three times as much fennel and cook them together in water, and then let him drink the water cold.

Si multis et diuersis cogitationibus scientia et sensus cuiuspiam hominis euacuantur, ita quod ille in amentiam uertitur, **hun** accipiat et ter tantum feniculi et in aqua simul coquat et abiectis herbulis istis eandem aquam infrigidatam frequenter bibat.

1b. *Physica*, col. 1202D

But if someone becomes demented from thinking too much, let him take **balsam**, and three times as much fennel, and cook them together in water, and let him drink the water cold.

Si **autem** multis et diversis cogitationibus scientia et sensus cujuspiam evacuantur, ita quod ille in amentiam vertitur, **balsamitam** accipiat, et ter tantum feniculi, et in aqua simul coquat, et abjectis herbulis, eamdem aquam infrigidatam frequenter bibat.

In this first example, although the recipes are close, they are *not* identical, and surely this vitiates against the idea that an anonymous compiler simply *copied* texts from the *Physica*. (Moulinier has replaced the elegant Latin ending *ae* in Kaiser's edition with the *e* of the manuscript, and I have used her edition here, although Müller used Kaiser's. These changes have to do with how a modern editor expands medieval abbreviations, and are not relevant to the argument.)

Even more suggestively, the herb referred to in *Causes and Cures* is the German, *hun*; in *Physica*, it is referred to in Latin, *balsamita*. This would mean that the compiler had replaced the Latin *balsamita* with the German *hun*. But why would a compiler, putting together a text in support of Hildegard's canonization, replace the elegant and proper Latin *balsamita* with the German *hun*? The inverse seems more likely—a compiler would have replaced the German with the more elegant Latin. This example, then, does not support the contention that *Causes and Cures* was compiled from *Physica*.

For a second example, Müller claimed that 1211C of the *Physica* is also found in *Causes and Cures*.[28]

2a. *Causes and Cures* (232:22–25)

Thus if a woman suffers from heavy periods that come at the wrong time, then let her take a piece of linen dipped in **cold water** and often tie it about her thighs, so that she becomes cold inside; for the coldness of the linen and of **the cold water** will restrain the flow.

Igitur mulier, que iniusto tempore multa menstrua **inordinate patitur** lineum pannum in **frigida aqua** intingat et femoribus suis sepe circumponat, ut interius infrigescat, quatinus per frigiditatem linei panni et frigide aque iniustus fluxus sanguinis retineatur.

2b. *Physica*, col. 1211C

But if a woman suffers from heavy periods that come at the wrong time, then let her take a piece of linen dipped in **water** and often tie it about her thighs, so that she becomes cold inside; for the coldness of the linen and of **the cold water** will restrain the flow.

Sed et mulier, quae injusto tempore multa menstrua **patitur inordinate**, lineum pannum in **aqua** intingat, et femoribus suis sepe circumponat, ut interius infrigescat, quatenus per frigiditatem linei panni et frigidae aquae injustus fluxus sanguinis retineatur.

Again although the passages are almost identical, they do differ, and in telling ways. First, the passage in *Physica* does not come at the beginning but right in the middle of a paragraph. Second, in *Causes and Cures* the recipe makes the point that the water must be *cold*—it is this *physical* coldness that will stops the menses. By contrast, in the version in *Physica* this kind of coldness does not occur; rather it is the more sophisticated, abstract, qualitative "coldness" of water that is therapeutic.

More importantly, to accept that the version in *Causes and Cures* came from *Physica* would mean that the presumed compiler first lifted an extract from the middle of a section and then also changed it, *adding* the therapeutic phrase, "cold." It seems more likely that it was the *Physica* version, where "cold" was dropped, that is derivative.

For a last example, Müller claimed that *Causes and Cures* 264:11–20 was taken from 1266 A of *Physica*. (This overlap was not mentioned by Kaiser):

3a. *Causes and Cures* (264:11–20)

A person whom a worm gnaws in some part **of his body** should take clay and twice as much *cridum* and with **vinegared wine** make a kind of glue, and with a feather put it on the place, and he should do this every day for five days. After let him take aloe and a third of myrrh and grind them; then let him take some new wax and prepare a plaster, cover it with a hemp rag, and tie it on for twelve days. For clay is hot and *crida* cold, and the heat of the clay with the coldness of the *crida*, and with the heat and sharpness of the **tempered wine** will kill the worms; but the heat of the aloe augmented by the heat of the myrrh will extract the rottenness of the sore, and heal the spot.

Homo, quem uermis in aliquo loco corporis sui **comedit**, cretam accipiat et bis tantum de **cridum** et ex hiis cum **acetoso uino** uelut tenue cementum faciat atque istud loco, ubi uermem patitur, cum penna **inmittat** et hoc per singulos dies usque ad quintam diem faciat. Postea sumat aloe et secundum eius tertiam partem myrram et simul terat, atque ex his cum recente cera **emplaustrum** paret et **canibeo** panno superponat et ita super locum doloris per duodecim dies liget. Nam creta calida est et crida frigida, et sic calor crete cum frigiditate cride et cum calore et acumine uini temperatus uermes **mortificat**; sed calor aloe **cum calore myrre** augmentatus putredines eorundem ulcerum extrahit et eumdem locum sanat.

3b. *Physica*, col. 1266A

But a person whom a worm may eat in some place, should take clay and twice as much *criden*, and from these **with vinegar or with wine** make a kind of weak glue, and put it on the place with a feather, and this should be done every day for five days; after let him take aloe and a third of myrrh, mix them together, with new wax and make **eyn plaster**, and cover it with a hemp rag, and tie it on for twelve days. For the clay is warm and the **crida** is cold, and the heat of the clay, with the coldness of the **crida**, and the heat and sharpness of the tempered wine will kill the worms. But the heat of the aloe will increase that of the myrrh, and will extract the rottenness of the sore, and heal the spot.

Homo **autem**, quem vermis in aliquo **loco suo comederit**, cretam accipiat et bis tantum de **criden**, et ex his cum **aceto seu vino** velut tenue caementum faciat, atque istud loco, ubi vermem patitur, cum penna **emittat**, et hoc per singulos dies usque ad quintam diem faciat; postea sumat aloe, et secundum ejus tertiam partem myrram, et simul terat, atque ex his cum recenti cera **eyn plaster** paret, et **canabineo** panno superponat, et ita super locum doloris per duodecim dies liget. Nam creta calida est et crida frigida, et sic calor cretae cum frigiditate cridae et cum calore et acumine vini temperatus vermes **commortificat**. Sed calor aloe **myrrhae** augmentat, putredines eorumdem ulcerum extrahit, et eumdem locum sanat.

Again the passages do have significant differences; they were not simply copied. For instance, in the first, the Latin *emplaustrum* is to be made; in the second, the German, *eyn plaster*. Most tellingly, the recipe from *Causes and Cures* prescribes a *vinegared wine* but the recipe from *Physica* prescribes vinegar *or* wine, yet when the explanation of how the medicine works is given, in both passages it has been assumed that the ingredient of the *Causes and Cures'* version (vinegared wine) not that of *Physica* (vinegar *or* wine) was used. This must mean that the original recipe was for vinegared wine; and, therefore, the passage that used vinegared wine, from *Causes and Cures*, must have preceded the one in *Physica*.

Thus, while these selected examples are more or less "the same recipes," that is, from the same person or tradition, they are not identical—and their differences point from *Causes and Cures* to *Physica*.

It seems, then, that although the facts of text and manuscript are not in doubt, their interpretation certainly is. Clearly one of the biggest problems in reaching a definitive conclusion has been the absence of manuscripts. Can any helpful information be gleaned *simply* from the physical manuscript, Ny kgl. saml. 90b, alone?

Unlike many medical miscellanies, the manuscript of Ny kgl. saml. 90b is not, itself, the product of an ongoing compendium of different texts added to in different hands at different times. Nor, given its lack of signs of use such as glosses and marginalia, was the manuscript ever used by a medical practitioner. It was copied by two or three different scribes in a Benedictine monastery from a single prior manuscript in the mid-thirteenth century;[29]

therefore, at least one other copy of the text must have existed by 1250.[30] Consequently, if the text as we have it was *not* written by Hildegard, the original text was indeed edited and excerpted very early on, probably in the first or second quarter of the thirteenth century. How could this be?

As we have seen, Moulinier proposed that *Causes and Cures* be attributed to one or more "continuers," "connoisseurs of Hildegard's work and desirous of increasing it" (2001, "Hildegarde ou Pseudo-Hildegarde," 144)— Theodorich, Gebeno, or Guibert—perhaps in view of her canonization. But to accept this, we have to accept a scenario that just does not seem plausible.

First, since no one has yet found the exact passages in Hildegard's extensive writing that correspond to the "theological" parts of *Causes and Cures*, the "continuer" must either have had access to otherwise unknown "notes" of Hildegard, or was able to invent a credible imitation of her style. Second, he must have been able to add surprising pieces to her theology, including practical instructions on bleeding, cupping, and the treatment of farm animals, as well as vivid passages on male and female sexuality whose linguistic and imagic flavor are perfectly Hildegardian. Third, he must have extracted hundreds of passages scattered throughout the manuscript of the *Physica*, rearranged and consolidated them, literally creating the series of treatments that make up Books Three and Four. Finally, unbeknownst to the nuns (who believed, as we have seen, that *Causes and Cures* had been written by Hildegard) the unknown compiler placed his manuscript in the library at Rupertsberg, where it was thenceforth taken as an authentic work of Hildegard.[31]

But the proposal that *Causes and Cures* came from *Physica* implies even more—the hypothesized compiler not only changed words and word order, occasionally substituting a German name for a Latin name or a Latin one for a German one. He (or she) actually *created* single recipes for *Causes and Cures* from chapters scattered over the whole nine-book text of *Physica*.[32] Doing so would not have been an easy task: *Physica* is large and clumsy, and it would have required an intimate knowledge of its contents. Is it reasonable to suppose that a literate compiler with no medical training would have done this?

What would have been the motive? Moulinier has suggested that it might have been done in order to promote Hildegard's canonization (Moulinier 1995, 44). But would such a medical text have promoted Hildegard as a saint? To the contrary, it could have presented serious impediments. In the first place, by the thirteenth century, monastics were not supposed to practice medicine, and *Causes and Cures* is an eminently practical text. In the second place, some of its contents, having to do with sexuality, pregnancy, fertility, and even, in some readings, abortion, would certainly have been problematic for a nun. There would seem to have been every reason to dispose of *Causes and Cures*, not to create it.

An alternative scenario has been put forward that accounts for all of the facts: that Hildegard only mentions one text; that *Causes and Cures* is not tightly ordered; that all of it is linguistically Hildegardian and that her nuns swore to the Pope's emissaries that she had written it. In this hypothesis, Hildegard had some kind of "ongoing" text about natural science and medicine that she worked on and added to over many years. Then, as Eliza Glaze noted, Hildegard herself would be Laurence Moulinier's "interpretative compiler," assembling a lengthy text from numerous sources; and the *Causes and Cures* we have today would be a "notebook," fundamentally in the state and form it enjoyed at the time of Hildegard's death.[33]

What is my own conclusion? Medieval authorship is notoriously difficult to pin down, since manuscripts resemble today's student notebooks more than they do the fixed texts of printed books, and this may be a question that cannot be answered.[34] For the purpose of this inquiry—to use as a practical manual from a particular time and place—whether Hildegard composed *Causes and Cures* precisely as we have it today is not critical. However, after working with the text for many years, I believe that Hildegard did write the text pretty much as we have it.

What seems most likely is that *Physica* and *Causes and Cures* had always been completely separate treatises, since they come out of very different textual traditions. The former clearly derives from the encyclopedic tradition of the herbal/lapidary and bestiary, and the latter comes out of the practical medical/antidotary tradition. But *Causes and Cures* was not written at once as a whole.

Instead Hildegard probably wrote Books Three and Four first, for the new infirmarian at library-less Rupertsberg, c. 1145–1155. She may even have begun it much earlier on, in the late 1130s, when she became *magistra* of the women and had to give up being an infirmarian. (Linguistic arguments for this hypothesis will be found in Chapter Five.) Books One, Two, and Five were written later on, to provide the medical—that is, theoretical (Book One), diagnostic (Book Two), and prognostic (Book Five)—background necessary for the nun-infirmarian at Rupertsberg to practice medicine thoughtfully.

It is even possible to imagine that the nun-infirmarian, like Volmar, kept a clean copy of Hildegard's notes and, perhaps, even of lectures or casual discussions, as well as of suggested drugs and practical techniques, and that *this* is the *Causes and Cures* we have today. Such a hypothesis would resolve some of the issues with the messy state of the text and still preserve the unique window onto medical practice and theory that *Causes and Cures* provides.

In any case, it is clear that the implication for Hildegard studies is that practical medicine needs to be taken into account when interpreting Hildegard's work. It may even explain some of the most puzzling of her images and metaphors.

For instance, Peter Dronke analyzed her odd use of images and gave as an example the following (in his translation):

> Today the closed *gate*
> has opened for us
> what the serpent *throttled*
> in a woman
> through this, there glistens in the dawn
> the *flower* from the maiden Mary (312).[35]

Barbara Newman also used this same text in a similar way, to illustrate the difficulty of understanding Hildegard's metaphors. "The closed gate of the temple signifies Mary's virginity . . . but . . . although the general drift is clear the exact sense [of suffocation] defies explanation" (1988, 62–63); she translates *suffocatio*, therefore, as "slammed." Dronke argues, instead, that the closed gate, the serpent, the throttling, dawn, and flower are all examples of Hildegard's unique style of mixed metaphor. Because he could find no clear, single image to unite this imagic profusion, he suggested that Hildegard might have been a synesthete, who experienced sounds for smells, colors for sensations. This is what can explain, for instance, the images in this hymn.

And yet, there *is* a unifying and comprehensible image behind this poem, but only if the text is placed in a medical context. An alternative translation would be:

> Today the closed door has opened
> What the serpent suffocated in a woman
> Whence at dawn
> The Virgin's flower appears.

Behind the poem is a *medical* syndrome—suffocation of the womb (*suffocatio matricis* or *uteri*), clued by the use of the word "*suffocatio*"—suffocated, not throttled. Manifested by episodic fainting, the cause of uterine suffocation was a blockage of the uterus, usually diagnosed by the absence of menses, often called a woman's flowers.[36]

Thus there are, perhaps, not many metaphors in this Christmas hymn (the day that a Virgin gave birth), but only one—a medical one. The *clausa porta* is the shut-up womb of the world[37] that was closed to fruitfulness by the "serpent"—the serpent in the Garden of Eden, of course, but also the serpent of sex which entered the womb of the world with the Fall. The serpent, then, is not "throttling" or "slamming the door," but causing the "suffocation" of the world, that is, causing its sterility, its infertility. It is on Christmas Day that the flower of the Virgin appears; that is, her womb is opened, and her flowers (menses) appear. (Of course, the flower is also the dawn star of Venus and the birth of Jesus.)

In other words, placed in a medical context, the text does not illustrate Hildegard's luxuriant use of mixed metaphors, but, rather, how thoroughly her medical practice had infused her entire worldview. Indeed, this very image—the world as womb or vagina—is explicit in Vision 3, Part I of *Scivias* (frontispiece). That is, not only is the illumination an image of the cosmos (and, as we shall see, a radial diagram of Book One) but it is also an anatomically correct, surprisingly detailed picture of a vulva.[38]

There are also implications for the history of medicine in Hildegard's authorship of *Causes and Cures*, since it means that we now have a practical medical text that can be attached to a particular person, place, and time. As noted in the Introduction, texts do exist that can be attached to particular people, but they are later texts, written after the great professionalization of medicine in the thirteenth century, and by university professors. It is not at all certain whether what is written in those texts reflected what was actually done, even by the professors writing the texts, as Cristina Alvarez-Millan has demonstrated in "Practice Versus Theory."[39]

At the same time, early texts that *do* reflect practice are all anonymous, despite their attribution to ancient authors.[40] Only Trota, and, perhaps, Bartholomew of Salerno, were peri-professionalization and also described a "method" of what seems to have been practical, performed medicine. But essentially nothing is known about either author. For instance, only an educated guess identifies the writer of the *Practica secundum trotam* with a twelfth-century woman named Trota in Salerno.[41] The same can be said for the putative author of Bartholomew's *Practica*, a twelfth-century text ascribed to a Bartholomew of Salerno, who may or may not have been the teacher of Maurus, and who may or may not have written a letter on health practices to King Louis.[42]

By contrast, much is known about Hildegard, as we have seen. Her authorship of *Causes and Cures* (within the limits of medieval authorship discussed above) means that we now have a text that can be used to illuminate the practical application of the ancient medical tradition within a particular context. It also means the obverse—the current understanding of medieval medicine can be used to illuminate *its* context.

HILDEGARD AS PRACTITIONER

Her Places of Practice: The Infirmary, the Hospice, and the Garden

Since the women's part of Disibodenberg was a small Benedictine monastery, it seems most likely that Hildegard learned her practical medicine as the nun-infirmarian for the women at Disibodenberg. Therefore, although information about medical practice in the twelfth-century Rhineland is scant, some educated guesses about her training and practice can be made. Let us begin with her likely places of practice—the infirmary and the hospice.

Indeed, the *only* fact known about medical practice on Disibodenberg is that there *was* an infirmary and a hospice, because their chapels were dedicated on May 27 and May 28, 1142.[43] The hospice (*hospitium*), a monastic structure derived from the Byzantine *xenodochia*, was where sick pilgrims who came to Disibodenberg to see Jutta would have stayed. Its function was to care for travelers and the poor, old, and disabled; and it usually provided medical and nursing care along with food, clothing, and shelter (Risse, 117–166).

As for the infirmary, almost all monastic Rules established some kind of place for the ill apart from the rest of the monastery. The most famous Rule was the sixth-century Benedictine Rule (which applied to Disibodenberg and Rupertsberg); it legislated that a place in the monastery be assigned to the sick far from the noise of the active monastic life. Other Rules also legislated such a special place for the sick. For instance, the Rule of St. Isidore required that a place for the sick be set up, far from the church and the cells of the brother so that no noise or clamor might disturb the ill;[44] the Rule of the Fructuosis, that the ill of whatever disease sleep together in one room.[45] The *Regula solitariorum Grimlaici* specified a room with a sign so that the brothers would know to visit.[46] In a Rule specifically for nuns, which might, perhaps, be applied most reasonably to the small women's monastery on Disibodenberg, the infirmary was prescribed as a separate room of the convent with everything necessary.[47]

Archaeological investigations of monastic sites confirm that there was often more than one space set aside to care for the ill—a hospice for travelers and the poor, an infirmary for the monastics, and, sometimes, a Great House for noble guests. The most pertinent to Disibodenberg are the records and archaeological findings of the great Benedictine house of Cluny, which demonstrate the gradual evolution and expansion of its infirmary.

Thus, at the first monastery, Cluny I (910–927), there was only a small infirmary. After Cluny II (950–981) was built, a hospice was eventually added (1015) next to the entrance gate, which was convenient for visitors.

By 1040 Cluny II had a larger infirmary to the east of the cloister. It had four sleeping rooms of two beds each, with an entrance on the west and a lavatory behind, and was situated beside the infirmary garden. In 1082 a still-larger infirmary was constructed on the north wall of the chapel, with twenty-four beds. By 1132 (at the height of Cluny's prominence), even that was too small, and the Great Infirmary was built, southeast of the old one. It was a large rectangular space with two separate fireplaces along each of the long walls, a canal running through the south end for lavatories, and a capacity of 80–100 beds (Jetter, 313–338).

This kind of setup was usual, since the infirmary was modeled on the church; it was typically a long hall where the patients' beds were placed perpendicularly under the windows. An open central nave ran down the middle, and often there would also be a chapel for the sick, a lavatory, and a cloister (Chazin). Indeed, this was a style that influenced hospital architecture until well into the 1920s.

More detail can be gleaned from the Plan of St. Gall, a ninth-century blueprint for the reconstruction of the then-current monastery (Horn). On the northeast side of the church, sheltered from the wind and isolated from the activity of the monastery, was to be built what amounted to a medical complex. The infirmary proper was to be situated between the house of the abbot (who was expected to take an interest in medicine) and the house for the oblates. In the Plan, the infirmary was set up like a miniature monastery, with large sleeping rooms for the sick surrounding a cloister; each room had its own lavatory. There was a special eating area for patients (the refectory) and a separate chapel.

Next to the infirmary complex was to be a four-roomed house for the doctors, the *domus medicorum*. The head physician was to have his own room, with fireplace and bathroom. From it a door would lead into a central room for the other doctors (*medici*); the area also gave onto a storeroom for medicines (*arma pigmentorum*), and a fourth room for the critically ill (*valde infirmorum*) with fireplace and bathroom. Behind the doctors' house were to be two specialized buildings for the practice of humoral medicine. The first, with eight beds and several fireplaces, was to be for purging and the "taking of medicine." The second building had one room for the physical application of medicines in the form of steam, sauna, and fumigation (the breathing of medicinal fumes), and another for bleeding and cautery. Finally, in front of the doctors' house was to be a large medicinal herb garden, for growing the opium poppy and the apothecary's rose, and other medicinal plants that were neatly labeled in the Plan. They included lily, sage, rue, rosemary, fennel, watercress, cumin, costmary, fenugreek, mint, and peppermint, among others. (There were also several other gardens at St. Gall, with many more medicinally active herbs.)

At Disibodenberg, archaeologists have located the infirmary away from the main monastic complex, to the west overlooking the mountains, with a view of the setting sun and protected by a wall from the north wind. The hospice was part of the same structure. Although nothing has been unearthed to help define its space, it probably had three or four beds on each of its long walls, with a fireplace at each end and a lavatory in a separate room in the back.[48] Since it seems unlikely that male and female patients would have been placed together, there may have been a separate infirmary and hospice for the women.[49]

Although nothing precise is known for sure about the structure of Disibodenberg's infirmary, it did have its own chapel and probably its own kitchen. (Hildegard does specify special food and drink for the ill.) There may also have been separate spaces for the bleeding, purging, cupping, steaming, bathing, and inhalations that infirmarians (Hildegard included, as we shall see) recommended; at least, they had been planned for St. Gall.[50] There would also have been a pharmacy (*arma pigmentorum*) for storing medicines, a place for their preparation, and, of course, a medicinal herb garden, with beehives for the fertilization of plants and the cultivation of medicinal honeys. What kind of medicinal plants would have been grown in the infirmary garden?[51]

To answer that we can turn to a remarkable resource—the 1100-word glossary of Hildegard's *Lingua ignota*.[52] Although its title is the *Unknown Language*, it apparently consisted mainly of nouns and a few adjectives. Its purpose is unclear.[53] Hildegard only rarely used words from it in her hymns. Scholars have suggested that it may have been used within the monastery, perhaps to create a secret or special atmosphere, like the robes she designed especially for her nuns. In any case, the *Lingua ignota* provides a window onto the physical objects in Hildegard's world, particularly onto her medicinal garden, since 181 of its 1100 words are names for medicinal plants. Here we have, presumably, Hildegard's own pharmacopoeia—the plants she used in the monastery, collected in the wild, and grew in the medicinal infirmary garden.

In the *Lingua ignota* Hildegard has words for ginger (barschin #812), licorice (diziama, #805), and springroot (gramko, #812); also for millefoil, centaury, plantain, papaver, marrubium, absinthe, rose, lily, agrimony, lavender, rue, tanacetum, garlic, parsley, viola, gladiola, sage, hyssop, wormwood, and dill. Indeed, there are all of the medicinal plants labeled in the blueprint for the gardens of St. Gall, and more. There are also names for the many foreign medicinal spices of *Causes and Cures*, including pepper, cinnamon, sugar, cardamom, galangal, gariofel, and zedoary. But mainly there are more than a hundred names for the native wild plants that Hildegard prescribes in *Causes and Cures*, and that can still be found behind the monastery at Eibingen, including, among many others, nettle, thistle, wildmint, and feverfew. Not only does this list give us a very good idea of her actual pharmacopoeia, but it adds confirmatory evidence to the idea that Hildegard was, in fact, the infirmarian for the women at Disibodenberg, with a strong interest in, and knowledge of, medicinal herbs.

Hildegard's Patients

Next, what do we know about Hildegard's patients and the kinds of diseases she treated? Records about patients are extremely rare during this period. Now and then there is the chronological account of an injury and its treatment (Risse, 88–106) and a few letters from physicians which have been analyzed as "medical consults,"[54] but infirmarians did not keep records of their patients, and Hildegard kept none. Medical practice by monastics was frowned upon, and no medical advice remains in her highly edited letters.

One approach to establishing disease prevalence in the period has been to use hagiographic accounts.[55] Diseases *are* mentioned in the *Vita*, the *Canonization* (also known as the *Acta Inquisitionis*), and the *Vita Juttae*, although such evidence must be approached gingerly, since translating medieval medical terms into modern medical terminology is not straightforward.[56] Moreover, the purpose of the *Vita* and the *Canonization* was to prove Hildegard's sanctity, and the miraculous nature of her cures was, therefore, highlighted.

Thus, in these hagiographical stories Hildegard is not shown using any of the treatments she prescribes in *Causes and Cures*; indeed, the only physical items she uses are those rendered powerful by her touch—water, hair, tears, and clothing. The stories also emphasize her cure of very particular diseases—epilepsy, leprosy, and insanity—partly because these were just the type of diseases not amenable to natural healing, but also because they were saintly specialties. In fact, when Hildegard imagined a clientele for St. Disibod she included a mute, someone with edema, and a leper (PL197, cols. 1105–1106). Given these caveats, what can we learn about Hildegard's patients from the *Vita* and the *Canonization*?

According to the *Vita*, Hildegard was a great healer: "Practically no one who came to her ill did not recover his health."[57] Some of the examples it gives of her successes include two patients with "tertian fever," a servant from Rupertsberg with a large and painful tumor of the chest, a man with edema, an infant who refused to eat, a person with a throat so swollen he could hardly breathe, a mute, a young woman afflicted with love-sickness, a woman with vaginal bleeding, and patients with prolonged labor, insanity, blindness, and epilepsy (*Vita* [1998], III, 186–200). The *Canonization* mentions her healing patients with fevers, epilepsy, blindness, and lovesickness. It also reports that she healed sick children, acutely ill adults, women in labor, and women with vaginal bleeding, infertility, and amenorrhea (*Vita* [1998], III, 248–264).

If we take these accounts as records of the kinds of diseases Hildegard did heal, or was expected to heal, then we can say that she was presented with what we would think of as routine medical problems, as well as with the most difficult diseases of the period—in modern terms, cancer, epilepsy, mental illness, and prolonged infections. This picture corresponds to what we might expect an infirmarian to see, since he or she was responsible for the care of everyone in the monastery, including monks, nuns, and their servants.[58] As infirmarian for the women's monastery, Hildegard was responsible for at least sixty women[59] and, perhaps, for the numerous ill visitors to Jutta.[60]

Again, the *Lingua ignota* can be mined for independent evidence of the kind of diseases that Hildegard expected to see at Rupertsberg. She made sure to have words for several types of patients, including the blind (*hochziz*), the deaf (*nosinz*), the cross-eyed (*hiszin*), the stutterer (*sciniz*), the wounded (*keliz*), the mute (*scarpinz*), the lame (*kolianz*), the eunuch (*pariziz*), the dropsical (*siliziz*), the paralyzed (*stragulz*), and the leprous (*pasizio*); and there are words for some diseases, including *lepra*, *ruga*, *eczema*, and scabies (Portmann, 3). She also provided words for many parts of the body—not only the visible but also the physiological—including the heart, kidneys, liver, lung, stomach, spleen, intestines, and bladder—and words for urine, blood, sweat, stool, saliva, and phlegm. Remarkably, she included two words for penis (*galich* and *creveniz*), a word for testicles (*virlaiz*), and a word for the female genitalia (*fragizlanz*).

From all of this, it would seem that Hildegard expected the nun-infirmarian at Rupertsberg to care for exactly the kind of patients her hagiographers, as well as the nuns interviewed by the Papal Legation for the *Canonization*, had documented that she, herself, *did* see, along with the usual physical ailments of the body detailed in *Causes and Cures*.

Her Training

Just as Hildegard left no records of her patients, so, too, she left no unambiguous record of her training. Because Disibodenberg followed the Benedictine Rule, it had an infirmarian, and probably two (one for its nuns, and one for its monks), and it seems likely that Hildegard received training in caring for patients and in preparing medicines from the monk-infirmarian. Since Disibodenberg's monks had recently come from the well-equipped and sophisticated monastery of St. Jacob's in Mainz, it is quite possible that its monk-infirmarian had had medical, as well as infirmary, training, although it is not clear how much medical training an infirmarian was expected to have.[61] Early monastic Rules emphasized the moral character of the monk who was to care for the sick, not his medical expertise. Thus he should be "a healthy man of holy conversation,"[62] "fit or predisposed for the job,"[63] and "careful and caring."[64] This was also true for nuns. For example, in the Rule of Donatus Vesontinus, the abbess was responsible for the care of the sick; she should make sure that they had whatever they needed, especially from the kitchen.[65]

How much of a medical practitioner was the infirmarian? Cassiodorus referred to "medical practitioners" (*medici*), not to infirmarians, when he wrote that certain monks should "learn the nature of plants and their mixtures"[66] by reading medical and botanical texts (see below). At St. Gall, too, there were *medici*—medical practitioners—on campus, not infirmarians; but whether these were monks with medical training or were outside practitioners is not clear. The Carthusian infirmarian was expected to administer medicines but not to cauterize or bleed, and so there must have been an outside practitioner to perform these functions.[67] Likewise, the Rule of the Camaldolese legislated that medicine be prescribed only by an outside practitioner "paid by wage, of good name in craft."[68] In fact, the first time any of the Rules name a monastic infirmarian (as such) is not before the early 1100s.[69]

The same chronology was probably true for the nun-infirmarian; only in the 1100s is the existence of a nun-infirmarian explicitly legislated. For instance, Gilbert, who set up a double monastery for both men and women, legislated not only for an infirmarian for the monks (*infirmarius infirmorum*, in Hostenius, 497) but also for "the same amongst the nuns" ("idem inter Moniales et Sorores," in Hostenius, 497). He called her the nun-infirmarian (*monial infirmaria*, in Hostenius, 535) or the *infirmaria*.[70] Although the Rule of the Cluniac Hirsau reform (which Disibodenberg is thought to have followed) did hold that the infirmarian was to be a priest (Baader 1973, 279), Gerhard Baader found that attached female cloisters still had evidence of medical activity (284).

As an *infirmaria*, what kind of training would Hildegard have received? Male infirmarians sometimes had formal medical training, either because they had been physicians before their entry into monastic life,[71] or because they trained in medicine at Salerno or Montpellier (although by 1130, this was frowned upon).[72] Jewish practitioners who converted (e.g., Petrus Alfonsi) had often been trained at home by relatives (Shatzmiller 1994, 22–24). The possibilities for Hildegard, who could have had no formal medical education, were fewer: she must have learned her medicine on Disibodenberg from the monk-infirmarian, other monastics, the lay staff, and in the library.

Indeed, that she had had some kind of hands-on training in phlebotomy, uroscopy, cupping, moxibustion, and the preparation of medicines is apparent in her detailed, idiosyncratic, but practical instructions in *Causes and Cures*. What her textual sources might have been are problematic, however, since she mentions no specific texts, and nothing is known about Disibodenberg's library or its infirmary.

Hildegard's Latin Sources

There were, however, general recommendations for the medical library of a monastery that can give us some insight into the likely Latin sources for Hildegard's medicine. The sixth-century monastic writer, Cassiodorus, thought it should contain "a Dioscorides' Herbal, which explains the wonderful properties of plants and has pictures . . . , Hippocrates and Galen translated into Latin, that is, the *Ad Glauconem*, and . . . a selection of anonymous authors, as well as Caelius Aurelianus, and Hippocrates' *On Plants*, and a number of other texts on medicine which you shall find in the library and which, God willing, I shall leave to the monastery."[73]

Can we assume that such were the texts, or the kind of texts, that Hildegard read? The frequency of surviving medical texts from the period pre-mid-twelfth century supports the argument that the kind of texts Cassiodorus recommended—herbals, recipe books, collections of theoretical texts, and texts on techniques—were, in fact, what monastic libraries did contain.[74] So-called "medical miscellanies" were usually collections of just such texts bound in a single manuscript.[75]

Surviving monastic library catalogues also indicate that most monasteries possessed a wide-ranging, if limited, collection of medical treatises on the order of Cassiodorus' recommendations. For instance, the abbey library at Weihenstephenense, not far from Disibodenberg, contained sixty-six books of which two were titled "Medicinalia," (usually anonymous compendia of several theoretical tracts) (Becker). The Benedictine monastery of Blaubern had 189 books, including Macer's *De floribus*, Isidore's *Etymologies*, and "some small books on medicine" (*medicae libellos*, Becker, 174). Although some monastic catalogues did not list any medical books, this was mainly because books with titles like "libri medicinales," "medicinalia," "antidota," or "botanika," (usually simply collections of recipes arranged in head-to-toe order), were kept in the infirmary (Nebbiai-Dalla).[76] Nothing is known about the books in Disibodenberg's or Rupertsberg's infirmary but *Causes and Cures* certainly resembles a twelfth-century infirmary medical miscellany, and perhaps that is what it was.[77]

In any case, it was usual for monasteries to borrow books from one another (Hildegard herself had lent St. Eucharius of Trier a copy of *Scivias*), so she did have access to books outside of Disibodenberg.[78] The Benedictine monastery of Prüfening, whose patron was Otto of Bamberg, (present at Hildegard and Jutta's entrance to Disibodenberg), produced a still-extant manuscript on cautery and bleeding.[79]

Although no catalogues survive for the monasteries with which Hildegard had most contact (Sponheim, St. Maxim's of Trier, St. Jacob's of Mainz), we do have the catalogue from the library of Bruno, Bishop of Hildesheim. He was a half-brother to the Bernard from whom Hildegard purchased Rupertsberg. Bruno owned twenty-six medical texts, among which were many of the translations and medical texts just coming out of the East. They included: *Quinque libri physici artis* (either Dioscorides or five separate books of anonymous authors bound into one text); the *Pantegni; Alexander Sarrocenus* (a widely disseminated, theoretical compendium of Alexander of Tralles); the *Passionarius* [Galieni] (probably Garipontus); the *Viaticum* of Constantine; two Antidotaries; a *Liber febrium* (probably either that of Isaac Judaeus, or of "Pseudo-Galen"); a *Liber urinarum*; a *Liber graduum* [of Constantine]; a *Liber cirurgie*; a *Liber cerebri* (probably another work of Constantine); an *Herbaria*; a *Liber melancholia* (probably part of the *Liber cerebri*); a *Liber Aureus* (probably of Constantine); a *Liber lepre*; a *Universales diaete* [of Isaac Judaeus]; a *Tegni Galeni*, newly translated; a *Liber stomachi*; a *Liber oculorum*; a *Particulare diaete*; an *Isagoge*; a *Glossae in aphorismus*; and a *Glossae in librum urinarum et in librum pulsum* (Sudhoff 1916).

But did Hildegard, in fact, know such texts? Based on the ingredients in her prescriptions, Hermann Fischer generated a list of her probable textual sources. Moulinier argues that she knew Isidore and Pliny, Ovid, Macer, Strabo, Quintus Serenus, and Palladius (1997, 435).

All we know for certain is that, in addition to its library and infirmary, Disibodenberg had at least two visitors with medical interests, and, possibly, with medical texts. There was Adalbert II, Archbishop of Mainz, who had studied at Hildesheim (which owned the medical library described above), and who had even visited the new medical school at Montpellier, where the revolutionary new translations of Arabic medicine were being copied and distributed. A second visitor to Disibodenberg was Siward, former bishop of Uppsala, who witnessed the translation of St. Disibod's relics in 1138 and probably stayed for two months. Of Siward, one of Hildegard's biographers wondered whether he "might have arranged for her [Hildegard's] scientific studies?" (May, 35–36), Siward did own medical books because he named them in his will: "an herbal and a lapidary in one volume, a *Gemma animae*, and an *Elucidarium*, as well as six other books of medicine."[80]

In sum, though we cannot say which *specific* Latin texts Hildegard had access to, we can say that if Disibodenberg had an "average" library, it would have contained several medical texts, including some kind of encyclopedia (Pliny's *De medicina*, and/or Isidore's *Etymologies* or *De rerum natura*, and, perhaps, one of the newer encyclopedias, like William or Honorius); some summary of ancient medical theory, such as the *Ad Glauconem*; an herbal, a recipe text, and a collection of tracts on techniques such as phlebotomy, uroscopy, cautery, and pulse-taking.

Hildegard's Folk Sources

In addition to practical training in the infirmary and Latin medical texts in the library, Hildegard had another medical source in folk tradition (Schipperges 1992 [1957], 41 and Pawlik, 12). Given Disibodenberg's many servants and its visitors, including aristocratic guests of the monastery, pilgrims, and the chronically ill seeking healing from Jutta, Hildegard certainly had an opportunity to become acquainted with oral folk medicine. In particular, during the years she was writing *Causes and Cures*, she had well-documented interactions with all sorts of persons, including the townspeople of Bingen and Rudesheim and even Jewish savants and merchants.

That she did gain a portion of her medical/botanical knowledge "in the field" is supported by the fact that she sometimes used the German rather than Latin name for a plant or disease, suggesting that she had learned its use from mono-lingual, German-speaking informants. For instance, in *Causes and Cures* forty plants are referred to by their German, not Latin, names, as are six diseases—*cramphe* (spasm), *freislicha* (erysipelas), *gich* (gout), *orfune* (scrofula), *slim* (livor), and *glewsuch* (jaundice).[81]

Is there any way of getting at the oral/folk content of her medicine? Two methods suggest themselves: first, using modern German folk medicine as transmitted orally; and second, using information from contemporaneous German texts, on the assumption that vernacular texts would convey more of what Hildegard would have heard from her sources than would Latin texts. There are, in fact, many parallels in style and treatment between modern folk remedies (and ideas!) and the recipes and ideas of *Causes and Cures*, but there are obvious methodological problems with this approach.

However, using vernacular texts from the period is not foolproof, either.[82] Although they have been used to get a sense of folk practice (Meany), often they are simply translations of Latin medical texts into English, and are not necessarily representative of vernacular practice at all (Getz 1990).[83] In fact, one of the few German texts we do have is simply a German translation of the *Capsulea eburnea* (a Latin version of Hippocrates' *Signa mortifera*).

Of the six known German-language medical texts from the twelfth century or earlier, four can be dispatched quickly, being simply short lists of recipes.[84] The *Prüller Steinbuch* (a lapidary copied 1143–1144 in the Benedictine monastery of Prüll near Regensburg) consists of twelve Latin lines describing the colors of precious stones and a few recipes in German.[85] The *Prüller* (also called the *Innsbrucker*) *Kräuterbuch* consists of two folios of German plant recipes;[86] the *Arzneibuch Hippocrates* contains German recipes for use in pregnancy.[87] The two-folio *Innsbrucker Arzneibuch* has Latin-German (macaronic) recipes ordered from head-to-toe, as well as a "Monatregeln" with directions for bleeding and for seasonal medicinal wine.

The most informative text is the twelfth-century German *Zuricher Arzneibuch*, which was copied c. 1140–1160 in a Benedictine monastery.[88] It, therefore, provides a good vernacular comparison to *Causes and Cures*.

There are similarities. It, too, is organized from head-to-toe; and it, too, is macaronic (i.e., it uses both Latin and German in a single passage). In its case, the titles of its recipes: "Against head pain . . . against eye pain . . . against pain in the teeth . . . for those unable to urinate . . . against dropsy"—are in Latin, and its recipes are in German.[89] For instance: "Against hair loss: Burn linseed, mix it in oil, and rub into the hair."[90] Or, for the eyes: "This collyrium is wonderful for the dimness of the eyes. Take good cinnamon and *caferan* and *milwez* and take ivy, *wrcunso* and honey; mix them together well and . . . drop the mixture in the eye using a feather."[91]

Following the recipes is a section on how to use medicinal wines according to the months and seasons of the year:

> These comforting drinks are very good and healing and balance the humors that overflow at certain times. In March, put one part sage, twelve peppercorns, feverfew, ginger, and honey into thirty measures of wine. . . . This should be drunk daily for health. In April, make the same drink but add wormwood; in May, add *lubestechil*; in June, betony; in July, others; in August, agrimony. [September is missing.] In October, add *simbrate*; in November, millefoil.[92]

At the end of the text are recipes for general problems, like insomnia, impotence, amenorrhea, and malaise, and a few random recipes for hiccoughs (vinegar) and cough (dittany).[93] The last few sentences seem to be nonsense syllables; perhaps they were meant to be recited as a way of marking time, a useful technique in a world without clocks.[94]

What does the *Zuricher Arzneibuch* tell us about Hildegard's potential folk sources? There are many similarities with *Causes and Cures* that suggest that both came out of the same kind of tradition. Its mixture of Latin and German, where the formal titles are in Latin and the recipes are in German certainly imply that its recipes came out of folk, empirical practice, but had been written down by someone who knew Latin and the ordering of a Latin medical text. Its recipes, also, are short and use readily available ingredients—native plants and kitchen herbs. Its measures, too, are from the kitchen—a handful, a palmful. Equipment is not technical—a feather serves as an eye-dropper. The tone is confident and has little recourse to magical healing practices. It most resembles practical herbal texts of the following thousand years, and even of today. As we shall see in later chapters, all of these same characteristics are also to be found in *Causes and Cures*. Most likely, Hildegard was, as has often been suggested, relaying the folk medicine, the "old-wive's wisdom" she had learned, not out of books, but in the field.

What the *Zuricher Arzneibuch* does not contain, however, is any explicit medical theory. Humoral theory is implied, certainly, for instance in the passage on medicinal wines, where "humors" "overflow" during certain seasons.[95] But these humors were not the named and differentiated humors of traditional Latin medicine; rather they were some kind of amorphous, flowing substance, like sap in spring or the rising water in winter's rivers.

Do we have here a simplified medical theory that infused vernacular practice? It is impossible to say. However, to the extent that vernacular texts do reflect what Hildegard *could have* learned from her oral sources, it would have not have been medical theory but the empiric use of medicinal plants.[96]

Putting together an idea of Hildegard's training from all of these sources: she acquired her medical theory in the library, her medical techniques from Disibodenberg's infirmarian, and her plant lore from informants and in the field. Behind each was a very particular, and very different model for practice: healer, infirmarian, medical writer. Can we get a sense of how Hildegard, herself, understood her practice?

Her Image of Herself as Practitioner: Hildegard the Pigmentarius

Hildegard never wrote of herself as a practitioner, nor of anyone else, for that matter. But she did use the various terms for medical practitioner as metaphors for the wise, or compassionate, bishop and priest. Using searchable versions of the *Patrologia Latina*, the *Cetedoc Corpus of Christian Latin Texts*, (which includes Hildegard's *Epistula*, her *Liber vitae meritorum*, her *Liber divinorum operum*, and *Scivias*), and *Causes and Cures*, we can analyze all the passages in which she uses a term for medical practitioner, including *infirmarius*, *medicus*, *physicus*, *fisicus*, *chirurgicus*, *apothecarius*, *hortulanus*, *herbarius*, and *pigmentarius* (in masculine, feminine and plural endings).

Surprisingly, Hildegard never used the term *infirmarius*, *physicus*, *herbarius*, *chirurgicus*, or *apothecarius*, although they were used frequently by other authors.[97] She did, however, use two terms for medical practitioner fairly often—*medicus* and *pigmentarius*, and, occasionally, a third quasi-medical term, *hortulanus*.[98]

How does she conceive of the *medicus*? In the (metaphorical) passages that contain the term, her *medicus* is usually concerned with the healing of wounds. Thus, of the twelve separate passages in *Cetedoc* and the *Patrologia Latina*, v. 197 that used the term, the treatment of wounds came up as the main reason to fetch a *medicus* seven times.[99] But Hildegard's *medicus* also gave medicine, either by putting medicinal ointments on wounds[100] or by giving bitter medicaments ("pigments," see below),[101] or medicine.[102] The *medicus* also purged the ill,[103] roused those who kept vigils, awakened those who slept, and even poisoned evildoers.[104] He treated the leper,[105] and gave medicine for pain.[106]

But there was a second figure in Hildegard's writing who was also responsible for caring and curing the sick—the *pigmentarius*. Now in general usage, *pigmentarius* meant the pharmacist, spicer, or herbalist—because he had to do with *pigmenta*, that is, spices and medicinally active herbs.[107] The *pigmentarius* was the provider of all kinds of medicine, including the bought, the imported, the prepared, and the grown.[108]

For Hildegard the *pigmentarius* was mainly responsible for making medicines; that is, for composing *pigmenta*, the medicinal juices of plants.[109] He does so by collecting medicinal plants in the wild,[110] by planting them in his garden where he takes care that they stay green,[111] by watering them properly,[112] and by weeding out useless herbs.[113] In fact, a good *pigmentarius* is known by his sweet-smelling, well-watered, and productive garden.[114]

Out of his garden's fruit, he will make his *pigmenta*, that is, his herbal medicines. The *pigmentarius* is he who "knows how to properly water his garden and collect the products from which to make many different medicines."[115] Not only is he responsible for collecting, growing, and fabricating them, Hildegard's *pigmentarius* is also in charge of administering them—that is, of curing: good monks "will have no need for the care and cure of the *pigmentarius*."[116]

Now usually it has been said that it was the monastic *infirmarian* who was responsible for formulating medicines and treating patients (although, as noted above, the word, "infirmarian," was not often used until sometime in the twelfth century). Typically, the infirmarian was also responsible for the medicinal herb garden attached to the infirmary, "a rather private place, with a wall outside, where the sick can be refreshed by the greenness of the garden."[117] Indeed, in the Plan of St. Gall, the medicinal infirmary garden was located just in front of the doctors' house, separate from all the other gardens. Although it is not known whether the infirmarian was always in charge of the medicinal herb garden, in Westminster Abbey the medicinal herb garden was known as the "infirmarer's garden."[118] Given the dual function in Hildegard's work of the *pigmentarius* as both gardener (producer of medicines), and medical practitioner (provider of medicines to patients), it would appear that in Hildegard's world, it was the *pigmentarius* who performed the functions of the *infirmarius*.

One piece of confirmatory evidence is that Hildegard used *pigmentarius* to designate bishops and priests, a "usage never satisfactorily explained" (Baird, 8). But it was common to speak of these same figures as doctors—*medici*—for the soul-sick, a usage found throughout the *Patrologia Latina* and the *Cetedoc*. Apparently Hildegard called these figures *pigmentarii* for the same reason—what she meant by calling bishops *pigmentarii* was that they were medical practitioners for the ailing soul. Indeed, according to Corvi, the *pigmentarius* was, in fact, the monastic infirmarian (1996, 36–39).

Given this evidence, it seems probable that Hildegard thought of herself not as an infirmarian but as a *pigmentarius*, the monastic responsible for the sick patient *and* for the medicinal herb garden. In fact, in both *Causes and Cures* and *Physica* Hildegard is careful to describe not only how to make a medicinal recipe but also where and how to acquire its required herbs—whether they should be wild or domestic, fresh or dried, collected in the morning, noon or evening, in spring or fall.

Did she supply a specific word in the *Lingua ignota* for the medical practitioner? Perhaps. She did come up with words for the tools of the infirmary, including knife and forceps, as well as for the tools of the garden—sickle, trowel, and garden knife, and she had hundreds of words for medicinal plants, as we have noted. *Garginz*, a position that follows *camerarius* and *cellerarius* (traditional positions in the monastery for those who cared for the physical comforts of the monastics) may be her unknown language's equivalent for *pigmentarius*.[119] *Garginz* is glossed (by a second hand) as *hortulanus*—gardener; and in Hildegard's work the *hortulanus* overlaps with the *pigmentarius*. Thus the *hortulanus* removes useless weeds[120] and collects herbs, good ones, and those that are the perfect ones for certain uses.[121] *Garginz*, although glossed as *hortulanus*, probably signified Rupertsberg's *pigmentarius*, who was in charge of the body of the garden *and* the garden of the body.

It seems, then, most likely that Hildegard thought of herself as a *pigmentarius* (or, perhaps, as a *garginz*). As such she would have had a dual charge—taking care of bodies and taking care of plants. Her medical works, *Causes and Cures* and *Physica*, would be, therefore, the products of a double figure, at home in the world of the infirmary and the world of the garden, and their concepts best approached as coming from a mixed experience of medical practice and gardening.[122]

Nor would such a professional doubling have been unique to Hildegard. Rather, in the time before the professionalization of medicine it was common, even usual to have more than one specialty—to be a blacksmith *and* a doctor, an herbalist *and* a midwife, an apothecary *and* a grocer, but above all, to be a gardener and a medical practitioner. Plants were the basics of medicine. Knowing where to find them in the forest, how to collect them, transplant them, cultivate them, grow, harvest, and prepare them was as much a medical art as knowing when and how to administer them.

Given such dual occupations, it was easy for concepts to cross fields, and be used to explain the body. Thus the ideas, even of the blacksmith or the architect, might end up in medicine as the fiery forge of the stomach or the building blocks of the body (Pouchelle). But the most congenial, popular, and common was the interplay between the concepts of body and plant, gardening and medicine. So the gardener might doctor his plants, or keep them in a nursery; the doctor might nurture the body; and the patient might husband his forces.

It was precisely *because* of her experiences of the garden—its seasons of greenness and dryness; its responses to sun and rain, wind and soil; its rising sap in spring and fallowness in winter—that Hildegard was able to grasp so vividly, the abstract concepts of ancient medicine. And this makes perfect sense because, as we shall see, it was these very experiences—of the effects of seasons, times, and climate on plant and body—that gave rise to the ancient concepts in the first place.

Chapter Three

Wind, Land, Rain, Sun–The Elements

The elements were the basic building blocks of the medieval and ancient universe, as they are the basic building blocks of ours, but the ancient elements were not those of the periodic table.[1] In Book One of *Causes and Cures*, Hildegard gives us an opportunity to explore the medieval idea of the elements, not from the point of view of the philosopher or the rhetorician, but from the practical point of view of the *pigmentarius*. The question for this chapter is: why were the elements important in practical medicine? That is, what made them so useful to the medical practitioner that they survived the fall of the Roman Empire, even though transmitted in so few texts? What did they provide that was, apparently, irreplaceable for more than 2000 years?

SUMMARY OF BOOK ONE—THE ELEMENTS

In Book One Hildegard sets up her medical manual by first presenting the basics of medieval science, although her material at first appears to be a jumble of texts, ideas, and genres. Thus it begins with theology—Hildegard's idiosyncratic version of Creation and the Fall—jumps to a brief description of the four elements, and then continues with dozens of passages on apparently disconnected subjects: on the sun, the weather, and the winds; on the firmament, Lucifer, the zodiac, the planets, and the moon; on the varieties of water; and ends with passages on plants and trees. Kaiser's edition added to this confusion by italicizing the rubrics in the manuscript, which were probably scribal additions and were often misleading.

However, if the rubrics are ignored (as they are in Moulinier's new edition) and the passages analyzed in and of themselves, it becomes clear that a strict, though invisible, structure undergirds Book One and binds its disparate pieces together. *Causes and Cures* begins with God because God began

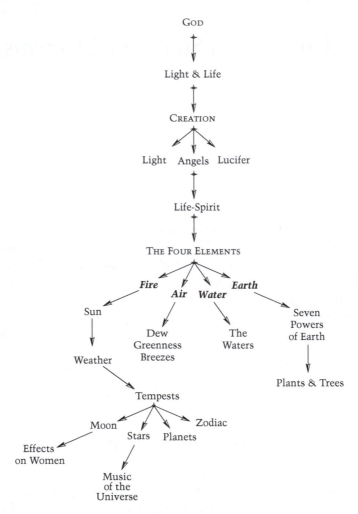

Figure 3.1. Implicit Linear Structure of Book One of *Causes and Cures*.

creation, as its very first line indicates: "Before the creation of the world, God was from the beginning, and is; and he was, and is, light and brilliance, and he was life."[2] The aspect of God accentuated here is life: God begins *Causes and Cures* because God is life, the underlying subject of medicine.

A passage on creation therefore follows, and then material on Lucifer. Next the four elements make their appearance, in the traditional order of their lightness of being—fire, air, water, earth; and they bring with them their embodiments in the sublunar sphere.[3] Thus *fire* gives rise to the sun, the moon, the planets, and the stars of the zodiac; and these are discussed. *Air* is known

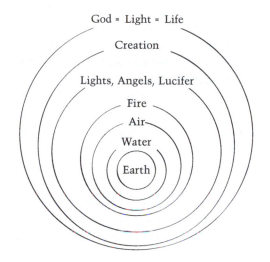

Figure 3.2. Implicit Radial Structure of Book One of *Causes and Cures.*

by the winds and so a lengthy passage on the winds follows. The element *water* is embodied by the different kinds of waters, and *earth* by its plants and trees. When all the passages of Book One are ordered according to this *actual* content instead of the apparent content signaled by the added rubrics, it can be deconstructed as a linear structure (fig. 3.1). But if this outline is portrayed radially (i.e., as circles within circles) instead of linearly, then it is clear that Book One was organized by the customary medieval image of the cosmos (fig. 3.2).

Moreover, if this radial outline of Book One is compared with an illustration of the universe from *Scivias* (frontispiece), it appears that the radial structure of Book One reflected the scheme in the illumination. For instance, in the illumination, Creation is within God and structured by the four elements. *Fire*, represented by the sun, moon, stars, planets, and constellations is at the top; *air* comes next, as the winds surrounding the earth; then comes *water* as rain-bearing clouds and streams; and last, at the center, is *earth*, with its plants and trees.

Book One reflects this very same cosmology but as text. Thus within God, Hildegard tells us, is creation, which is structured by the four elements on which the discourse centers. *Fire* is embodied in the sun, in fiery weather—tempests, thunder, and lightning—in the moon, stars, planets, and the zodiacal constellations (and they are discussed in this order.) *Air* is embodied in the winds. Under the rubric *water* are the medicinal effects of the many kinds of waters—the salty and sweet, the fresh and the marshy, those of underground sources and those of the well, the river, and the sea. The element *earth* ends Book One just as it punctuates the *Scivias* illumination. It produces *viriditas* (greenness) and *ariditas* (dryness), reproduces plants and animals, and bears all things.

The congruence of the implicit structure of Book One with Hildegard's contemporaneous illumination in *Scivias* supports the notion that Book One was not a random collection of excerpts (although this escaped whomever added the rubrics). But it also speaks to the question of Hildegard's authorship of *Causes and Cures*. To create this implicit structure, the hypothesized "continuer" of the last chapter would have had to arrange the text to correspond to what would have been a fifty-year-old illumination. More important, this internal organization of Book One indicates how thoroughly the concept of the four elements affected Hildegard's medical understanding.

And it allows us to ask: how would a student *pigmentarius*, say Hildegard's student at Rupertsberg, have understood this eruption of the elements into her medical manual? Would she have understood them as formal principles inherited from the past? As abstractions from the physical world? Or as the physical substances—fire, air, water, earth? What, that is to say, was the relationship for the practitioner between the element, water, and the water that one can touch and drink (Lorcin, 260).

Surprisingly, despite an enormous literature on the ancient theory of the elements, this question has not often been addressed, either generally, or with reference to Hildegard.[4] Thus, although several scholars have examined the elements in her work, it has mainly been with a theological focus, in her theological work.[5] Here the elements will be examined in *Causes and Cures* against a backdrop of practical medicine, not theology, as if we were its apprentice, to learn what it was that the four elements provided to the practitioner.

Air is Abstraction, Air is Stuff, and Air is Wind

At the beginning of Book One of *Causes and Cures*, Hildegard defines the elements explicitly:

> And God made the elements of the world which are in the body. . . . They are fire, air, water, and earth, and they are so mixed up together that in no way can they be separated, for each contains the other.[6]

Again:

> God tied together the four elements in such a way that none can be
> separated from the other; the earth could not exist if the elements were
> separated one from another.[7]

Here by the elements Hildegard certainly means something abstract, although
as she elaborates her ideas, concreteness and materiality quickly enter. Thus,
the preceding passage continues:

> Fire is stronger than air and tames it and kindles it, while like a fuller, air
> makes fire burn and so tempers it, because fire is like the body of air and
> air is like the insides and the wings and feathers of fire.[8]

Hildegard here is clearly thinking of the elements not as abstractions but as
their embodiments as physical fire and physical air. Indeed in Book Two she
seems to suggest that the "Elements"—the abstract organizing principles of
the world—were, in fact, also the "elements"—the elements of farmer and
gardener, as well as the "elements" that make up the body:

> Just as the world prospers when its *elements* are well-ordered, so that
> *heat*, *dew*, and *rain* come down moderately and temper the *earth*, while
> if they come down suddenly and all at once, it is torn up and its fruit
> not healthy; so, too, when the *elements* in the body work well together,
> they conserve and even return health, while if they are in discord they
> make illness and even death.[9]

By "elements," then, Hildegard meant several different things—the abstract
elements of the world, the basics of gardening (sun/heat, land/earth, and
rain/dew-water), and the building blocks of the body. (The same ambiguity
exists today, where "elements" can be the abstractions of philosophy, the elements of weather, or the building blocks of chemistry.)

To investigate Hildegard's understanding of the elements further,
then, let us take a single element—*air*—and trace it throughout *Causes
and Cures*. (This is not to suggest that "air" was privileged; the same analysis could be performed on any of the other three elements.) This can
be done by extracting and examining all of the passages that contain the
term, *aer*. As Hildegard's introductory material on the elements had suggested, air turns out to have had several meanings, from the most abstract
to the most physical.

At its most abstract, for Hildegard air was an heuristic—that is, a cat-
egorizing principle. For example:

> *Air* with its four powers is in the breath and in the power of thought.
> For *air*, by the living breath which is the *anima*,[10] serves the body
> because it bears it and is the wing of its flight, and by it humans
> breathe in and out, and so live. . . . It shows these four powers in the
> body: as dew in emission, *viriditas* in excitation, *wind* in motion, and
> heat in dilation.[11]

Also:

> Adam was made of earth and fire awakened him; air set him in motion;
> water moistened him.[12]

Hildegard also used "air" to classify disease. Thus the four kinds of fevers
were caused by the four kinds of air—hot, cold, wet, and dry.[13]

At the same time, it is clear, even in these examples, that by "air" Hildegard also meant some kind of physical stuff. For instance, air was breath,
the stuff of life:

> Humans are made of the four elements, and from fire have heat, from
> air, breath, from water, blood, and from earth, flesh. . . .[14]

Air carried the dew:

> And so the stars warm and strengthen the air with their heat, and the
> heated air sends its sweat, that is, its dew, onto the earth and makes it
> fertile.[15]

Air supported the birds above the earth and kept all breathing beings alive:

> Birds and animals and beasts, which can be and are used by humans,
> take their life from the air according to God's rule, whence they abide
> on the earth.[16]

Here by air Hildegard clearly meant that invisible but still material stuff that
fills the sky, supports birds in their flight, brings the dew, and through breath,
keeps all beings alive. This was, of course, the ancient conception: "Wind is
the famous breath that generates the universe."[17]

But, finally and most concretely, "air is wind"[18] and it is mainly as the
wind that air turns out to play its role in *Causes and Cures*.[19] But what was
the wind doing in medicine? How was it related to the abstract concept of

air? And how would a medical practitioner—Hildegard herself, or better, perhaps, her nun-infirmarian—have understood the reason for "wind" in her medical manual?

To answer these questions from within a *medical* context, another analysis can be performed on how Hildegard used wind, that is, *ventus*, in the thirty-one passages in *Causes and Cures* that contain the term. As it turns out, the wind had three meanings in *Causes and Cures*, and three corresponding reasons for its importance in medicine. First, when Hildegard defined the wind explicitly, it meant the traditional directional winds that stood for the *seasons'* effect on the body. But by *ventus* she also meant, occasionally, various named and local winds that encapsulated the effect of *weather* on the body. Last, by wind she was sometimes referring to an internal circulating wind that mediated and tempered the four qualities *inside* the body.

Hildegard's Explicit Winds

Early in *Causes and Cures* there is a very long passage on the winds.[20] The most important winds are the four cardinal winds, and each is associated with one of the four directions and one of the four elements:

> The four cardinal winds under and above the sun are present in the firmament and contain it and the whole universe; they cover it from the lowest to the highest like a cape. The *east* wind holds the *air* and sends the *sweetest dew* onto dry things; the *west* wind mixes with clouds and restrains their *water*, lest they break out. The *south* wind in its dominance restrains *fire*, lest it burn up everything and the *north* wind holds back the outside *darkness and blackness*, lest they exceed the usual. These four winds are the wings of the power of God; and if they should ever break out at the same time they would bring together all the elements at once and divide themselves and shake the sea and dry all the waters.[21]

The main function of the four cardinal winds, apparently, is to mark the four cardinal directions, but each is also implicitly connected to one of the four elements—east to air, west to water, south to fire, and north, to darkness and blackness, that is, to earth.

In the *Scivias* illustration we have already examined (see frontispiece), Hildegard portrayed these four winds and their relation to the four directions and elements.[22] At the top (which, as in most medieval world maps, signified east) is the east wind, colored bluish-white, (at least in the hand-painted, twentieth-century copy of Eibingen), as appropriate to its watery nature. It sits atop the sky under greenish clouds and within great, flowing, pink clouds; and it has three heads, representing the main wind and its two subsidiary winds.[23] The central head, which represents the cardinal east wind, blows into the moon; the northeast head onto clouds and the southeast into the firmament of stars; and each is embedded in *water*. On the right is the south wind; it is colored red and located in a layer of *fire*. On the bottom is the west wind, colored greenish-blue and located in the (dark blue) *airy* firmament. Finally on the left is the north wind, situated in an empty shell of darkness between fire and firmament; it is colored gray-green and blows into the fire and *darkness*.

However, the four cardinal winds are not the only winds in medicine. According to *Causes and Cures*:

> Each of the cardinal winds has two weaker ones, like two limbs into which each breathes certain of its powers. Thus these weaker winds have the same nature as the main wind, which is like their head, and they are like two ears that hear only one thing though they are separated. They receive breezes and strength from their main wind; and they can create much agitation, just as ill humors create turmoil for people when they become sick.[24]

That is, each of the four cardinal winds has two subsidiary, derivative winds, carriers of the same kind of forces but less predictable. These subsidiary winds derive their powers from their directions, that is, from their contiguity with the great winds. It is as if the cardinal winds stored their particular powers or qualities—of coolness, dryness, wetness, heat—at the four corners of the universe, from which all other dryness, wetness, heat, and cold derived proportionally.

But Hildegard has added something else—the winds in the universe can also act like bad humors in the body, agitating the world. Indeed, the main function of the winds is to maintain order by the perfectly-balanced tempering of the elements under their control. Should the winds fail in this task, the very elements of the world would fall apart, dis-integrate:

> Just as the *anima* pervades the entire body, so the winds pervade the firmament, lest it be destroyed. Like the *anima*, the winds are invisible and come from the hidden God. And just like a house without cornerstones will fall apart, so nothing can exist without the winds, neither the polestar, nor the earth, nor the abyss, nor indeed, the whole universe with every created thing, because all these things are composed and held together by them. Indeed, the whole earth would tear apart and break if these winds did not exist, just exactly as a person would fall apart

with out bones. The main east wind holds together the east; the west wind, the west; the south wind, the south; the north wind, the north.[25]

Here Hildegard has explained the macrocosm by means of the microcosm—the physical universe is like a body and the winds are like its *anima*. How so? On the one hand, the winds are like the bones of a body or the cornerstones of a house. Like bones, they hold the body together; like cornerstones, they structure the four "corners" of the universe and so contain it.[26]

On the other hand, the winds are also like the life-force—the *anima*—that pervades the body; they are invisible, organizing principles of the world. Just as when the life-force departs, the body decomposes (so proving that it was this *anima* that held the composed elements of the body together) so, too, the universe without its winds will decompose into its four constituent elements. Indeed:

> The principal winds since the beginning of time have never moved nor ever will move with their full powers until the Last Day and then they will reveal their full strength, sending out their breaths all at once. Then the clouds will break apart and the firmament collapse, just exactly like the body of person breaks apart and disintegrates when the *anima* is loosened from the body.[27]

Thus the cardinal winds maintain order by keeping the four elements separated, since they would mutually annihilate one another should they ever come together.[28] On the Last Day, this is precisely what will happen—all the winds will blow at once and the universe, consequently, will dis-integrate.[29]

Indeed, in *Scivias* when Hildegard envisioned the Last Day, the associated illumination portrays this very moment—all four of the cardinal winds blow at once. (See fig. 3.3 follwing page 124) Remarkably, the text in *Scivias* does not describe winds but simply a "horrible moment" when the mountains would fall, etc. So the illumination in *Scivias* does not depict the *text* in *Scivias* but the *text* in *Causes and Cures*. In fact, the image in *Scivias* illustrates the scene in *Causes and Cures* almost exactly, and the words used to describe the moment are almost (but not quite) the same: "Quia tunc omnia purgabuntur" (CC, I, 24:23) and, "Et omnia elementa purgata sunt" (*Scivias*, col. 604:37).

This similarity supports the notion that Hildegard authored the text, and at about the same time that she wrote *Scivias*. What we see in the illumination is that the four winds are located at the four cardinal directions, and the three heads of the main and its two subsidiary winds blow into the center of the cosmos. Humanoid but amorphous, frightening in their vague and dissolved quality, the fundamental characters of these universal props blur, melt, dis-inte-grate. Unstoppable forces collide, opposites unite, and there is mutual assured destruction. Hildegard is saying that just as the body dis-integrates at death when its animating force departs, so too, on the Last Day the universe will dis-integrate when its ordering principle, the winds, annihilate one another. That is, in the macrocosmic universe, the winds are the ordering life-force.

Thirdly, and most significantly, is that in *Causes and Cures*, the winds are important because they *also* define that critical premodern structure, the zodiacal ecliptic:

> The east wind has two wings with which it attracts the whole earth, so that one wing from the top to the bottom [i.e., from east to west] holds the sun in its course, and the other does the opposite, preventing the sun from moving lower than it should. And this wind moistens all wet-ness and makes all seed germinate.[30]

This description of the special powers of the east wind—moistening and ger-minating—hints at how its direction was originally derived.

What Hildegard means is that the east wind, which by definition is due east (as defined by the equinoctial sunrise), establishes the center of the zodiac. Its two "wings" delineate the width of the zodiac, that is, the winter and summer course of the sun. Were it not for the confining power of this east wind, the sun would not stay in its orbit around the earth but would "fall down" the sky too far in winter, too high in summer. Were it not for the east wind, time, therefore, as reckoned by the zodiacal turning of the sun during the year, would be out of true.[31]

The *east* wind, then, is not simply a wind from the east. It is the very power of the direction—due east—and its primary quality is its moistness. It is, therefore, identified with the element, *water*. Finally, since it causes seeds to germinate, it is identified with the season, spring, as defined by the spring equinox, also in the east. As for the west wind:

> The west wind is like a mouth that blows away clouds and dissipates and disperses water; and this wind *dries* all green things.[32]

By definition, the west wind is directionally opposite the east wind, but it is also its opposite qualitatively—where the east wind moistens, the west wind dries. It is identified with autumn because it marks the place—due west—of the autumn equinox. Now the south wind:

Can be thought of as if it held an iron staff with three branches and a sharp point. Strong as iron, and like the heart of a person, it holds the poles apart lest they expand. Its three branches are its three powers of tempering the sun's heat in the east, in the south, and in the west. It is sharp because it cools on the bottom, lest wetness and cold come up more than normally. This is the wind that makes all things mature and ripen—the leaves in the woods, and the seeds, grain, fruit, vines and all the products of the earth.[33]

That is, the south wind by its coolness keeps the icy poles from expanding into the temperate sector that belts the earth. It controls the cold, which would take over were it not for the south wind's ability to temper the heat with a cool wind. And, since it is responsible for the maturing of fruit, it stands for the season of summer. Finally:

The north wind has four columns through which it keeps together the whole sky and the whole abyss. And if it should pull them back, then it would fold up the sky and the abyss. These four columns hold and separate the four elements which are mixed up in the north, as if the columns were empty, lest they fall. And when it shall blow on the Last Day, then the sky will fold up like a trestle table. The north wind is cold and brings cold and with its cold it restrains everything and prevents them from expanding.[34]

The north traditionally signified darkness and evil partly because there are so few stars in the north. (Perhaps this is one of the reasons for the myth that Lucifer took the angels, that is, the stars, with him into Hell). But for Hildegard, the north was also dry and cold, a bitter cold that caused tension and so maintained "constriction," that is, individuality.[35]

To sum up what has been learned so far from *Causes and Cures* about the winds and their significance for medicine. The four cardinal winds maintain the structure of the world, which depends on a tension of opposites. Each is associated with an element and a quality—east with moistness and water, west with dryness and air, south with heat and fire, and north, with cold and earth. Each blows from a cardinal direction from whence it brings its weather; and each has an agricultural function—the east wind germinates, the south wind matures; the west wind disperses, and the north wind constricts. The primary function of the winds is of holding back, of balancing and compensating for one another; indeed, like the humors in the body, they cool what is too hot and moisten what is too dry. If the winds did not exist,

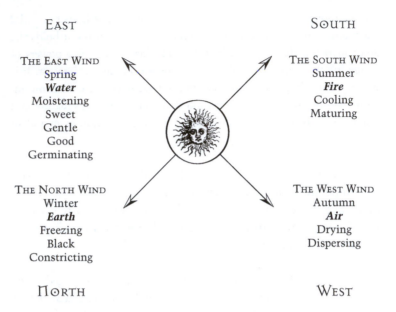

EAST   SOUTH

THE EAST WIND   THE SOUTH WIND
Spring   Summer
Water   *Fire*
Moistening   Cooling
Sweet   Maturing
Gentle
Good
Germinating

THE NORTH WIND   THE WEST WIND
Winter   Autumn
Earth   *Air*
Freezing   Drying
Black   Dispersing
Constricting

NORTH   WEST

Figure 3.4. The Characteristics of Hildegard's Explicit Winds. East is at the upper left as a compromise between the usual direction of medieval maps, where east is at the top, and the physical direction of east, on the left, for an observer in Europe who typically faces south to make celestial observations.

the qualities in the cosmos would become distempered—hot would be too hot and cold, too cold. Thus the winds function as the instruments of a universal allopathy.

Despite their oppositions, the winds also have much in common—they provide the animating, i.e., the moving, force, for the macrocosm, just as the internal *anima* moves the pulse and breath. The *anima* maintains the structure of the body by holding tensions in opposite, and it is this very tension that allows life to continue.[36] Like the body, the universe is an instrument of oppositional powers kept in balance by a mysterious force—in the body, known as *anima*, in the universe, *ventus*, wind. Only if *ventus* comes out of balance does the universe get ill—with floods, fires, droughts, and harsh winters. Only on the Last Day will the winds finally let go and when they do, the sky itself will fall apart, the universe disintegrate.

A diagram can summarize the main characteristics of Hildegard's explicit winds—their directions, seasons, elements, qualities, and powers (fig. 3.4).

THE ORIGIN OF HILDEGARD'S EXPLICIT WINDS

The Winds in Isidore

Do Hildegard's connections between winds, directions, elements, and qualities tell us something about the ancient system of winds, directions, elements,
and qualities? Or were the connections that she makes as idiosyncratic as the
prose with which she describes them?[37]

This is a relatively easy question to answer, since so few texts on the
winds were available at the time she was writing Book One (c. 1150).
They included: two ancient poems that gave the Greek names for the
classic twelve winds in pentameter; an obscure text by Nimrod; a few
wind diagrams scattered in manuscripts; and short passages in the ency-
clopedists, Pliny, Bede, and Isidore.[38] The two poems pass on very limited information on the ancient wind system; simply the Latin and Greek
names for the winds and short descriptions of their individual properties.[39] Nimrod reveals even less. But Isidore in his *Etymologies* and his
De rerum natura, and Bede, who mainly copied Isidore, did pass on a
summary of the traditional material originally formulated in Aristotle's
Meteorology and Theophrastus' *De ventis*.[40] Both would have been easily
available to Hildegard. Isidore's winds, therefore, allow for a telling comparison with Hildegard's.[41]

Isidore begins by defining the wind: it is "moving and agitated air, as
shown by a fan when it is used to chase off flies."[42] The wind we feel on earth
is generated, therefore, by movement—of the sky or of bodies;[43] and its purpose is "to ripen fruit and temper the heat of the sun."[44]

In addition to this generic wind, however, there are also twelve
named winds (these are the winds of the two ancient poems). They include
the four cardinal winds located at the four cardinal points of the compass (north, south, east, west) and the eight minor winds that flank them
(northeast and northwest, southeast and southwest, etc.)[45] Each one brings
a characteristic weather. For instance, the north wind when it blows pres-
ages cold and dry weather, but with clouds; while the northeast wind brings
clear weather. Isidore classifies the weathers mainly by their qualities of
humidity (wet or dry) and temperature (cold or hot); and he also associ-
ates each wind with a season. For example, the west wind is a spring wind
that "pushes away winter" and the north wind blows in winter.[46] A similar
diagram to that formulated for Hildegard's winds can be put together for
those of Isidore's (fig. 3.5).[47]

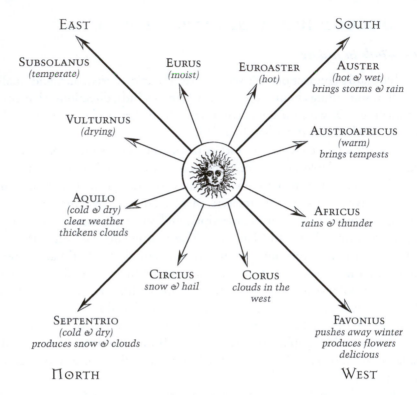

Figure 3.5. The Characteristics of Isidore's Winds.

As we see, like Hildegard, Isidore also associated the four cardinal winds with the four cardinal directions, and like her he connected the weather and the seasons to the winds and the directions. But, and this is telling, the *particular* seasons and weathers that Isidore connected to the winds and their directions were *not* the same as Hildegard's. For instance, while Hildegard's south wind was cool, Isidore's was hot; while Hildegard's west wind was associated with autumn, Isidore's was associated with spring.

Why should this be so? It is impossible to imagine that Hildegard had spontaneously recreated in a vision the ancient tradition which Isidore, Bede, Pliny, Nimrod, and even the short poems on the winds, convey. But if she were merely copying an authoritative text (and we know which texts were available when she was writing, and none of them expresses the system exactly as she does) then why would there be differences between her wind system and theirs?

The answer lies in *how* and *why* the winds had initially been assigned to their traditional directions. This turns out *not* to have been due to any actual blowing of any actual winds but, instead, was a way of spatializing the temporal effects of the equinoxes, solstices, and zodiacal ecliptic, as Hildegard had hinted. It was because of their connecting *time*, in the form of the year and its seasons, with *space*, in the form of the directions, that the winds, and through them, "air," achieved such an elemental role in health.

Aristotle Explains the Origin of the Wind System

In his *Meteorology*, Aristotle explains how the wind system evolved. The first of their characteristics to be established was the four directions. East was determined (or defined) first: it was the location on the horizon of the sun at dawn on the spring equinox. Due west was next defined, as the position on the horizon where the sun was setting at twilight on the autumn equinox. Thus "east" and "west" winds were *defined* by the cardinal directions, located by the spring and autumn equinoxes.

Next, Aristotle explains, north and south were located by dropping a perpendicular to the line of the equinoxes; so the north and south winds were thus determined (or defined). The two minor winds that flanked the east and west winds were derived from the position of *sunrise* and *sunset* at the winter and summer *solstices*. That is, the position of the sun at dawn of the winter solstice defined the direction, "southeast," and, therefore, the southeast wind. Its position at *dawn* of the *summer* solstice determined the direction northeast, and, therefore, of the northeast wind (fig. 3.6).[48] This is why there were eight main winds, each of which was correlated *ipso facto* to a direction, a season, and, eventually, to the weather characteristic of that season.[49]

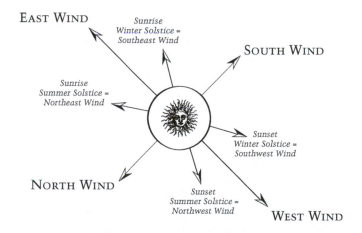

Figure 3.6. The Eight Winds are also Defined by Time.

To grasp these connections it was not necessary to have read Aristotle's *Meteorology*, however. This knowledge was well-diffused in the ancient world. For example, in Athens an ancient clock could be found well into the twentieth century that displayed not only the hours but also the times of the solstices and equinoxes. It was an eight-sided tower with the winds pictured on each of its faces, and a weathervane at its top (Robinson). "Compass cards" in Vitruvius' *De architectura* (a copy of which existed as early as 800 A.D.), also conveyed this ancient relationship between time, the directions, and the winds, and so did Pliny's instructions to farmers in *Natural History* (Obrist 1997, 43). Nimrod passed the idea along when he explained that the four cardinal directions had to do with the rotation of the sky over the earth, inclining north and south.[50]

Because of this derivation from the equinoxes and solstices, the original eight winds did not signify *local winds* that predicted weather but the *seasonal* forces that belonged to the eight directions. The winds acquired their characteristics *not* from an experience and observation of actual fluctuating winds but, rather, from a constructed relationship to the *seasons*. Thus the east wind was always a spring wind *because* it was defined by sunrise at the spring equinox; and it had the characteristics of spring—*whatever they were.*

But this meant that the characteristics of the winds were almost always fungible, not codified. In Germany, for instance, Hildegard's east wind "was"—warm, moist, temperate, and fructifying, like the German spring. The character of the west wind was similarly determined by its association with the autumn equinox and with sunset. In Southern Europe and the Mediterranean, autumn was defined by the rains that ended the summer drought—and so the west wind was known as a cool, wet wind. In the same way, the north wind was assigned to winter and the south wind to summer not because of their observed characteristics of cold or heat, but rather because the night sky moves southward in winter and northward in summer (as Hildegard noted).[51] *This is to say that the reason that the winds were connected to the physical qualities of humidity and temperature, to weather, and even to the directions was because they served as stand-ins for the seasons.*

Airs, Waters, Places

This connection between the winds and the seasons (and, eventually, with the weather brought by the seasons) was a main reason that the element, air, entered medical theory in the first place. How and why "air" did so is explained in one of the earliest Hippocratic texts, *Airs, Waters, Places*. This happens also to be one of the few Hippocratic texts available in Latin in the early Middle Ages. It was translated from Greek to Latin as early as the sixth century, and conserved in several ninth- and tenth-century manuscripts. It was retranslated in the eleventh century and preserved in a twelfth-century manuscript from Munich (Jouanna 1996, 115), so in theory Hildegard might even have read it. In any case, what it conveys about the body is known, about plants, to any gardener.[52]

"Whoever wishes to pursue properly the science of medicine must . . . consider what effects each *season* of the year can produce . . . and also the effects of the hot *winds* and the cold—of those that are *universal* and also of those *peculiar* to each particular region."[53]

The reason for this is that "things that grow in the earth assimilate themselves to that earth"(Jones 1984, 137),[54] and, therefore, "you will find assimilated to the nature of the land both the physique and characteristics of its inhabitants" (137). That is, "the character and the physique of people parallel the region in which they are nourished" (137).[55]

According to the text, there are *four* distinct types of regions, and these are determined by the *direction* each faces and the *winds* to which each is exposed. Consequently, there are, precisely, *four* distinct human physiologies (or temperaments), and these four are the main determinants in health and medicine.[56]

For example, a town that faces *north* gets the cold, dry, north wind, and its climate, therefore, will be winter-like—cold and dry. Its citizens, having "assimilated themselves to the region" will, therefore, also be cold, dry, and bilious.[57] An east-facing town gets the mild, temperate east wind. It has a spring-like climate and its citizens will be well-tempered (Jones 1984, 81). "A city that faces *east* is healthier and produces better bodies. . . . There the people . . . are . . . more blooming than elsewhere . . . just as all things growing there are better" (81). A *south*-facing town gets the hot, dry, south wind; it has a hot, debilitating summer climate, and its people will be phlegmatic (75).[58] Finally, a town that faces *west* gets the cold, wet, storm-bearing west wind. It has an autumn-like weather and its people will be unhealthy (83). A city that faces west "is precisely like autumn in respect of the changes of the day" (83). "With the seasons, men's diseases, like their digestive organs, suffer change" (73).

In *Airs, Waters, Places* then, a medical connection was made between directions and seasons, climates and weathers, and these connections were understood to be through the winds. For comparison, the Hippocratic material can also be diagrammed (fig. 3.7).

That these are not hypotheses but observations about the effect of weather and climate on people is supported by the fact that the hot, dry south wind was not said to make people bilious (which will, later, be the inferred humoral temperament) but "phlegmatic"—slow in thinking and

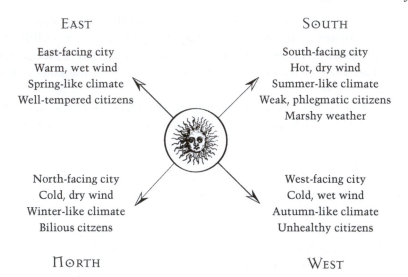

EAST   SOUTH

East-facing city   South-facing city
Warm, wet wind   Hot, dry wind
Spring-like climate   Summer-like climate
Well-tempered citizens   Weak, phlegmatic citizens
                                                 Marshy weather

North-facing city   West-facing city
Cold, dry wind   Cold, wet wind
Winter-like climate   Autumn-like climate
Bilious citzens   Unhealthy citizens

NORTH   WEST

Figure 3.7. The Hippocratic Directions, Regions, Winds, Temperaments, and Weathers.

movement. It is no news flash, of course, that heat makes people sluggish and cold weather is stimulating (up to a point), but what is important to see is that these early connections were local, experiential, and, as we shall see, fungible. It is their fungibility—their relationship to observation—that will explain the idiosyncrasy of Hildegard's own wind system.

TRANSMITTING THE WIND TRADITION

The Winds in Agriculture

Not surprisingly, this set of notions about weather, climate, the winds, and health was passed along in agricultural, as well as in medical texts, but to a lesser degree.[59] For instance, in his *De re agricultura*, Palladius explained to the (gentleman) farmer that his success would depend on the healthiness of the "air" of his farm. He should take care, therefore, to situate his farm with attention to the *direction* it faced and to the prevailing winds (Rodgers 1975, 7). "Air" here takes on the meaning of local climate.

Varro, another agricultural writer, suggested that in cold regions the farms should face the warm south and east; in hot regions the north and west, to counteract the maleficent imbalances of qualities (that is, hot/cold and wet/dry) likely to prevail. Indeed, he is explicit: the farmer, like the doctor, had to take care to balance the qualities of his land (Heurgon, 22, 24, 35).

Winds brought not only the seasons. Local winds brought local weather,[60] and the farmer had to take local and particular winds into account as well.[61] So Pliny suggested that the farmer should not gather fruit or plant seeds when the northeast wind blew;[62] when the dry west wind blew in spring, he should attend to his trees and plants.[63] Palladius warned that certain local winds might stir up the bees (Rodgers 1975, 46). As late as the fourteenth century the English horticultural writer, Godfrey, would point out that the sudden changes of weather brought by the (local) winds (through their abrupt changing of the qualities of hot and cold, wet and dry) could turn wine sour.[64]

On the whole, though, of the surviving agricultural literature, the material on the winds, while present and clearly conveying the effects of seasons, weather, and climate on plants, is sparse. Perhaps such connections were obvious to anyone who, unlike the gentleman readers of agricultural treatises, actually took care of the farm.[65]

The Winds in Medicine

Hildegard, then, was not idiosyncratic in her placing of the winds front and center in medicine. Was she expressing an understanding that was not only, perhaps, common, but also usual? Did the winds have a similar place of importance for other contemporary medical practitioners? Can we take it that her explanation and understanding was typical?

There are only two medical texts that allow for a comparison with *Causes and Cures* c. 1150: the *Practica* of the Salernitan Bartholomew and the *Medicine* of the Jewish writer, Asaph Judaeus. Each was a very different kind of practitioner from Hildegard. Bartholomew was writing and practicing in twelfth-century Salerno, a professor associated with the medical school who had access to the revolutionary new translations coming out of Salerno. Asaph was Jewish, writing in Hebrew, perhaps as early as the ninth century, and in the Middle East.[66] While providing, therefore, very imperfect comparisons, their texts do, at least, demonstrate what place the wind had in two other (more or less) contemporary medical texts.

Wind had very little place in Bartholomew's *Practica*. Except for a few references to the wind-producing effects of beans, Bartholomew had nothing to say about wind or air. For Asaph, however, wind was a very important concept: "When the seasons and the prevailing winds are not in harmony with the temperament, then diseases arise and the doctor must consider the prevailing atmospheric conditions."[67]

However, and this is highly significant, the *particular* season, weather and direction that Asaph associated with each wind differed, not only from *Causes and Cures* but also from Hippocrates and Isidore. In Asaph, the south

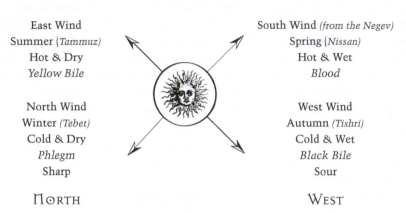

EAST	SOUTH
East Wind	South Wind *(from the Negev)*
Summer *(Tammuz)*	Spring *(Nissan)*
Hot & Dry	Hot & Wet
Yellow Bile	*Blood*
North Wind	West Wind
Winter *(Tebet)*	Autumn *(Tishri)*
Cold & Dry	Cold & Wet
Phlegm	*Black Bile*
Sharp	Sour
NORTH	WEST

Figure 3.8. The Winds, Directions, Seasons, Qualities, Humors, and Tastes in Asaph.

wind was warm and wet, not hot and dry; the east wind was warm and dry, not warm and wet; the west wind was cold and wet, not cold and dry, and the north wind was cold and dry, not cold and wet.[68] So Asaph connected the east wind with hot-dry bile; warm-wet blood went with the south and cold-wet phlegm with the north, not with the west, winds. Asaph's winds, and their connections with medicine were those of the Middle East, not of the Mediterranean or of Northern Europe (fig. 3.8).

The "wind schema" then, was not traditional and authoritative only; it was also fungible and heuristic. It was an ancient tried-and-true structure that organized many observations of the body's connections to its environment.

HILDEGARD'S IMPLICIT WINDS

The Winds of the Rhineland

Now it turns out that wind comes up in *Causes and Cures* not only explicitly but also implicitly, because Hildegard also frequently used "wind" as a metaphor. When the winds in the thirty passages where she used wind as a metaphor are analyzed, it becomes clear that Hildegard had not just "wind" but many winds in mind—cold and hot, wet and dry, sweet and bitter. Indeed, using details in these passages, her metaphorical winds can also be categorized according to the traditional rubrics of direction, season, weather, taste, humor, quality, and temperament.

Hildegard's most frequent use of wind was as a metaphor for sex or passion, and this was a wind that was hot and dry.[69] Located in the loins, it was strong, hot, and blew into the groin.[70] It brought joy and passion[71] but also dried up the dew.[72]

However, there seems to have been also a second kind of wind; it was, by contrast, cold and dry and blew in winter. It brought frost and stripped the trees of their leaves.[73] A third metaphorical wind blew in spring; it was temperate,[74] sweet, and gentle.[75] It brought forth seeds,[76] and within the body it was the filling wind of joy.[77]

A fourth wind was drying but gentle;[78] and a fifth, temperate and cool.[79] Since it scattered the summer dust, it was an early autumn wind.[80] A sixth wind was strong and powerful; it brought storms, lightning,[81] and thunder.[82] It was a dangerous wind. It was an autumn wind, too, because it harmed plants and fruit (which are found in late summer and early autumn).[83]

In other words, when Hildegard used wind as a metaphor in *Causes and Cures* it was not generic wind as moving air. Rather, she evidently had in mind *specific* winds, with defining characteristics of direction, season, weather, taste, humor, quality, and temperament. In fact, there are enough details that a comparable, though incomplete, diagram of her metaphoric winds can be put together, as can be seen in figure 3.9.

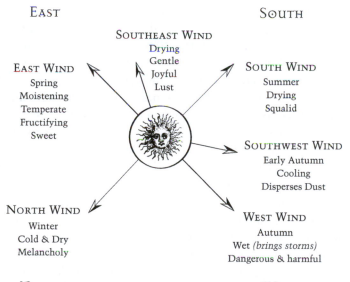

Figure 3.9. The Winds, Directions, Seasons, and Emotions in *Causes and Cures'* Metaphors.

Clearly, the winds that Hildegard had in mind for her metaphors were nearly identical to the winds in her explicit passages—except for a few telling particulars. First, when she used the wind as metaphor, she often associated it with a specific emotion—the north wind to melancholy, the east wind to joy, and the south wind to passion. Second, when she used winds explicitly and traditionally, the south wind was cool and temperate, the west wind was dry, and the north wind, cold and wet. But in her metaphors, the south wind was hot, dry, and burning, the west wind wet, and the north wind cold and dry. There are at least two explanations for this discrepancy. She may simply have made a mistake, since one of the collateral winds of the north wind *was* cold and dry; one of the collaterals of the west wind *was* wet, and one of the collaterals of the south wind *was* hot and dry.

But a second, perhaps more likely, and certainly more interesting possibility is that when Hildegard used wind as a metaphor she was actually thinking of the real local winds that she knew from experience. Traditionally, local winds were also accepted as having specific effects on the body. For instance, Isidore remarked that, in addition to the twelve named winds, there were "innumerable winds that take their names from rivers, places or regions" and affected the body through their effect on local weather.[84] Even today, local winds are named, known by their directions, the weather they bring and their specific effects on the body.[85]

In fact, to this day there are a number of named winds in the Rhineland.[86] Of most importance is the Föhn, a hot, dry south wind held responsible for migraines and murders, and the Bise, a cold, dry north wind responsible for colds.[87] Indeed, Hildegard did ascribe one of her serious illnesses to the south wind. (See Chapter One.) We have already seen in Pliny and Palladius that local winds were responsible for weather, and that they had horticultural implications; and, of course, weathervanes were a usual part of the premodern environment.

All of this suggests that Hildegard's use of winds as metaphors may have derived from her own experience (and local, oral tradition) of the winds in the Rhineland and that the medical observation of real, actual winds played a role in the medical longevity of the wind scheme.[88] Even more significant, her use of the winds as metaphors showed that the winds were not only stand-ins for the *seasons*, but also for the local *weather* typical of the seasons in a specific locality. Winds were in Hildegard's medicine, therefore, not only as signifiers for the effect of the seasons on the body, but also for the effects of local weather.

An Internal Circulating Wind

Hildegard's metaphors of the wind not only imply that the winds were local as well as cardinal; they also suggest that a wind moved around *inside* the body, bringing the qualities of hot and cold, wet and dry, to the organs. Thus she observes that:

> Sometimes in an idle person, a hot wind arises which fills the chest, and makes that person happy. From the chest it ascends to the brain and fills the head and all the veins with heat; then it goes to the lungs and heart, and thence to the umbilicus in a woman and the loins in a man.[89]

And:

> When something is pleasing, either good or bad, then this wind goes out of the bone marrow to the thigh, then to the spleen and fills its veins. From the veins it goes to the heart and then to the liver, and suddenly the person bursts out laughing.[90]

Wind, it would seem, was not only outside but inside the body; and it circulated in the veins to the major organs—brain, lungs, heart, liver, spleen, and genitals.

Once again, though, Hildegard's notion that a wind inside the body circulated was not idiosyncratic. It was, in fact, explicit in the oldest text on the winds that we have, the third-century B.C. *De ventis*.[91]

According to *De ventis*, air was not only outside the body as *pneuma* but inside it as *phusa* (wind)[92] and, in a way, it could be said that this internal wind was the reason for all internal diseases.[93] Because, as everyone knows, it is poor *regime* that causes internal, constitutional disease. But poor regime does so only by its effect on this internal wind, which is agitated by wrong foods and sex and emotion, just as the sea is stirred up by the winds. This internal wind causes not only general but also local disease, because whenever it cannot circulate freely and becomes trapped in a part of the body, it causes specific, localized disease.[94]

That this idea of an internal circulating wind did become an accepted idea in medicine is shown by Asaph. Indeed, he described four separate internal winds that "reflected the four cardinal winds outside the body." They include:

> A cloud-bringing wind that corresponds to the desire to eat; a rain-bringing wind that stores and uses food and drink; a clearing wind that kindles and inflames the entrails in the kidneys; and a wind like that which brings forth plants from the earth. It corresponds to the lower,

expulsive gastrointestinal system. There is a fifth wind as well; it dwells in the brain, heart, and abdomen and is the spirit of life, governing the body all the days of its life upon the earth until its very end.[95]

Indeed, the traditional medical technique called "cupping" in English was known as *ventosa*—the wind technique—in Latin and French.[96] It was based on this very notion—that an internal wind did circulate in the body and could cause disease when its circulation was impaired. As a matter of fact, Hildegard described *ventosa* in a number of passages in *Causes and Cures*.[97] For example:

> If someone has pain in his eyes or ears, or even in his whole body, put a horn (*cornu*) or a *ventosa* on his neck and back. If he has pain in his chest, put *ventosas* on the scapula; for pain in the side put them on the opposite arm and wrist. For thigh pain, on the legs; for pain in the legs, between the buttocks and on the back of the knees. Leave the *ventosas* or the horns on for about a quarter of an hour, so that the blood comes to the surface.[98]

For gout she gave more detail about this "wind technique":

> If a person has gout [*podagra*] in his thighs or in his feet, and the pain is new, then put a number of horns or *ventosas* on the legs, starting with the heels so that the humors are attracted into them. Then take the *ventosas* off and reapply them somewhat higher. Do not break or cut the skin, and continue until the buttocks are reached, then put a tie around the upper part of the knees, so that the humors cannot flow backwards, and let them out of the buttocks by incising the skin. When this is done, the pain of the *podagra* will cease.[99]

What is going on here? It seems that for Hildegard *ventosa* had to do with drawing some disease-causing stuff, which might be in the blood or in the humors, out of the part of the body into which it had lodged into another, more accessible part from which it could be removed. But was this stuff wind?

The answer would seem to be yes, at least in the earliest surviving Latin text that describes the technique of *ventosa*. According to the late-antique writer, Celsus, *ventosa* can be performed with cups (*aenea*—open at one end) or horns (*cornea*—open at both ends). In both cases air (*spiritus*) has to be withdrawn so that the cups will adhere to the skin. When cups are used, the practitioner lights a flame inside them so as to use up their air; when horns are used the practitioner sucks the air out of the horns and then closes the end with wax.[100] There are two reasons to perform *ventosa*, Celsus explains; first, to remove *spiritus*, in which case, the cups are simply applied. Second, if the purpose is to remove small amounts of blood locally, then the skin is cut, and the cups are placed.[101] *Ventosa* was useful whenever there was pain or swelling of an extremity or whenever "there was rotten matter or an evil *spiritus*,"[102] because it removed disease-causing air from the body.

Clearly the early idea is that some substance in the body—bad blood or bad air—caused disease and should be removed. This early description supports the idea that the passages in *Causes and Cures* imply: that there was an air in the body that moved (i.e., a wind) and could cause illness.[103]

Indeed, the eleventh-century Arabic surgeon, Albucasis, provides more evidence that a circulating wind existed inside the body. Translated into Latin only in the 1200s, Albucasis was not a source for Hildegard and his list of cupping points was much more detailed than hers (Spink, 656–672). *Ventosa* was to be used mainly for removing disease-causing humors from the body,[104] he writes, but also, by placing the cups on the abdomen, the technique could be used to remove air that had congested internal organs (Spink, 668).

This internal circulating wind was not quite the same thing as *ventositas*—windiness—what we usually think of as "internal wind." According to the twelfth-century medical writer, Bartholomew of Salerno, *ventositas* occurred in the body for two reasons—from food (chickpeas, fresh meat, and beans)[105] and from unused "seed." Such windiness caused distinct diseases in men and women: in men, prolonged erections;[106] in women, uterine suffocation.[107] Thus, in addition to a circulating wind in the body, there was also wind (*ventositas*) that was, apparently, the flatulence caused by certain foods. As a kind of internal wind, it could injure the body because it came from corrupted humors, by analogy, perhaps, with the miasmic external winds that brought epidemics (401).

Now Hildegard does not mention *ventositas*, which seems to have been the physical flatulence we generally experience with a variety of foods. Nevertheless, these two pieces of evidence from *Causes and Cures*—its metaphorical use of wind, and the wind technique of *ventosa* (along with additional evidence from other sources)—indicate that air as wind was not only outside the body as the seasons and as weather but inside it as well.

Homo Ventosus—The Permeable Airy Body

But how did this system of external winds and internal wind work as a way of understanding, and even treating, the body?[108] That is, what was the relationship between the external winds of seasons and weather and this internal wind? The earliest text on the winds, *De ventis*, implied that internal and external winds were connected by the orifices of the body; they were the channels by which the winds carried weather and climate (i.e., the qualities of hot-cold/wet-dry) into the body.

Indeed, that air entered and left the body through its openings (not only as air through the nose but also as air through orifices such as eyes, ears, mouth, anus, and vagina, pores, and open sores and cuts), was commonly accepted, at least by the fourteenth century.[109] In fact, this was how "bad air"—*malaria*—injured the body, and how good air as herbal fumigation, worked. In both cases bad or good air got into the body through its pores and orifices. Certainly the technique of *ventosa* implied that there were certain sites on the body that allowed for the egress (and presumably the ingress) of wind. In *Causes and Cures* there is the suggestion that the body was permeable to the wind through its orifices, too, since Hildegard explained that hearing got into the body because air passed into the ears; and odors were carried into the body by the wind.[110]

But *Causes and Cures* alone does not present any overarching conception of the permeable, airy body. Only with Hildegard's last work, the visionary *Liber divinorum operum*, did her idea of the windy body become explicit.[111] In it she explains quite clearly that the external winds affected the internal body by entering it:

> All the winds, whether naturally or by God's disposition, penetrate man's body, so that by their breaths he is either strengthened or made destitute.[112]

The winds enter the body and then, they, somehow, make the body stronger, or weaker. How? By changing the humors:

> The different qualities of the winds and of the air go together, and they change the humors in the body, which take up the qualities of the winds.[113]

The medium through which this occurs is the *anima*, which seems to be identical with the internal circulating wind:

> This makes man somewhat mutable in his humors; for when he takes on this changed air, and his soul [*anima*] transmits it to the interior of

his body, his humors are changed, and often induce sickness or health
in him.[114]

To sum up: the body of *Causes and Cures* was permeable. It was
affected by the external winds of season and weather, and also, as we shall
see, by food, which itself was affected by the winds of season and weather.
All of these influences could enter the body through its pores and openings,
to modify the final common pathway in the body—the internal circulating
wind. Wind, that is to say, was the tie that bound the unseen; it invisibly
connected macrocosm and microcosm, the fabric of outside and inside.

Would this analysis of the elements in *Causes and Cures* have led to the
same kind of connections had it been focused on one of the other elements
instead of air? On fire as sun, moon, and stars? On water and the waters that
we drink? On earth as land and soil, clay, sand, or rock? It is impossible to
say. But fire, water, and earth were certainly not left out of the premodern
system. They turn out to be, at the very least, the basic connectors between
the wind outside the body and the wind inside the body. This was because
sun, rain, and earth were the main determinants of the second basic premodern concept—the humors.

Chapter Four
A Fluid Concept—The Humors

"Four are the winds, four the corners of the sky, four the seasons, . . . four are the humors in the body."[1] So began the wise author of the eighth-century *Art of Medicine*. What were these humors he wrote about? What did they have to do with the winds, the directions, and the seasons? And how did Hildegard understand them?

We know from Schöner's diagram in the Introduction (fig. 0.1) that humors—of some kind—were an important part of the premodern schema; that they somehow connected and linked winds, seasons, ages. But how? What made the humors so perfectly adapted and adaptable to serve as the linkage between the abstract cardinal winds, the real winds of weather, the seasons, the ages, and the temperaments?

The humors were a key concept in Hildegard's medical thought, especially in Book Two of *Causes and Cures*; and a detailed examination of that text will demonstrate that she had in mind three separate kinds of humors. First, she had an idea of humors as an (idiosyncratic) version of the traditional four "qualities"—hot and cold, wet and dry. She also used the usual bodily humors of blood, phlegm, bile, and melancholia. And last, she used humor to mean the essence or "juice" of anything, but especially the medicinal juice of plants. Did this multivalence of the Hildegardian humor reflect common usage? Or was it entirely idiosyncratic, a result of Hildegard's unusual education, training, and mind?

When compared with the humors in contemporaneous popular texts, Hildegard's humors will turn out to be paradigmatic. In the ancient tradition she inherited, there were, indeed, three distinct meanings for "humor." Her synthesis reflected an odd but traditional amalgam of a medical system where the humors were bodily fluids, a cosmological system where the humors were the carriers of the four qualities, and a horticultural system, where the humors were the essences of plants. All three meanings were indissolubly linked because they derived from the four-cornered, four-directional cosmos that first began to take shape with the four winds.

BOOK TWO OF *CAUSES AND CURES*

Summary: A Physiology of Humors

Book Two of *Causes and Cures* is the richest and longest of its chapters, making up more than half of the total text. It is made up of hundreds of passages on many subjects, ranging from sex, conception, and reproduction, to disease and its physical treatment. As previously noted, this apparent disorderliness has been a leading reason why some scholars have rejected Hildegard's authorship of the text. But it is mainly the rubrics and paragraphing that make the text seem without order; when they are set aside, Book Two, like Book One, does have an indistinct but appreciable structure. Here it is the concept of "humor."

Thus near the beginning, Hildegard explains that, while other creatures are formed from a single *element*—reptiles and worms from earth, fish from water, and birds and beasts from air[2]—human beings come from a mixture of all four *humors*. Just as the four elements outside the body temper one another, so the four humors inside the body are there, in part, to temper one another:

> But humans could only have been made from the four humors—not from one or two or even three—so that each may temper one another, just as the universe is made of four elements in harmony with one another.[3]

Now before Adam's sin and the Fall from Paradise the human body was perfect—unchangeable and incorruptible; and so it never became ill and disease did not even exist. But at the moment of the Fall, human blood soured and spoiled in an instant. Bodily imperfection became possible and so did disease:

> God made human beings so that all animals would serve them, but when they disobeyed Him, they were changed as much in the body as in the mind. For the purity of the blood was soured, and seminal ejaculations became necessary for cleaning. Had humans, instead, remained in Paradise, they would have stayed in a perfect and unchangeable state, but after the transgression, everything was turned into another and bitter mode.[4]

This is how sex and the ever-changing life of the world came into being.

So we are led into the first main theme of Book Two—reproduction—which Hildegard explains mainly with the concept of *humor*. Reproduction takes place by means of the humors: every being produces seed that replicates its *humoral* nature. Just like wheat, rye, or barley, the nature of the seed planted by one's parents determines one's nature:

> [On reproduction.] Thus the *flecmata* and the *humors* increase according to the nature of the semen, just as wheat, barley, or rye are produced according to the type of seed which is sown.[5]

This is how strong, virile men in union with robust and faithful women produce healthy progeny and how weak men produce sickly offspring.[6]

Not only does the concept of humor account for some aspects of reproduction, conception, sexual predilection, and inborn tendencies towards certain diseases, but it also helps explain human physiology—the workings of heart, lung, liver, spleen, and stomach (CC, II, 137–139). For instance, Hildegard explains that one function of the liver is to receive and then recirculate the humors to the other main organs:

> Like a vessel put into a fountain accepts water and then sends it out again, so the liver is in the body. The heart, lung, and stomach pour juices into it, and it pours the juices back. But if the liver is rotten and perforated, then it cannot receive the good juices and so the juices and humors return to the heart, lung, and liver in a flood.[7]

In fact, Hildegard explains much of pathology on the basis of the humors (CC, II, 137–144). She also uses the humors to explain constitutional diseases, such as paralysis:

> Often it happens that storms of bad humors fall into a limb and by their unhealthiness they close up the veins so that the humors cannot flood the veins, which dry up because they lack an opening for blood; and so a person begins to limp.[8]

Or gout:

> And it happens often to those who use many different kinds of foods that they easily become ill. . . . This is because the bad humors overflow and increase so that they cannot be contained, whence they go

hither and yon, up and down . . . eventually settling in the limbs, from which comes gout [*podagra*].[9]

That is, although in theory disease results from the corruption of the humors that followed the Fall, in practice disease is more complex, a consequence of how a person's inherited complexion interacts with his environment, especially with the humors in food and drink. Indeed, it was by the intermediary of the humors that the environment, in the form of the traditional non-naturals—food and drink, rest and exercise, waking and sleeping, sexual activity, emotions, and climate—affected the body.

For instance, Hildegard explains that food and drink affect the body mainly because of their effect on his "*juices*," a concept that she uses almost synonymously with "humor":[10]

> When someone eats and drinks, the pull of consciousness draws the taste, subtle *juice*, and odor of the food to his brain, which heat its veins. But the rest of the food and drink that arrives at his stomach heats the heart, liver, and lung, and its taste, subtle *juice*, and odor fill and nourish the veins, just like a dried piece of intestine puffs up when it is put in water. By eating and drinking a person replenishes his veins by the *juice* of food and drink, so that it warms the *sanguis* [blood] and *tabes* in his veins.[11]

Also waking and sleeping (another pair of the "non-naturals" although Hildegard does not refer to them in this way), affect the body through the concept of *succus* (juice). For instance:

> After eating a person should not sleep until the taste, *juice*, and odor of the foods eaten reach their proper places. . . . Otherwise, they can go to the wrong place like dust blown hither and yon.[12]

Similarly, it was through the humors that the other non-naturals—sex, exercise, perspiration, and emotions—affected the body (although, again, Hildegard does not refer to them as the non-naturals.) For example: "Humors . . . are expelled from the body by *sweating*."[13] Also:

> Thus those humors in the body, like a bitter smoke, through *sadness* can come out of the heart and scatter around it.[14]

And:

> In these people various humors of *flecmata* are excited, so that those
> same *flecmata* (from immoderate food and drink or excessive joy, sad-
> ness, anger, or desire) are stirred up. . . . [15]

Hildegard also explains the physiology of menstruation, fertility, and lacta-
tion using the concept of the humors. In fact, health itself is the result of a
proper equilibrium of all the internal humors, through proper diet, or, if
necessary, through medicinal herbs. Indeed, herbs work in the body *because*
they balance the humors. For example:

> For the heat of agrimony and galangal and the strength of storax and
> the heat of polypoidy and the heat of chelidony overwhelm the cold
> humors from which *flecma* is born, and the coldness of fenugreek and
> the coldness of storkesnable dissipate their coldness. [16]

That is, Hildegard understood her recipes as working through their effect
on the humors, although she was not perfectly consistent. For instance, here
"cold humors" differ from the humor of phlegm, which they produce.

In fact, it was because the urine mirrored the internal humors that it
could be used for diagnosis. In a healthy person the urine should be a perfect mixture of all four humors; and upon standing it should layer into red,
black, yellow, and clear, reflecting the colors of the four humors. Any deviation from this is a sign of illness. For example:

> If the urine is pure and clear like water, then the patient will die. This is
> because his blood has become cold and, therefore, the other humors all
> flow together like coagulated milk, because they have lost their heat and
> blood. The reason that the urine is clear is because the humors are no
> longer mixed in it and no longer perform their functions. [17]

All of the physical techniques she describes in some detail—bleeding (Sweet,
390–391), cupping (*ventosa*), moxibustion (*coctura*) (CC, II, 167:17–170:9),
and scarification (CC, II, 165:26–166:10)—also work mainly by restoring
the humoral balance of the body. Weather, too, acts on the body through the
medium of the humors. For instance, "The heat and brightness of the sun
causes the humors in the brain to increase past their proper balance. . . ."[18]

At the end of Book Two, Hildegard describes several constitutional
diseases, including *lepra* (probably some kind of skin disease, not necessarily leprosy), *gutta* (probably arthritides, not necessarily gout), and,
especially, fevers, to which she devotes several pages. In each of the long

passages explaining these problems, she uses the humors to describe their pathophysiology. For example: "Fevers arise from the movement of harmful humors . . . ";[19] or "for those who are of soft flesh from too many drinks, the bad humors which they contain can suddenly fall into a limb and disturb it. . . ."[20]

In sum, although Book Two does seem, at times, to be a jumble of medical passages, like Book One, it, too, has a recognizable heuristic thread. In Book One, it was the elements. Here, it is the concept of good and bad liquids—*succus, liquor, tabes, flecmata, livores,* but, especially, *humor.* It is the humors in the blood that became corrupted at the Fall of Man, and it is the humors that account for many of the diseases in the body. The humors also provide Hildegard with a way of rationalizing the efficacy of various physical treatments and the workings of her medicines. What is not clear, however, from this summary, is what Hildegard meant by "humors."

Hildegard's Explicit Humors

Now it *should* be clear what Hildegard meant by humors because she defined them near the beginning of Book Two.[21] Nevertheless, even her explicit discussion is complex and confusing. She writes that just as there are four elements in the external world, so there are four humors in the body; they are called *siccum* (dryness), *humidum* (wetness), *spuma* (foam),[22] and *tepidum* (warmth).[23]

All these are unusual terms for the humors, but *spuma* is the oddest. Hildegard uses *spuma* to signify foam or scum—the foam of turbulent waves for instance, or the froth on something when it boils.[24] But she also uses it to mean some stuff specific to the body, which is neither blood, nor water, *livor,* nor *tabes.*[25] *Spuma* is watery;[26] it is the liquid part of semen[27] or sexual lubricant;[28] but it is also in the blood like frothy milk.[29] Apparently by *spuma* Hildegard meant *anything* cold, wet, and foamy in the body: including the saliva of the epileptic,[30] the semen or sexual secretions of men and women, and the white creamy layer of lipids that sometimes settles on the top of blood.

After naming her humors she goes on to explain in a particularly confusing way that the two most abundant of the humors are called the *flecmata* and the least two, *livores*[31] (although she also writes that there are three *flecmatas—siccum, humidum,* and *tepidum*). For the body to be healthy, there must be a specific ratio between all four humors—each should be related to the one that follows it by being a quarter and a sixth more than it, and the one that comes after should be twice and five-sixths more:

But the humor that is highest is greater than the one which follows by 1/4 part and 1/2 of the third part. And the next should temper two parts and the rest of the 1/3. Whichever is the first humor according to this measure exceeds the second, which are called the two *flecmata*, and in this way the second exceeds the third, and the third the fourth, and they are called *livores*. And when it is this way a person is tranquil. But when some humor exceeds its measure, then there is danger.[32]

From this it would seem that, on the one hand, Hildegard's definition of the humors—the dry, the wet, the tepid, and the foamy (or dryness, wetness, tepidness, and foaminess)—was a reification of the four Galenic *qualities* of dry and wet, hot and cold. As in the Galenic system (McVaugh 1975, 103), for Hildegard disease will arise whenever the humors are in a "wrong" proportion. But, on the other hand, her prescription of a seemingly precise mathematical ratio amongst her four humors is unique, without historical precedent and without a clear physiologic correlation. Some authors have hypothesized that the ratio she prescribes might have derived from Plato's *Timaeus*, but the passage that has been cited shares with *Causes and Cures* only the concept of ratio, not the actual proportion (to the extent that can be calculated from her ambiguous formula.)[33]

It is difficult to imagine how Hildegard came up with these ratios, or what she actually meant by them, physiologically, for medical practice. It was not unusual for blood to be examined; indeed Hildegard explains how to use blood for this purpose. After it is removed it should be examined for color, turbidity, waxiness, and for black specks (CC, II, 164:1–5). She does not, however, explain how to determine the ratios that she describes. Other medical writers did explain how to examine settled blood, so perhaps her proportions were meant to suggest the physiological proportions of the colored fluids seen in settled blood. If so, how would the hotness, dryness, warmth, or foaminess of blood be determined?

I can only conjecture. The hotness and coldness, wetness and dryness of foods and medicines were often determined by their tastes, as wine is today. For example, a sweet taste meant that a food or medicine was warm and wet. If it had a bitter taste, it was warm and dry; if sour, it was cold and wet; if sharp, it was cold and dry (see below and see Chapter Five). Perhaps what Hildegard meant then is that the practitioner should examine the blood or urine (or both), by look, by feel, and especially by taste, to determine its proportions of coldness, dryness, tepidness, and foaminess.

In any case, her notion has unusual implications for understanding disease as a humoral problem, because it privileges some kind of optimal *order* of the humors. In the Galenic model as customarily described, it was the *balance* or *proportion* of humors or qualities—in a given person, for a specific organ—that determined health, and their *imbalance* that caused disease.[34] Perfect balance meant an equality among the four humors, so that a perfectly tempered complexion had an equal amount of each of the four humors. Proportion was understood as some optimal *combination* of the four humors for a particular complexion. But for Hildegard, it was the *order* of the humors in a complexion that defined the proper humoral mix.[35]

This dependence of her proper humoral ratios on the order of the humors meant that two apparently similar humoral mixes were, in fact, radically different. For example, in the complexion she names *siccum/spuma/ humidum/tepidum*, the humor, *siccum*, must be, according to her definitions, approximately one and a half times as much as *spuma*, two and a quarter times as much as *humidum*, and about three and half times as much as *tepidum*. The humoral makeup of what would seem to be a similar combination—*spuma/siccum/humidum/tepidum*—is radically different—for *spuma* is one and a half times as much as *siccum*, two and a quarter times as much as *humidum*, and three and half times as much as *tepidum*, and gives a very different temperament.[36] The former produces a person who is dull, irritable, and robust;[37] the latter a person who is gentle, tender, and does not live a long time.[38]

Thus in Hildegard's humoral system it was the *order*, not merely the *combination*, of humors that determined character and the propensity to disease, since it was the *order* of the humors that produced different humoral makeups.[39] Because of this, the number of possible basic complexions was much larger than in traditional Galenism, as there are *twenty-four* permutations of four objects, while only five or nine basic *combinations* of four. (There are five in the physiology that relied on humors—the sanguine, the choleric, the melancholic, the phlegmatic, and the well-tempered—and nine in the one that relies on qualities—purely hot, cold, wet, or dry; mixed hot/wet, hot/dry, cold/wet, cold/dry; and one that is perfectly balanced). But, since there are *twenty-four* (4x3x2x1) permutations of an ordered set of four, Hildegard's humors allow for *twenty-four* types of basic physiologies. And indeed, she does provide twenty-four descriptions of the kind of characters and propensity to disease that accompanies each possible humoral type.[40]

In sum, Hildegard's explicit humoral system, while bearing a superficial resemblance to the Galenic (see below) was not simply derivatory, and yet it was, somehow, related to the traditional qualitative system. Perhaps Hildegard had re-intuited a kind of Stoic medicine, since, according to Max Wellmann, in Stoic philosophy the four qualities, not the four elements, made up every-thing, and they were regarded as physical substances (Wellmann, 146–148).

Hildegard's Implicit Humors

In addition to these *explicit*, defined humors, however, Hildegard also referred to blood, phlegm, yellow bile, and melancholy; that is, to the traditional four humors of ancient medicine. This bivalence has been noted before. For instance, Sabina Flanagan observed that Hildegard's humors were contradic-tory, being both qualities and bodily fluids (1996, 15). Although Hildegard does not define the traditional humors directly, she does refer to each of them frequently, so it is possible to get a good idea of how she regarded them by looking at each passage where she mentions *fel, melancholia, flecma,* and *san-guis.* For each, she refers, sometimes obliquely, to a variety of characteristics, including color, quality, taste, and pathological effect.

For example, *fel,* which, as we shall see, usually stood for what became, in English, yellow bile or bile, was for Hildegard also, yellow, since she writes that "*pestis,* also called jaundice, comes from too much *fel.* . . ."[41] It was bit-ter[42] or sour in taste;[43] it came from bad, weak humors and increased with certain fevers and with certain emotions, especially anger. "*Fel* comes from the weaker humors, from fevers, and from great, frequent anger. . . ."[44]

Melancholia was black[45] and also dry, since "feces turn black and are dry from too much *melancholia.*"[46] It was bitter, because "*melancholia* decreases with sweet foods and increases with bitter ones."[47]

Flecma (phlegm) was tepid and sticky, since it was "neither wet nor thick but tepid, and like *livor,* it is sticky. . . ."[48] Also, *flecma* was cold because "if someone eats very cold foods in a hot summer, *flecma* will appear."[49] It was more wet than dry.[50] *Flecma* waxed and waned and so, more than the other humors, was easy for the practitioner to deal with.[51]

Finally, *sanguis* was hot, since wine as the *sanguis* of the earth, "carries its heat from the bladder to the bone marrow. . . ."[52] It was dry, because "*sanguis* is essentially dry, and would not flow without wateriness. . . ."[53] *Sanguis* was distinct from the "water," (that is, probably the serum), in the blood, which was called *tabes*[54] and differed from all the other humors.[55]

In other words, despite Hildegard's careful and explicit definition of the humors as *spuma, tepidum, humidum,* and *siccum,* implicit in *Causes and Cures* were the traditional four humors of the bodily fluids. Although they *had* qualities, they were clearly not *identical* to her explicit humors of *spuma, tepidum, humidum,* and *siccum.* Indeed, it is noteworthy that in the *Lingua ignota* Hildegard did not provide words for her explicit humors, but did pro-vide them for some of the traditional ones: for *fel* (idiez), *sanguis* (rubianz), and *tabes* (scirinz). This suggests that her explicit humors of *spuma, tepidum, humidum*, and *siccum* may have belonged to a more theoretical and abstract understanding of the body, while her implicit humors, which corresponded to the four traditional humors of the bodily fluids, were actual physiologic liquids for her.

These extracted passages even imply a physiology for these bodily fluids that roughly corresponds to traditional physiology.[56] For instance, they suggest that Hildegard assumed most of the major organs—heart, lung, liver, stomach—stored and excreted humors:

> But the humors that *come out of the heart, liver, lung, stomach and other organs*, if sometimes they turn into a wrongful difference or superfluity, then they become sticky, slippery, and tepid, and if they stay inside, will cause illness.[57]

Or:

> Whoever is too fat or too thin often has too many bad humors, since they do not have a proper temperament, and sometimes the bad humors that cause *melancholia* flow out of the heart, liver, lung, stomach, and other organs. . . .[58]

That is, the humors were inside the heart, liver, lung, and stomach, and could be excreted by them. In fact, Hildegard seems to have assumed an orderly flow of humors—from stomach to spleen,[59] from spleen[60] and other organs to the heart;[61] from blood to brain,[62] from brain to lung,[63] from uterus[64] and kidneys[65] out of the body. Not only did the humors move in the body, but in illness they moved in a disorderly way: "Healthy people do not have mobile humors, that is, those that move hither and yon."[66]

In sum, in addition to the explicit humors of *spuma, tepidum, siccum*, and *humidum*, Hildegard also made use of what has traditionally been called the humors—blood, bile, phlegm, and melancholia. And these humors had a physiology; they traveled from stomach to liver, from liver to the rest of the body, and were excreted from the body as urine by the kidney.

A Third Kind of Humor

But a close reading of all of the extracted passages in *Causes and Cures* that contain the term "humor" reveals still a third meaning. *Humor* also signified a group of substances that could be bad or good, temperate, or delectable. Thus Hildegard very often refers to "bad humors"—to evil,[67] wrong,[68] harmful humors.[69] For instance, "bad humors can cause anxiety in a human being when they make him ill."[70] Or, "just as a storm arises and then quiets, and just as in wine the must boils up and then falls, so also the bad humors in humans sometimes arise and then become quiet. . . ."[71] Bad humors can cause specific illnesses, such as back pain: "Many times pain in the back or kidneys will come from these unjust humors."[72] Or they can cause paralysis: "Often it happens that tempests of bad humors fall in a certain limb, and block the blood from getting to the veins, which dry up and cause the person to be lame."[73] Bad humors can also cause fistulas,[74] pustules,[75] and rashes.[76] Bad humors can cause testicular swelling,[77] dullness of mind,[78] facial swelling,[79] and fevers.[80]

Not only are there bad humors, there are also good humors, as well as delectable humors, temperate humors, cold, hot, mobile, and acid humors. For instance, immoderate laughing destroys the good humors, which can be replenished by drinking wine heated with nutmeg and sugar: "And when wine is tempered with heat, it restores the good humors that have been lost by immoderate laughing."[81] "A certain delectable or delightful humor is in some people, which is not depressed by sadness or bitterness, and which flees and avoids the bitterness of *melancholia*."[82] Also, "Some people have a temperate body and are neither too fat nor too thin, nor too muscular, and they have temperate humors."[83] There are also sharp, acidic humors,[84] and cold humors,[85] hot humors,[86] and right humors.[87]

Hildegard, then, used many different adjectives with *humor*, leading to two possible interpretations. First, she could have meant that some of her "explicit" humors or some of the traditional humors were sweet or acid, mobile or stationary, bad or good. Or she could be referring to yet a third kind of humor, a fluid or liquid of some sort.

A number of authors have opted for the first interpretation—that the humors had theological significance for Hildegard and were, therefore, divided into good or bad. For instance, Danielle Jacquart held that blood and breath were two (good) spiritual substances, and phlegm and *livor* were two (bad) material ones (1998 "Hildegarde et la physiologie," 128–130 and 133–134). Sabina Flanagan also argued that Hildegard's humors should be understood as hierarchically ordered, like all of Hildegard's world, into good and bad (1996). Hildegard certainly implied that *fel* and *melancholia* were the cause of disease when she wrote that "if a person lacked *fel* and *melancholia*, he would never get sick."[88] Indeed, as shown below, there was often a sense in premodern medicine that some humors (particularly *sanguis*) were intrinsically good and others (usually *melancholia*, but sometimes also bile) were bad.

Here, however, I argue for the second possibility: that, in addition to her explicit humors of *spuma*, *tepidum*, *humidum*, and *siccum*, and the four bodily humoral fluids, Hildegard also meant that there were general fluids, mostly plant essences, that were also humors. Thus she never specifically refers to bile and *melancholia* as the "bad humors," or to *sanguis* as the "good humor;" and she always refers to bad and good humors as if there were more than two harmful ones and more than two delectable, good, and tasty ones. In this sense humor meant something different—it meant juice, liquid, or essence, especially of ingestible things like plants, animals, and stones. For instance, humors are in plants:

> Heat them again in the water while the patient is in the bath, so that through the *humors of these herbs* the skin and flesh on the outside, and the uterus on the inside, will be softened and so that the closed veins will open.[89]

Even stones have humors—"Stones have fire and many humors in them."[90]

In fact, sometimes Hildegard seems to use *humor* as a synonym for "juice" (*succus*). For example:

> A person, whether old, or young, if he has weakness in the eyes should gather the leaves of this tree [*affaldra*] in spring before the fruit appears, because then it is sweet and healing, just like young women before they bear children. And he should rub the leaves so that their juice [*succum*] is expressed, and anoint his brows and eyes with a feather. . . . But if in spring *affaldra* has already fruited, then let him cut a single branch so that it becomes wet with its juice [*succum*], and when you sense no further wetness, then with a knife cut that eruption with small hits, so that more of its juice [*humor*] can come out. . . .[91]

Here Hildegard first uses *succus* to mean what we would call the sap of the tree, and in the next sentence refers to the same substance as *humor*.

Humor, then, could be the juice or sap of a plant, but it could also be anything liquid that comes out of the body. For instance, "sometimes humor can come out of ulcers."[92] Or humor can be something that comes out of the earth. For example: "Sometimes this earth no longer works against illnesses, because its *humor* and the juice of this tree have already produced fruit."[93] *This* kind of humor can even be equated with the dew.[94] Even the earth has "humor," and its humor produces useful plants.[95]

Can it be that Hildegard meant that such "humors" were simply mixtures of the basic four humors, whether the set of *spuma/siccum/tepidum/humidum* or *sanguis/flecma/melancholia/tabes/fel?* Apparently not, because she explains that the liver is damaged when there are too many *different* humors from foods:

> If someone immoderately eats many different foods and his liver is injured and hardened from the different humors of these foods, then let him take a little lettuce . . ."[96]

Since it is the *variety* of humors in foods that is damaging, this must mean that there are different humors in different foods.

What seems to be going on is that *each* distinct thing has a distinct "humor" unique to it. At least, this is what she is suggesting when she writes that:

> When a person eats a new and unknown food or drinks a new and previously unknown drink, then from these new humors, other humors in the body are moved and flow from the nose, just like a new wine in its bottle ejects its dregs."[97]

This must mean that every food or drink has its own peculiar humor, which the body has to get accustomed to. Also, each peculiar humor seems to have its own particular odor and taste. For instance, "after eating, one should not go to sleep until the flavor, juice, and odor of the foods reach their proper places."[98] *Succus* (juice) is used as a synonym for humor, in the broad sense; I infer that it is the food's plant *humor*, which has its own taste and odor.

It seems to be through these *specific* humors that medicines work in the body. For instance, it is the unique humors in food that, after being processed by the body, help the body to grow and to heal:

> From these different humors, both good and bad, the flesh and veins of the body are formed; just the way flour rises from yeast.[99]

Hildegard, then, used "humor" in *three* different ways. In the first the humors corresponded roughly to a reification of the four qualities; and in the second, humors meant the traditional four bodily fluids. In the third, she used "humor" generically, to mean liquid, juice, or essence, but, especially, sap. Each of these three kinds of humors implies a somewhat different and characteristic model for disease causation to which she obliquely refers. To the humors as qualities, disease mainly occurs because of an *imbalance* (or dis-order). To the humors as the four bodily fluids, disease occurs whenever they are in the wrong place, at the wrong time, or for the wrong temperament. To the humors as plant essences, disease occurs mainly as the result of toxins—bad humors.

How are we to understand this multivalence of Hildegard's humors? Was it idiosyncratic, her own personal synthesis of medicine? An inspired vision? Or does she tell us something about the commonly-held understanding of humoral theory transmitted to the Middle Ages from the ancient world?

HUMORAL THEORY AND HILDEGARD

To answer this question, we must first establish some way to determine what the "commonly-held understanding of humoral theory" was. This is not easy, because humoral theory stood for different things at different times. It developed and regressed; was complicated and simplified; was taken over by different sects, modified, and even changed.

Moreover, how it did develop and change is almost impossible to ascertain, as noted in the Introduction. What we do know is that the concept of four humors—blood, phlegm, bile, melancholy—did not come from philosophic speculation but from medical observation—scrutinizing settled blood[100] and observing the role of bodily fluids (red blood, black feces, yellow urine, white pus) in a variety of disease-states.[101] When these medical observations were combined with the philosophic theory of the elements, humoral theory came into being (Klibansky, 10).

This was in the fourth century B.C.; it took centuries to develop into the organized tetradic system of Galen (130–220 A.D.), who fused the medical model of the humors with the Stoic philosophy of the Pneumatics. For Galen the four qualities, not the four elements or humors, were the building blocks of the body.[102] Seasons, ages, temperaments, and sexes resulted from fluctuations in the body's mix of hot and cold, wet and dry.[103] Galen's model, it is usually maintained, is what passed into the Middle Ages as the *summa* of ancient medicine, and, presumably, what we should use to establish "the commonly-held understanding of humoral theory" against which to measure Hildegard.

Unfortunately, although this is the accepted history, none of Galen's own texts on humoral theory was available when Hildegard was writing *Causes and Cures*. Instead the theory was passed on in short summaries in the encyclopedias or as letters attributed to Hippocrates or Galen. Thus it was not Galen *per se*, but this schematic "Galenism" that was transmitted to the Early Middle Ages and that represented both the common understanding of humoral theory, and what Hildegard would have had to teach her humoral theory.[104] Only in the early 1100s did more sophisticated summaries of Galen begin to appear (the *Isagoge* and *Pantegni*). Galen's works *per se* did not become available in Latin until after 1150, that is, after Hildegard had formed her medical understanding. Galen, then, cannot exactly be used to determine this "common understanding."

Another impediment to determining the background of Hildegard's humors is that, while our canonical view of humoral theory is straightforward, humoral theory itself was never unitary. For example, although the humors *do* appear early on as four fluids linked to the four qualities and the four seasons,[105] nevertheless, this was not *only* how they appeared. For instance, Aristotle referred to three humors, not four,[106] and Praxagorus to ten (Smith 1979, 188); while the author of *On Ancient Medicine* "scoff[ed] at the idea of four qualities, positing a multiplicity of continuous powers that would include not only hot and cold, but sweet and sour and astringent and insipid" (Nutton 1995, 284).

Indeed, even as late as Galen there were medical theories that did not use humoral theory at all. For instance, the Methodists considered the body to be a structure of canals and corpuscles ruled by the two physiological responses of "laxity" and "stricture"; and the Empirics focused on the actual effects of particular drugs on particular symptoms.[107] (In fact, a Methodist text that survived into the Early Middle Ages probably did provide one of the strands in Hildegard's own humoral theory; see below.)

Thus there was more than one approach to the humors in the ancient world, and Galen, despite his subsequent importance to medical history, was not its only representative.

It is no wonder then, that confusing, and sometimes even completely opposite, conclusions about Hildegard's humors have been reached. Thus Margret Berger used a Galenic schema of qualities and explained that Hildegard's use of the qualities as humors was "Galenic" (1999, 15–18). But Sue Cannon used a definition of ancient humoral theory that understood the humors as bodily fluids and concluded that Hildegard's use of the qualities as humors was idiosyncratic. Danielle Jacquart concluded that Hildegard's humors are ultimately confusing because her use of them was confused (1998, "Hildegarde et la physiologie," 127).

Texts and Contexts for Hildegard's Humors

An anthropological approach might be more organic. Suppose we imagine ourselves as Hildegard's student, her nun-infirmarian, perhaps, trying to use and understand the humors in *Causes and Cures* from the point of view of actual practice. The question then becomes: What textual background would such a person have had access to for understanding the humors of *Causes and Cures*? That is, in the Rhineland c. 1150, what texts about the humors were available? Once this canon has been determined, it can be used to put together the "commonly-held understanding of the humors" against which to measure Hildegard.

But putting together such a canon is surprisingly difficult, since complete texts for most of the important ancient writers do not exist (Lloyd, 75–90). Even so straightforward a task as learning what ancient texts survived into the twelfth century, in what state, and what they have to say about the humors, is difficult. To be sure, humoral theory was implied in many texts (such as herbals and even theology),[108] but it was *explicitly* transmitted in only two kinds of texts—general encyclopedias and medical texts. Certainly Hildegard and her nun-student had access to at least one of the encyclopedic versions of humoral theory—to Pliny, Isidore, Bede, Hrbanus Maurus, and/or those of their contemporaries, Honorius Augustodunensis, William of Conches, or Pseudo-Bede.[109]

But "Pliny" (actually, Plinius Secundus, the usual form in which Pliny's medicine was known in the Early Middle Ages), had no specific passage on the humors; and Hrbanus Maurus, Bede, and Honorius Augustodunensis mainly copied or echoed Isidore, just as they did with their material on the winds. We can, therefore, use Isidore of Seville, William of Conches, and Pseudo-Bede to get a good idea of what kind of humoral theory such "general encyclopedias" passed on.[110]

Next, which *medical* texts had material on humoral theory and were also theoretically available to Hildegard and her student infirmarian?[111] Of the most ancient Hippocratic texts, only a few were generally available in Latin c. 1150: the *Aphorisms*, ("the best known and most widely disseminated" according to Kibre 1985, 29); *Airs, Waters, Places*; the *Prognosis*; and a fragment of the *De natura hominum*. Although each text does imply some kind of humoral theory, (e.g., in their linking of winds, seasons, qualities, and humors), none of these texts, as they were available to Hildegard and her student, had anything *specifically* to say about the humors.[112]

Establishing which Galenic texts were available in Latin c. 1150 is even more difficult. For one thing, modern editions of Galen give exemplary, scholarly texts that were translated into Latin from Greek texts only available in the Renaissance or even later. These were not, necessarily, representative of the 1150 versions. For another, they (in particular, Kühn) sometimes include spurious texts written after 1150, such as the *De compositione medicamentorum secundum locos*, as well as texts not yet translated into Latin in 1150.

Indeed, although there were a number of "Galenic" texts c. 1150—that is, texts that proffered a version claiming to be derived from Galen's works—it is not clear whether any actual text by Galen on humoral theory was popularly available pre-1150.[113] There *were* a few copies of *De sectis*, *Ars medica*, *De pulsibus*, and the *Commentary on the Aphorisms*. However, none of these contained anything *explicit* on humoral theory.

What scholars have called "Galenic" texts—i.e., texts not by Galen but conveying "Galenic theory"—can be divided into two types. First, there were the newly translated (from Arabic) summaries of Galen's ideas; they began to be available in Latin in the late eleventh century. Indeed, by 1150 a collection of such texts, the so-called *Articella*, was used in Salerno. It consisted of the *Isagoge*, the Hippocratic *Aphorisms* and *Prognosis*, Theophilus' *Urines*, Philometus' *On Pulses*, and Galen's *Tegni*.[114] Although Hildegard's historians have not yet agreed on whether she had access to this new material,[115] it seems reasonable to include in our canon a representative example (Johannitius' *Isagoge*).[116] In addition, there is a passage on the humors from the Greek text of Nemesius, which was paraphrased into Latin by Alfanus in the eleventh century,[117] and it seems reasonable to include it as well.

Second, there were other "Galenic" texts on humoral theory that were widely diffused and readily available *before* the new translations.[118] Hildegard, and our stand-in student practitioner, would very likely have had access to one or more of these. Unfortunately, the historiography on these texts is even more confusing. Some texts carry a variety of titles, while in other cases different texts have been given the same title.

To give the reader an idea of the problem. On the one hand, according to Erich Schöner, Galen's humoral theory went into the Middle Ages *via* four popular anonymous tracts, which he named: the *Ad Maecenatem*, the *Ad Antiochum*, the *Letter to Diokles* and the *Isagoge saluberimum*. On the other hand, Pearl Kibre wrote that the subject of humors, an important element in ancient and medieval medicine, was covered for the Middle Ages in *Ad Antiochum: De quatuor humoribus* (171). According to Augusto Beccaria, material on the humors was transmitted in numerous popular texts, as established by the number of surviving manuscripts that contain them: in the *Ad Maecenatem* (10 manuscripts), the *Ad Antiochum* (20), the *Ad Pentadium* (or Vindician) (20), the *Epistle Ipocratis et Galieni* (2), the *Sapientia* (10), the *Epistula ex quatuor humoribus (sanitas est integritas corporis)* (3). Gerhard Baader added the *Ad Glauconem* (18)[119] (1985, 253). There was also the *Liber Aurelius*, probably a Methodist, not a Galenic text.

Of these, six were popular and widely disseminated medical texts with explicit material on humoral theory: the *Ad Maecenatem*,[120] the *Ad Antiochum*,[121] the *Ad Pentadium*,[122] the *Sapientia*,[123] the *De quatuor humoribus*,[124] and the *Liber Aurelius*.[125] This is, indeed, a meager selection, probably deserving Baader's anachronistic comment that "the state of medical knowledge in late antiquity and in the early Middle Ages in Western Europe . . . is deplorable" (1985, 251).

Finally, were any other sources of humoral theory available to a student practitioner like Hildegard or her apprentice? There were Greek works by Alexander of Tralles, Rufus, Oribasius, and Paul of Aegina, but they were not available in Latin. The Latin medical text of Theodore Priscian was popular, but it had no specific passages on humoral theory. Herbals and recipe books were common and they assumed humoral theory, but again, they did not present *explicit* material. Although urine physiology was based on humoral pathology and might have provided some insight into the humors, none of the elaborate compositions on urines had yet been translated during the early period (Baader, 1985, 255).

Putting these all together, we have eleven texts that were popular and available c. 1150 which contain passages on humoral theory. These include: texts from the three encyclopediasts—Isidore, William of Conches, and Pseudo-Bede; the late eleventh-century medical summary known as the *Isagoge* of Johannitius; the late eleventh-century translation *De humoribus* by Alfanus; and six early (pre-tenth century) texts: *Ad Maecenatem, Ad Antiochum, Ad Pentadium, Sapientia, De quatuor humoribus,* and the *Liber Aurelius*. Although they are neither a complete set of every single thing available in 1150 nor are they precisely what Hildegard or her student did have, they do supply a reasonable canon of texts on humoral theory. Our approach will be to examine them as if we were their students and had them, and them alone, to explain to us the ancient tradition. From them we can try to reconstruct the notion(s) of humoral theory that would have been available to a practitioner like Hildegard c. 1150 and would have informed her text.

This is the background that will help us decide whether Hildegard's multivalent humoral system was representative or idiosyncratic. This is not as impossible as it sounds because, although Danielle Jacquart rightly concluded that "it is difficult to make out in the whole of the medical literature of the High Middle Ages . . . the expression of any theoretical positions or of any clearly defined medical thought" (1998, "Medieval Scholasticism," 201), there was still a consistent tradition that agreed on the basics. Confusion has arisen because there were not one but three meanings for "humors," and Hildegard's multivalent humoral system was the norm.

THE PRIMARY TEXTS

The Humors Were Bodily Fluids

In these eleven texts the predominant meaning of humor is as one of the four bodily fluids—blood (*sanguis*), phlegm, *choler* (or *bili*), and *melancholia* (also called *fel*, black *coler* or *choler*). The Latin names for the various four humors vary from text to text and prove the fungibility of the scheme. Certainly the variability in names, spelling, and, as we shall see, even characteristics of the four humors implies that this material was not simply copied from one text to another. Rather, a common, but not identical, understanding of the humors surfaces in each text.

For instance, so-called "red *choler*" could be yellow or even green, and could be written as *choler, coler, colera,* or *bilis.* Black *choler* was also known as *fel, melanchtha, melancholia,* or *atrabilis;* and *phlegma* was sometimes in the plural, *phlegmata.* In all of the texts that explicitly transmitted humoral theory, however, each humor had a distinct color, taste, qualitative ratio, and medicinal effect, although these characteristics might differ from text to text. For example, in *De quatuor humoribus*:

> There are four humors in the human body—*sanguis, phlegma, cholera rubea,* and *cholera nigra.* . . . The blood dwells in the arteries and veins, *phlegma* in the brain, red *fel* in the liver and black *fel* in the spleen. A part of the blood is also in the heart, and part in the liver. Red *choler* is on the right, and black *choler* dominates in the spleen. Part of *phlegma* is in the bladder, and part dominates in the chest. Blood is red, *phlegma*, white, red *fel*, pink, and black fel, *spissa.* . . . Therefore, blood is by nature bitter, *phlegma* is salty, and sweet; red *fel* is tart, and black *fel*, strong and sharp.[126]

Although this seems straightforward, even within this single text there is some confusion, with different names for the same humors. When we turn to a different text, although the same ground is covered, it is with still more variability. Thus, the anonymous author of *Ad Pentadium* writes that:

> The power of the humors in health are that blood is hot, wet, and sweet; yellow *choler* is red, bitter, *viridis,*[127] fiery, and dry. Black *choler* is black, acid, cold, and dry, and *flegma* is cold, salty and wet.[128]

Notably, Vindician's humors *Ad Pentadium* have somewhat different characteristics from those in *De quatuor humoribus*: blood here is sweet, not bitter;
choler is red, not yellow, and it is bitter; *flegma* is not sweet but salty; and
black *choler* is sour, not sharp. Vindician must represent a variant tradition.
In addition, here each humor predisposes to a specific emotion—*sanguis* to
good humor, red *choler* to irritability, black *choler* to sadness, and *flegma* to
introversion ("intra se cogitantes," 489:1). Also, each humor has a characteristic pulse.[129] Last, with the changing seasons each humor cycles through
periods of increase and decrease.[130]

Also, according to Vindician, it is the *flavor* of each humor that is the
clue to its particular effect [*virtus*] in the body. For instance, *sanguis* is sweet
and, therefore, wet and hot; *flegma* is salty and, therefore, cold and wet. Red
fel is tart, and, therefore, hot and dry; and black *choler* (or *fel*) is acrid, and,
therefore, cold and dry. This means that the *tastes* of a humor signal its *qualitative* effect in the body, or, conversely, that the qualitative effect in the body
of each humor can be known by its flavor.[131]

In the later, more sophisticated *Isagoge* all of these humoral characteristics
are elaborated on. According to its eleventh-century Arabic author, Johannitius,
each of the four humors has sub-types, owing to their mixing with the other
humors. For instance, there are five varieties of *phlegm*, which come from adding one of the other humors to the basic *phlegm*, and each sub-type is identified
by its taste—salty, sweet, acrid, glassy [*vitreum*], or neutral (315). There are also
five kinds of red *coler*, and they are distinguished by their color—clear, yellow,
red, white, and green (317). There are only two types of black *coler* (319). The
first is dark and seen when blood is allowed to settle, and it is known as the
dregs [*faecium*] of the blood. The other type of black *coler* is a true black and
very toxic (325). Humoral complexion reveals itself in hair color: black hair, red
hair, yellow hair, and white hair roughly correspond to the four humors—*coler*,
heat [sic], *melancholia*, and *phlegm* and so, perhaps, by implication, does race
(although race is not, *per se*, singled out).

In short, in humoral theory there was both consistency and variability.
Many characteristics—tastes, colors, complexions—were connected to, and,
therefore, somehow caused by each humor, but how those effects took place
is somewhat of a mystery.

These pre-1150 sources on humoral theory also agree, in a general way,
on a kind of humoral physiology, which resembles what we have already discovered in *Causes and Cures*. For example, the humors enter the body in two
distinct ways: through the orifices (pores and senses) and through food. Thus
black *coler* enters the body through the eyes and red *coler* through the ears,
sanguis through the nose and *flegma* through the mouth.[132]

But the humors can also enter the body in food; in fact, this is the most important way they *do* get in the body. After food is eaten, it goes to the stomach where it undergoes a third cooking (*pepsis*—digested, cooked, or ripened). (The first cooking is by the sun, and the second is by fire; the importance of all this cooking is why raw food was looked upon so skeptically.) This well-cooked food next goes to the liver where it is transformed into blood (*sanguis*)—a mixture of all four humors, as is proved by phlebotomy.[133]

These newly created humors then travel in the veins until they reach the various organs—brain, heart, spleen.[134] Each organ then extracts the humors from the blood in proportion to each organ's own peculiar humoral makeup and each humor's concentration in the blood. Each organ is dominated by a humor—red *coler* controls the liver on the right; *sanguis*, the heart; black *coler*, the spleen; and *flegma*, the brain and bladder.[135] Each organ, therefore, is made up of all the humors in characteristic proportions, and each extracts the proper ratio of humors from the blood and transforms them into itself.[136] Excess humor is either excreted in the sweat or returns to the liver, where it is cooked again, goes to the bladder and exits the body as semen.[137]

This is to say that the humoral theory transmitted in pre-1150 texts describes a progressive transformation of the humors of food into the bodily humors, into the organs of the body, into its semen, and into its excretions. Hildegard's use of the humors as bodily fluids, then, which are refined from food, cooked, transformed by the liver, and sent to the main organs, reflected this early model and probably came from it. Here she was traditional. Most likely she put together her own understanding of this system, which was not exactly congruent with any of the writers available to her, as we do here, from fragments and pieces.

Also, just as *Causes and Cures* hinted of two different ways that humors could cause disease—one where they injure by disproportion (there are right and wrong ratios for the humors) and another by toxicity (there are intrinsically bad humors)—so, too, in this early material, there are hints of two different etiologies for disease. Thus humors can be (more or less) neutral and disease, a matter of imbalance and disproportion. For instance:

> These humors are mixed together, and if they did not exceed their proper measure, people would not get sick. But when they are too little, or too much, or if they are too thick, or too weak, or fall off from the natural, or too bitter, or leave their proper places or go to the wrong ones, then various illnesses emerge.[138]

Here health comes from a proper mix of humors. If the humors are too abundant or too scanty, too hard or too soft, too bitter [*acerbiores*] or too wayward, (abandoning their proper place for that of other humors), then a person will get sick. To establish this balance, as well as the state of the digestion and the general health of the body, the texts tell us, the urine is the easiest and best tool, just as we saw in *Causes and Cures*. For instance, in a healthy person the urine will be white in the morning, red after meals, and pink before bed.[139] This is because when everything is functioning properly in the body, what gets excreted in the urine is the appropriate humoral excess from blood.

This, then, is a balance model, where humors come into the body not necessarily in the proper proportion, and the body's organs must "rebalance" intake by excretion. Therefore, treatment means helping the body rebalance the humors by nourishing the humors, or by purging them, by adding to them, or by tempering them:

> What must be done is to nourish some of the humors, to diminish others, some to strengthen, some to balance, and others to temper.[140]

In this model, that is, the humors are neutral fluids; it corresponds to the balance model of Hildegard's explicit humors of *siccum, humidum, tepidum,* and *spuma*.

But there seems to be a second, less explicit model for the humors hinted at in these texts, where the humors, or at least some of them, some of the time, are not neutral but toxic. This is best appreciated in the *Liber Aurelius*, a translation or paraphrase of a lost Greek medical text by Soranus of Ephesus, a Methodist near-contemporary of Galen:[141]

> There are acute diseases that are born from blood and red *fel*, and they are called acute because they either disappear or kill quickly; and there are chronic diseases, which arise from the *phlegmata* and black *fel*.[142]

Indeed, even in Isidore's *Etymologies*, the hot humors—*sanguis* and *fel*—cause acute illness and the cold humors—*melancholia* and *phlegmata*—cause chronic ones.[143] In this kind of model the four humors do not cause disease because they are in the wrong place, in the wrong proportion, or out of balance, but because they are toxic in themselves, when they are in excess. Thus Aurelius holds that:

> Illnesses are generated from the four humors of which bodies are made. . . . And this happens mainly when there is too much of a humor in the body.[144]

In *Ad Maecenatem* also, humors cause disease when they are in excess. In it, however, each humor has a particular appropriate place in the body, where it tends to cause disease. Each of the four parts of the body (head, chest, abdomen, and bladder) houses a particular humor, which causes illness when there is too much of it. For example:

> If an illness arises from the head, it appears as pain, and the temples pound, the ears ring, the eyes water, and the nose cannot smell. When this occurs, the brain must be purged; for this, use hyssop or oregano or oxtongue with water so that *pituita* flows.[145]

Pituita, (which here seems to be another word for phlegm), belongs in the head (or brain), and when there is too much of it, the head aches, the ears ring, the eyes water, and the nose loses its sense of smell. These symptoms can be resolved when excess *pituita* is purged.[146] Thus, like any drug, *pituita* in small quantities is necessary to the body, but in larger quantities, it is toxic.

Bilis acida et amara belongs in the chest. When it is in excess, there is sweating, a bitter taste, inflamed tonsils, yawning, restlessness, and a dry cough; and the patient should purge himself by vomiting, or, at the very least, should fast:

> When an illness arises in the chest, then the head sweats, the tongue thickens, the mouth is bitter and the throat hurts. . . . The body should be made to vomit its bile, which we, therefore, call the mother of illness.[147]

Ad Maecenatem hints, then, that the humors could be toxic in themselves, and this is confirmed by its treatments, which have to do with various methods of *removing* the humors (by bleeding,[148] cutting, cautery, or cupping[149]), or countering their *toxic* effects with specific antidotes. For instance:

> If the cause [of illness] is blood, which is sweet, wet, and hot, then it must be countered by its contrary; that is, whatever is cold, bitter, and dry.[150]

Although this may sound just like the proportional idea about humors, here the cause of an illness *is* blood—not too much or too little blood but blood itself—and the medicine chosen does not balance but counteracts its effects. It is an anti-dote.

To sum up: Hildegard's implicit use of the humors as actual bodily substances with typical tastes, colors, and characters, was widespread. Moreover, the ideas implicit in her text, that the humors entered the body through the pores, through the senses, or in food, was common. Her notion that humors were cooked in the stomach and again in the liver, circulated in the blood, and eventually distilled to form semen, a kind of humor of humors, was also traditional.

Although the humoral information in the early texts was reasonably self-consistent, no single text laid out the entire system. Rather, to the extent that the ancient medicine was passed along in texts, a practitioner like Hildegard could only have inferred the ancient system from many texts, as we have done here. This may partially explain why the much richer text of *Causes and Cures* conveys a humoral theory that is similar, but not quite consistent with, that transmitted by the early texts.

The Humors Were Also Internal Qualities

As we have seen, in *Causes and Cures* there was also a second idea of the humors as reifications of the traditional four qualities. This model, too, is hinted at in the early texts. For instance, in *Ad Maecenatem*, although there are four substances in the blood—*sanies, pituita, bilis acida et amara,* and *bilis atra*—humans are *also* said to be made up of the four qualities. "Every body, whether human or animal, but especially human, is made of four types of thing (*genus*—kind, type, or category; Lewis and Short, s.v. "genus"): hot, cold, dry and wet."[151] In particular:

> The lungs by which we breathe are cold; the *anima* by which we live is hot; dry are our bones, and wet the blood that nourishes life.[152]

Here the basic organs are not made up of mixtures of humors in various proportions but of four qualities. Perhaps this idea was a source for Hildegard's identifying the humors *as* the qualities.

What the early texts make clear, moreover, is that these concepts—of the humors as fluids and the humors as qualities—were not entirely distinct; rather they were connected by the cycling of the year through the seasons. The humors as bodily fluids increased and decreased in the body with the seasonal and climatic variations of hot and cold, wet and dry:

> The four bodily fluids rise and fall according to the seasons. In spring, *sanguis* rises; in summer, red *choler*; in autumn, black *choler*; and in winter, *flegma*.[153]

Why do the bodily fluids wax and wane with the seasons? This is because of the inevitable heating and cooling, wetting and drying that naturally occurs in each season. This is why the humors (as bodily fluids) were linked to the humors (as reified qualities). The eighth-century monastic writer, Bede, explained the connection most clearly:

> In winter, the sun is distant from the earth and, therefore, the earth is *cold* and *wet*. In spring, as the sun moves closer to the earth, the earth warms up, and spring, therefore, is *warm* but (still) *wet*. By summer, the sun is very near the earth and the earth has become *hot* and *dry*; in autumn, as the sun moves away, the *dry* earth *cools*; in winter, with the sun far away, *wetness* increases.[154]

In other words: *the four qualities on earth are due to the ever-varying distance of the sun from the earth, as conceptualized in a geocentric cosmos.* The four *qualities*, in turn, give rise to the *elements* that are associated with each season—the cold-wet *water* of winter, the wet-hot *air* of spring, the hot-dry *fire* of summer and the cold-dry *earth* in autumn:

> The earth, therefore, is dry and cold, and water is cold and wet; air is wet and hot and fire is hot and dry. They are linked, respectively, to autumn, winter, spring, and summer.[155]

Bede goes on to explain that humans are made from these same four separate but continuous qualities, which are embodied in the body as the humors, *sanguis, cholera rubea, cholera nigra*, and *phlegmata*. This is why *sanguis* increases in spring and is warm and wet; why red *choler* increases in summer and is warm and dry; why black *choler* increases in autumn and is dry and cold; and *phlegmata* in winter, which is cold and wet.[156]

The qualities, then, are embodied as the humors and the humors are the internal *counterparts* to the weather, temperature, and humidity *outside* the body. They increase and decrease in the body in accordance with seasons precisely because of this homeopathic relationship. For instance, when spring turns into summer, its wetness dries, and so do the corresponding humors inside the body. Hildegard's explicit model for the humors, as *spuma, tepidum, siccum*, and *humidum*, expresses this same idea of the humors as inevitably, inextricably linked to the qualities of weather and season. She is unique, however, in having merged the four qualities with the four humors, at least explicitly.

Plant Saps Were Also Humors

Finally, Hildegard's notion that humor can also mean *any* liquid but, especially, the sap of plants, also occurs in these pre-1150 texts. In fact, this was the original meaning of humor. In Greek *chymos*, from which the Latin, *humor*, derived,[157] originally meant any fluid or juice, but especially the sap in plants and the blood in animals.[158] *This is the key to how all three meanings of humor are linked.*

Thus the *Sapientia* recounts that, just as the seasons cause the sap and roots of plants to rise and fall, so, too, the seasons cause the waxing and waning of the humors:

> Starting on February 1, all the humors, diseases, *reuma* [liquid][159] and roots of plants surge. The body is vexed by this and can swell. This morbid matter should be extracted by phlebotomy and purged by *catarticum* [green moss, *Bracyfolium radulare*]. All of the humors dominate until July. After the eighteenth of July, the physician should stop phlebotomy and *catarticum*, because the seventy-three dog days are come. After the eighteenth of September there are sixty-four days of autumn, during which red *coler* and *flegma* increase along with the body's heat. Do not engage during this time in coitus, which weakens the body. Just before November 27 black *coler* begins to surge, which is salty and wet and permits coitus. In winter distilled grapes should often be drunk; *flegma* and *reuma* increase until the kalends of March and it will be rather *rheumy*.[160]

What this text tells us is that the humors in the body wax and wane with the seasons, just as sap does in plants. *Ad Antiochum* provides more specifics:

> At the beginning of the year, at the *winter solstice*, *humor* starts to increase and does so until the *spring equinox*. During this time very hot and excellent foods should be used, with moderate coitus. Starting with the *spring equinox*, *fleuma* begins to increase, and it is a thick, cold humor, until the rising of the Pleiades, on the fourteenth of May. Hence it is good to take fragrant but sharp and bitter foods. After the fourteenth of May, *sanguis* reinforces more hotly the increase of fever and *humor* is less, until the *summer solstice*, on which day one should fast and refrain from coitus. After this, *fel* increases until the autumn *equinox*, during which time sharp, tasty, oily foods should be taken. Then after the *autumn equinox* and until the setting of the Pleiades, *acres biles* rises; sharp and acrid food should be taken, and little coitus. Finally, around November 11, *sanguis* begins to decrease as the Pleiades descend; and light foods, some wine and moderate sex should be used until the return of the *winter solstice*.[161]

What all this means, mainly, is that during the year there was an orderly rise and fall of the bodily humors; and the effects of the seasons on the body were to be managed with compensatory food, drink, and activities.

It is remarkable that in these two texts the *number* of humors and the times of their rising and falling were not the same. Thus, on the one hand, in the *Sapientia* there were four fluids—*reuma*, red *choler*, black *choler*, *flegma*—which were correlated to the four moments of spring and fall equinoxes and summer and winter solstices. On the other hand, in *Ad Antiochum*, there were *five* bodily fluids and *six* time periods, because there were *two* distinct celestial phenomena—not only the four seasons but also the rise and fall of the Pleiades. The first created a cycle of four seasons and four fluids—*humor*, *fleuma*, *fel*, and *acres bilis*. The second was associated with the rise and fall of a fifth fluid, *sanguis* (fig. 4.1).

Thus there were two *separate* humoral cycles. Cycle #1 represented the seasonal increase and decrease of light, warmth, dryness, and dark, cold, wetness. Four bodily fluids—the humors—responded by increasing or decreasing. Humoral excesses, therefore, could be expected as the seasons came and went. They were to be managed by using foods with prescribed flavors because flavors signified the qualities of hot and cold, wet and dry

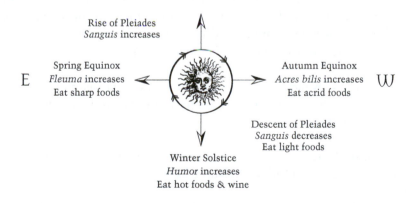

Figure 4.1. Waxing and Waning of the Humors with the Seasons.

and therefore, of the compensatory humors. In particular, too much *humor* was to be treated by the hot and acrid; too much *fleuma*, by the fragrant and sharp; too much *fel*, by the tasty and oily; and too much *acres bilis*, by the acid and acrid. This first cycle assumed that the humors in the body responded to the qualities of the seasons and could be treated with food whose *qualitative* effects—hot and cold, wet and dry—were known by their *tastes*—acrid and oily, sweet, and sharp.

Cycle #2, the rise and fall of *sanguis*, was linked to the rise and fall of the Pleiades, probably because this coincides with the life-cycle of plants and their *humor*, since flowers appear in May and the last fruit in November. Indeed, the Pleiades marks the traditional beginning and end of the planting season in many cultures. Thus inside the body, while four of the fluids vary with light, warmth, and wetness, the fifth, *sanguis*, rises and falls in conjunction with plant growth, maturation, and death. Cycle #2, then, suggests that there was a second kind of humoral physiology related to plant physiology where "humor" meant sap.

When Cycle #1 and Cycle #2 are combined with Bede's seasonal cycle in which the qualities are connected to the humors by the changing distance of sun to earth, we can begin to see that all three of the meanings of humor were linked. As the sun moved closer to and farther from the earth, the seasons changed, the qualities changed, and the humors of both plants *and* bodies rose and fell in accordance.

A Fluid Concept: Well-Aged and Not Too Dry

Humor provided even more than that, however. For it was not only a fluid *concept*. It, itself, *was* fluid—a fluid stuff that flowed from earth into plants, from plants into animals and humans, and back to earth again. In the next chapter this process will be examined in detail; for now, we can find hints of how *humor* connected plants and bodies. For instance, Pseudo-Bede notes that:

> Plants grow by sucking juice [*succus*] from the earth through their roots;
> it flows to their external parts and is transformed by their natural heat
> into their strength.[162]

That is, first plants transform earth-juice into themselves. Next: animals and humans eat plants and Isidore explains what happens then:

> By its heat the liver turns the *succus* which has been extracted from food
> into blood. The *liquor* made out of food and cooked by the body dif-
> fuses through the veins and bone marrow to the kidneys, where it is

thickened and ejected in coitus, and received by the uterus, and, through heat and menstrual blood, is hardened in the body.[163]

Once again, we see that the stomach "cooks" food. Here it turns from *succus* into *liquor*, which eventually flows through the veins and bones into the kidneys. The kidneys then transform this *liquor* into semen. It also goes into the uterus, which adds blood, cooks it still again, and creates the matter for a fetus. This *liquor* also goes to the liver, which cooks it, and turns it into blood. Thus *liquor* from food is the cooked *succus* of food, and it is this liquid that becomes the bodily fluids.

This, then, is another humoral cycle: plants transform earth-juice (*succus*) or *humor*, into themselves; bodies transform these plant humors into bodily humors. The bodily humors are, at their most basic, the transformed humors of plants.

Now, these plant humors were classified by the way in which they affected the body. Thus:

> There are two types of foods. There is a type of food that engenders good humor, and a type of food that engenders bad humor. Good food is good because it creates good blood, balanced in mix and power, like bread. Bad food works in an opposite way; for instance, certain types engender red *choler*, such as nasturtium, mustard, or garlic; others engender black *choler*, such as lentils; and others engender phlegm, like pork.[164]

How was the way in which plant humors affected bodily humors ascertained? Mainly by the way a plant tasted. Tastes signaled the quality or complexion of a plant—what it brought into the body. Of course, each plant humor was complex, made up of its own very unique and individual flavor, but a flavor that could be parsed and classified. The only remnant we have left of this system is, perhaps, the tasting of wine, which both picks out the very particular flavors of the grapes, as well as its terroir, its sweetness, dryness, etc.

Thus the anonymous early author of Zurich MS C128 explained that:

> The complexions of medicines, foods, and drink, can be known by their tastes. Because all things are made from hot and cold, wet and dry. Which qualities are signaled by their tastes. There are eight tastes: sweet, oily, sour, salty, sharp, bitter, styptic, and *ponticus*. Certain—sweet and oily—are pleasurable, the others are unpleasant. Sweetness is balanced, well-tempered. Oily is hot and wet in the second degree. Bitter is hot

and dry in the third degree. Styptic is cold and dry in the fourth degree, and *ponticus* similarly. Those things that have no taste, which are insipid, are cold and wet in the first degree.[165]

The twelfth-century author Marius wrote that there were nine basic tastes: sweet, bitter, sharp, oily, tart, salty, astringent, pungent, and insipid; and to each there was a specific complexion—temperate, hot, cold, hot-temperate, etc.—because each was made up of a varying amount of the four elements (Dales, 169). Asaph observed that "any sweet taste is hot and wet; any bitter taste, hot and dry; any sour taste, cold and wet; and any sharp taste, cold and dry."[166]

Food of course, was made up of many ingredients and so was even more complex. Nevertheless, its complexion, its mix of medicinal qualities, was signaled by its flavors. As Albert Magnus later explained: "It is taste [*sapor*] that gives the most certain experience of the power of a plant."[167] Taste meant flavors, and flavors meant the qualities.[168]

So the humor/taste of a plant or medicine could be parsed into its flavors (which equal the qualities), but each plant also had a unique identifiable taste as well, its own *specific* humor. "Thus early medieval medical writers . . . had come to conceive of medicines as acting to restore health almost exclusively by means of their individual, specific properties (8). . . . Most practicing physicians made constant use of an entirely different set of properties, ones that Galen had not really explained(16) . . . , their individual, and specific effects" (McVaugh 1975, 18–19). Isidore called this *particular* power of a plant its *dynamidia*; and so did the author of *Ad Maecenatem* ("dynames herbarum").[169]

This is the root of Hildegard's (and others) confusing and multivalent use of *humor*. Plant humors (saps) affected bodies in two different ways: *indirectly*, by means of their qualities (tastes); and *directly*, by means of their particular essences.

Thus humor twice linked cosmos, plant, and body in a cycle. First, by analogy. Humors in bodies and humors (saps) in plants were subject to the same forces of season and weather; all humors rose and fell as the year turned. Second, by practical effect. When ingested, the humors of plants turned into the bodily humors and affected the body through their qualities. And they also directly affected the functioning of the body through their unique, specific, medicinal humors.

Hildegard's humors have confused scholars because she had in mind not one but three different kinds of humors, and all three were connected. First, she had an idiosyncratic set—*tepidum*, *spuma*, *humidum*, and *siccum*— a reification and identification of the four humors with the four qualities.

Second, she had blood, bile, phlegm, and melancholy, although she did not identify them specifically as humors. Last, she used *humor* to mean the individual essence of a plant, an animal, or even a stone.

All of these meanings were traditional and are found in the early texts. Usually the humors were the four bodily fluids, although in early texts and even in Galen sometimes the four qualities seem to be the real humors. *Humor*, especially in texts with herbal recipes, could also mean the juice or sap of plants. Although these three meanings of humor explained three different kinds of observations—the horticultural, the physiologic, and the climatologic—they were not separate. *Humor* itself was not only a concept but a pluripotential *stuff*, which was acted on by the external environment of winds and seasons and weather, and the internal environment of plant, animal, and human bodies. *Humor* absorbed all of these influences, was transformed by them, and transmitted them. *Humor* linked earth, plant, and body in a cycle.

How did Hildegard manage to put together these scattered hints of an early medical system? We ourselves have had access to databases, microfilm, and the Internet, to trains, planes, and cars, to the libraries of America and Europe, and to the work of a century of scholars and wise librarians. She, by contrast, lived out her life on Disibodenberg and Rupertsberg with, at most, a few manuscripts. Yet, somehow, she put together a private understanding of this buried system from its fragments—from what she read, but also, probably, from what she heard—old wive's tales and proverbs, other practitioners—and from her own experience of plants in the garden and bodies in the infirmary.

It would have been her work as *pigmentarius* that allowed her with such scant resources to unearth the ancient sensibility. That plants, like bodies, had humors; that bodies, like plants, cycled with the seasons; that both responded to climate, weather, and place; and that hot and cold and wet and dry were very useful terms: all this, she could see for herself. This was her vision—not her spiritual but her earthly vision.

Chapter Five

The Green Humor

"The life of plants is called *viriditas*," Gregory the Great wrote. "They live, I say, not through *anima*, but through *viriditas*."[1] Greenness—*viriditas*—has been taken to be a defining term of Hildegard's thought but, as we shall see, it was instead an ancient concept which, like the elements and the humors, originated in horticultural observation. Hildegard did, however, do something different with viriditas;[2] she found it not only in plants but also in bodies. How she used viriditas, how she came about discovering it, formulating it, and how she applied it uniquely to understanding the life of the body—this is the story of this chapter.

HILDEGARD'S VIRIDITAS

Books Three, Four, and Five: A Medical Manual Based on Plants

When examined as a unit, Books Three, Four, and Five of *Causes and Cures* function as the therapeutic kernel for Hildegard's medical text. *With* them, it would be possible to practice as an infirmarian; *without* them, *Causes and Cures* provides little more than a basic understanding of the elemental macrocosm and the humoral microcosm.

Twenty pages in print, Book Three is organized according to the traditional head-to-toe format (*capite ad calcem*) of most anonymous medical recipe books of the early medieval period. As with Books One and Two, the structural organization of Books Three, Four, and Five is clearer when the added rubrics are ignored. Thus Book Three begins with a recipe for preventing hair loss, goes on to the treatment of headaches, weak vision, ear pain and toothache, and continues into the chest (lung pain), abdomen (hard liver, spleen pain, stomach pain, poor digestion), and genitalia (hernia, kidney or testicular pain, urinary incontinence). It ends with problems of the knees, feet, and skin. Once again Hildegard seems to have had a topography—here, it is the homunculus.

Each passage is short, to the point, and formulaic, consisting of an indication, a recipe, and an explanation for Hildegard's choice of plants. For example, to treat a *cancer*:[3]

> Take violets and press their juice and put it in a pan; add 1/3 as much olive oil and then take as much lanolin as violet juice and heat the mixture in a new pot until it becomes an ointment. On the area of *cancer*, or where worms have eaten, smear the ointment around and on the *cancer*, and they will die when they eat it.[4]

Most of the prescriptions are just as practical, resembling nothing so much as modern culinary recipes. Rough but reproducible measurements are recommended—two coins' weight[5] or the amount an average egg might hold.[6] In each recipe, ingredients are few and readily-available; and they can be parsed into ingredients from the garden, from the wild, from the kitchen, and from the market (fig. 5.1).[7] It is notable that most of the garden herbs are referred to by their Latin names, while the wild plants are often in German. This suggests that Hildegard knew about garden plants from a Latin (e.g., monastic, textual) tradition, and about wild plants from a vernacular (e.g., oral) tradition.

GARDEN	WILD	KITCHEN	MARKET
dill	laurel leaves	butter	nutmeg
fennel	marrubium	olive oil	ginger
fenugreek	bertram	bear grease	sugar
rose (water)	hun *(tanacetum)*	water	pepper
absinthe	sverthilem		galangal
sage	calamino root	honey	myrrh
aloe	fig leaf	flour	
poppy *(papaver)*	lungwort	egg-white	
ebech *(apium)*	feverfew	salt	
oregano	lehesticla	faba flour	
rue	hechernezzil		
verbena	dactilosa	liver	
cerifoil	millefoil	cow gallbladder	
cinquefoil	ringula *(calendula)*	spleen	
aristologia	stichwurz	uterus	
pimpinella *(anise)*	winda		
white pisa			
parsley			

Figure 5.1. The Medicinal Ingredients in Book Three of *Causes and Cures*.

Generally, each prescription consists of several (one to three) plants—wild, grown, or purchased—which are mixed with a carrier (usually flour), and a liquid (olive oil [sic], water, wine, honey, or vinegar). Given that sugar, ginger, galangal, nutmeg, pepper, and myrrh are its modest spices, and common weeds and garden plants its recommended herbs, *Causes and Cures'* pharmacopoeia corresponds to the likely contents of a (monastic or domestic) household. Even with her conservative use of household ingredients, however, Hildegard ends Book Three with a warning against polypharmacy: "Eating too many even of noble plants will not profit but injure the body."[8]

At forty pages, Book Four is twice as long as Book Three and uses almost twice as many ingredients. It, too, is made up of short, formulaic passages that provide an indication for treatment, a prescription, and a short explanation for Hildegard's choice of plants. For instance:

> For nausea: take cinnamon and 1/3 as much pepper and 1/4 as much *bibinella* and pulverize the mixture; then add the same amount of pure flour, an egg-white, and a little water. Make little cakes that can be cooked in the oven and have the patient eat them. The coldness of the

GARDEN	WILD	KITCHEN	MARKET
dill	salix	butter	nutmeg
fennel	febrifuge	olive oil	ginger
fenugreek	millefoil	vinegar	sugar
rue	tanacetum	wine	pepper
absinthe	dittany		galangal
sage	viola	honey	licorice
basil	hyssop	flour	thuris [*myrrh*]
aristologia	agrimony	eggs	storax
pandoniam	saxifrage		cinnamon
bibinella	polypoidy	liver	sulfur
verbena	laurel	gallbladder	garifoil
cerifoil	bachminz		
ebech=apium	lungwort		
	ashelock		
	helun		
	nettle		
	malva		
	storksnabel		
	plantain		
	mustard		
	merlinsen		

Figure 5.2. The Medicinal Ingredients in Book Four of *Causes and Cures*.

cinnamon, *bibinella*, and egg-white will be tempered by the heat of the pepper and flour; and the sweetness of the water and the furnace heat all together soothe nausea, which comes from a mixture of too much heat and too much cold.[9]

The tone and structure of Book Four are identical to those of Book Three, and its ingredients overlap, as can be seen in a comparable table (fig. 5.2).

In both lists what is perhaps most striking is the modesty and practicality of Hildegard's ingredients—no eye of newt or tongue of frog here. Nor, from the point of view of modern therapeutics, do these medicaments have only placebo effects. For instance, she recommends the highly active, anti-bacterial sulfur, as well as the pharmacologically active organ products of liver (with its iron and B12), gallbladder (with its bile acids), and uterus and testicles (with their active hormones). This supports the notion that Hildegard's pharmacopoeia was based, at least in part, on empiric, not simply magical or authoritative concerns, as has sometimes been argued.

The underlying structure of Book Four is more difficult to appreciate, however. Its organizing theme is, perhaps, what might be called the "pathophysiology of organ systems," if organ system is taken to mean a part of the body through which fluids can be excreted. Thus Book Four covers, in order, problems of the uterus, nose, penis, eyes, gastro-intestinal tract, skin, kidney, and joints. It begins with treatments for illness due to the under- or over-excretion of the uterus, then of the nose (bleeding, discharge), penis (nocturnal ejaculation), eyes, gastro-intestinal tract, skin (pustules, rashes, jaundice) and kidney (stones). It concludes with chronic skin disease (leprosy—perhaps as the body's failed attempt to get rid of toxins), fevers (also a generalized attempt to get rid of toxins), and "worms." Its last pages provide practical recipes for treating farm animals—horse, ass, pig, goat, cow, and sheep.

At the end of Book Four are a few recipes that lack Hildegard's customary format. That is, they are more like typical infirmary recipes, where a problem is followed by a recipe, rather than by an explanation. Perhaps these are recipes added later, as was often the case with infirmary manuscripts. The ordering principle for Book Four may be the decreasing "naturalness" of the excretions, given that the primary function of uterus, nose, and penis, *is* excretion, while eyes and gastro-intestinal tract excrete only secondary to their primary functions (of vision and absorption). Disorders of the skin (leprosy and fevers) would be the least natural way for the body to rid itself of toxins and so would end the Book.

Last, Book Five is a twenty-page text focused on the techniques of prognosis. Again, its underlying structure is (more-or-less) head-to-toe. Thus it begins with the prognostic value of the eyes. For example:

When a person is healthy, if he has eyes that are pure and clear (of whatever color), it is a sign of life. Thus if his eyes are clear like a white cloud through which another can be seen, he will live and not soon die. For the *anima* in the eyes of this person is powerful, since his eyes are pure and clear, and his *anima* sits strongly and has many more things to do in this world. For the eyes are the window of the *anima*. But whoever has cloudy eyes even if he is healthy, it is a sign of death, like those clouds that are thick; he will soon sicken and die. In his eyes, his *anima* is weak, as if it had only a few things left to do, and they are cloudy, like a person who cannot decide whether to go or to stay.[10]

Here the *anima* is not the abstract "soul," but something more like physical fire; yet it also has personality.

The text goes on to explain the prognostic uses of facial color, voice tone, the brachial and femoral pulses, and urine and stool. For instance:

When a person is healthy and has moderately red cheeks, then it is a sign of life, since the red color can be seen just under the skin like in an apple. For the red color under the skin is there because the fiery breath of life is the *anima*, which so shows in the cheeks because it sits securely in the body and will not leave soon. But if, when a person is healthy his cheeks are so red that no skin can be seen under the redness, then he will die, because the redness cannot be seen like the red of an apple; this redness shows that the *anima* is about ready to leave.[11]

As we saw in Chapter Four, the urine is an especially important prognostic and even diagnostic indicator, because whatever a person eats or drinks shows up, in some form, in the urine. For example:

[On the significance of the urine.] Whatever someone drinks, whether wine, beer, mead, or water, his urine will show his health or illness.[12]

Hildegard recommends that the urine should be observed only after it sits for a while, and the humors in it have had time to layer out:

If the humors in someone are heated too much, his digestion cannot be good and natural, and the humors will be mixed in the urine and a bit thick. Although this person will be sick for quite a while, he will not die, because his humors are still mixed together and not yet separated. But if a new urine is taken, and if it is pink and then, after a while, it pales, and if strands of different colors—pink, watery, and turbid—are seen, then he will die.[13]

For Hildegard, then, as for most premodern practitioners, the urine revealed the overall humoral state. In a normal urine the humors are mixed together; in a state closer to death, they begin to separate from one another, because the ruling force of the body, the *anima* (the physical fire of life), is no longer functioning properly.

A section on the medicinal effects of water in bathing and sauna follows, and then comes a section on character prediction using eye color. Thus:

Those who have gray eyes are light and lascivious, and in morals rather distractible; although whatever they do, they do properly. Those who have fiery eyes like black clouds near the sun are prudent and sharp, but quick to anger.[14]

The very last section of Book Five in Kaiser's edition, and Book Six in Moulinier's, provides predictions of character that are based on the age of the moon at the day of conception. This has been called a "lunary," but the classic lunary used birth, not conception, for its predictions. Hildegard's is much more interesting. Since the day of conception (but not of birth) is controllable by its participants, her text may be less about prediction than it is about manipulation. After all, it implies that parents can take an active role in the kind of children they conceive.[15]

If they are seen as a unit, Books Three, Four, and Five provide a kind of manual of practice, similar to that of many other infirmary texts—they provide the tools for humoral diagnosis, medicinal treatment, and prognosis. Although, as we have seen, it has been argued that Books Three and Four were merely later rearrangements of passages from *Physica*, and that Book Five was by a different author entirely, the foregoing analysis shows that for all their apparent confusion these Books, taken together, were most likely the product of a single author. Each has an internal structure based either on "anatomy" (head-to-toe) or on "physiology" (of excretory systems), and the three Books complement one another. That is, Books Three and Four provide the recipes (using a nearly identical pharmacopoeia) and Book Five the diagnostic/prognostic tools (observation of the patient, examination of the urine and stool). It seems likely that a single author wrote them. But was its author Hildegard?

Arguing now from *within* the text, several observations can be made. First, the author was likely German, even Rhenish, because the wild plants and weeds of the text are those found in the Rhineland and are called by their Middle High German names. Second, its garden herbs are those typically grown in monastic infirmary or kitchen gardens; and its ills are those found in the monastery or the village. Third, its equipment is typically that found in any kitchen—nothing more complicated than fire, pot, mortar, and pestle. Last, its instructions for the preparation of recipes are simple and explicit (so much so that Hildegard's prescriptions are being formulated today).

It is, that is to say, a kind of manual for practicing medicine, probably written by a German practitioner but in Latin; and this points to a German monastic infirmarian as its author. Although there were other monastic, even female, writers of medical texts in the Rhineland,[16] the literary style of Books Three, Four, and Five is unmistakably Hildegardian, with her straightforward sentence structure, biblical vocabulary, and rich imagic stream of mixed metaphors. Given the mutual consistency of their ingredients, their complementary structure and their Hildegardian style, it seems likely that Books Three, Four, and Five were written by Hildegard as a manual of practice or infirmary text.

Indeed, Books Three, Four, and Five, taken as a unit, were probably one of the *first* things that Hildegard wrote. Hildegard would most likely have had to train a nun to replace her as infirmarian when she became *magistra* at Disibodenberg in 1138, and certainly when she moved to Rupertsberg in 1150, and this new infirmarian would have needed a medical manual of some kind. Books Three, Four, and Five contain precisely the kind of information that such a trainee would have found useful; and they read like a typical infirmary text. They are straightforward and practical, with prose which is Hildegardian but unpolished. Certainly the text corresponds far more to that expected from an infirmarian than a mystic, and would have been an uncontroversial first text for a woman to have written.

Moreover, although the Hildegardian motif of greenness (*viriditas*) *is* found in this text, it occurs less frequently in Books Three, Four, and Five, taken as a unit, than it does in Books One and Two, and much less often in *Causes and Cures* as a whole than in Hildegard's other texts. Indeed, viriditas was a notion that Hildegard used more and more often as she matured. Thus if we examine its frequency in her *dated* texts, *Scivias* (1140–1150), *Liber vitae meritorum* (1158–1163), the *Epistolarium* (1147–1179), and the *Liber divinorum operum* (1163–1173), we can construct a bar graph of incidence

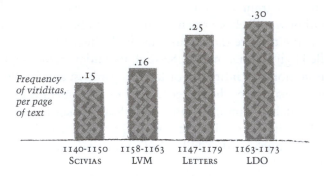

Figure 5.3. Viriditas Plotted Chronologically in Hildegard's Dated Works. Total hits in *Cetedoc*: 247 passages contain some form of the term, viriditas. Incidence per page (or column) of text in dated works.

per page, over time (fig. 5.3).[17] It clearly shows that, for the known dates of composition, the frequency of viriditas in Hildegard's works increases with time.

Where do *Causes and Cures* and *Physica*, whose dates are not known precisely, fit into this scheme? Despite the similarity of their content, in *Causes and Cures* and in *Physica* the frequency of viriditas is quite different. The frequency in *Causes and Cures* is .16 and thus closest to that in *Scivias* and the *Liber vitae meritorum*, while in *Physica* it is .21, putting it somewhere between the *Liber vitae meritorum* and the *Epistolarium*. A bar graph with these frequencies suggests that *Causes and Cures* was one of Hildegard's earliest works (fig. 5.4).

If we now plot the incidence of viriditas *within* the Books of *Causes and Cures*, we find that in the "manual" of Books Three, Four, and Five, it occurs the fewest number of times (fig. 5.5).

While admittedly not scientific, this graph does support the notion that the unit of Books Three, Four, and Five, with the least incidence of viriditas, may be the earliest text by Hildegard.[18] If so, then these Books allow us to see the earliest formulation of her conception of viriditas, which occurs in these Books only five times. In its first appearance, viriditas is simply the green color of grass:

> If the eyes become weak because their blood and water are thin due to age or illness, then the patient should go to a field of green grass and stare at it until the eyes tear. The *viriditas* of the grass will take away the haziness and make the eyes again pure and lucid.[19]

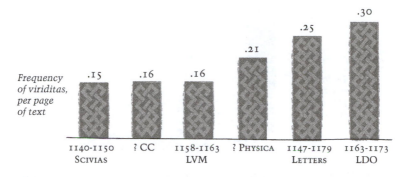

Figure 5.4. Viriditas Plotted Chronologically to date *Causes and Cures*.

In its second appearance, viriditas is something more substantial, more phys-ical, affected by such physical things as dew, and similar to sap:

> The heat of *ficus* and the cold of *erlen* (alvio or avo) are of such a nature
> that they attract moisture, but only dew should moisten them, so that
> its sweetness will temper their *viriditas*; and only the sun should heat
> them, so that their sap becomes less harsh, lest it hurt the eyes.[20]

Here viriditas is not just a color but a quality of rawness or youth, which is somehow related to the juice or sap (*succus*) of the herbs. In its third occur-rence in Books Three, Four, and Five, viriditas is something that increases a plant's medicinal effect:

> But in winter, let him powder the herbs and eat them with food; because
> at that time the *viriditas* of the plants is not available.[21]

That is, some plants do not have to be used green; they work without viri-ditas. On the other hand, some plants *must* be used green, "because their power particularly flourishes in their viriditas."[22]

Hildegard does not say that the plants to which she refers must be used when they are green (*viridis*); rather, she prescribes their viriditas. This indi-cates that viriditas was not simply a color or even a quality, but a thing or substance that augmented a plant's effect in the body.[23] There may be an observational basis for this usage, since the newest leaves of medicinal herbs do have a milder flavor and a paler green color. When they mature, they become not only larger but greener and "stronger"—that is, tastier, more odorous, and more medicinal. When the leaves of medicinal herbs dry up,

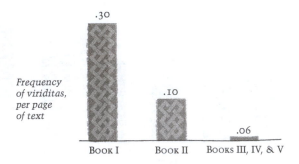

Figure 5.5. Viriditas Plotted Chronologically within *Causes and Cures*, to date the Books.

they lose taste, odor, and green color, that is, their viriditas, all at the same time. Their viriditas signals their medicinal effect.

If the hypothesis is correct that these Books recorded Hildegard's first writings, then for Hildegard viriditas was first a quality, power, or substance in plants that indicated or signaled their medicinal effect. This suggests that it may have been Hildegard's experience as *pigmentarius*, practical keeper of plants and bodies, rather than her experience as mystic or Benedictine nun that informed her understanding of what historians now regard as a key concept in her works.

A History of Hildegard's Viriditas

It was a scholar of the history of medicine, Heinrich Schipperges, who in 1957 argued for the first time that viriditas was a key concept for understanding Hildegard (1992 [1957]).[24] Previous researchers had noted her use of the color, green, but not her use of viriditas.[25] For instance, in glossing Vision I/3 of *Scivias*, Hans Liebeschütz suggested that the earth had been colored green in order to symbolize its fruitfulness (1930, 65, fn. 2), and Hermann Fischer noted, though without further comment, that in her portrait Hildegard was wearing a green robe (1927). The first translator of *Causes and Cures*, Hugo Schulz, translated viriditas simply as "zu grünen" without remarking on any special place for the concept in Hildegard's medicine or theology (1953).

But Schipperges translated viriditas as *grünekraft* (84)—greening power—in his translation and gloss of *Causes and Cures*. Indeed, he gave it pride of place, with more than twenty references in the index and a ten-page section explaining its role in her thought. According to him, although viriditas was something substantial, like flesh, phlegm, melancholia, or other humors (23), it did not come from Hildegard's medical practice. Rather it came from her spiritual thought, especially from her familiarity with the daily round of prayers (44); it was a religious concept from the first. It stood for vitality ("Lebenskraft," 210) or vigor ("Lebensfrische," 24). It was the image of Life ("das Bild des Leben," 301), the vigorous Force of Nature ("lebensfrischen Naturkraft," 304), even the Principle of Life itself ("vitalistische Prinzip," 304).[26] Its greenness was crucial to its meaning, Schipperges argued, because greenness meant life: "Under the image of green Hildegard gave a comprehensive idea of the vital power of nature, as she puts it before our eyes as a vegetative and animal Life-Force. It is the color of all sprouts and shoots, of all flourishing and growth."[27] However pagan this sounded in its evocation of fertility and vegetation, however, Hildegard's viriditas was grounded in God (307), and its significance was moral and spiritual (301).

Subsequent scholars agreed that viriditas was a key and particular concept for Hildegard,[28] which she used to signify a vital force or nature power that stood for God. For instance, Roland Maisonneuve defined Hildegard's viriditas as "vigeur, verdeur de la jeunesse, verdure" (259); Regine Pernoud, as that "internal energy that makes plants grow and people develop" (1994, 106); Laurence Moulinier, as life and vigor (1995, 267); Irmgard Müller, as "a creative principle" (1987, 31); and Renate Craine, as "the living life-force" (69). Miriam Schmitt called it a polyvalent symbol for "greening life energy" and Margret Berger explained that it was "the principle of life . . . [implying a] generative energy, growing and greening power, verdure, and fertility" (128). "No other writer prior or contemporary to her time uses greenness to convey the meanings Hildegard gave to it" (Schmitt 1999, 256). It was "a nature-power postulated by Hildegard,"[29] which did "not occur in any other treatise on healing. . . . [She] uses it to refer to the energy of life, which comes from God, the power of youth and sex, the power in seeds. . ." (Strehlow 1988, xxvi). In sum, it was "Hildegard's singular idea . . . with God as source" (Schmitt 1999, 258).

So historians hold that, while viriditas certainly had its origin in green nature, it was essentially religious: viriditas meant the acting of God within nature. Thus it was "the earthy expression of the celestial sunlight" (Dronke 1972, 84); the "fecundity of . . . God" (Maissonneuve, 257–258); "a mirror for the vigorous basis of eternity" (Müller 1987, 31), evoking "all the resilience and vitality of nature and its source, the Holy Spirit" (Newman 1988, 38). Viriditas was "a living power in which the working of God's word is made manifest" (Müller, 1987, 31); a "Christic dynamic of dying and rising in everyday life" (Craine, 75); one of the "manifestations of God's power"

(Baird 1994, 7). Although its linguistic origin was doubtless the green of leaves, it was really "the invisible principle of this unity."[30]

In short, scholars have agreed that viriditas, although probably derived from Hildegard's "forty year experience of the vibrant greenness of the vineyards gracing the Rhineland hills of Disibodenberg, and the verdant grasslands of the Glan and Nahe river banks" (Schmitt 1999, 257), meant the divine principle.

And yet, what Hildegard herself did mean by viriditas has not been closely examined. That viriditas was "*vis*, the basis of the life-force of the world" (Lautenschlager 1997, 206) is suspiciously reminiscent of nineteenth-century vitalism, with its claim that a unique quality differentiates life from non-life.[31] Indeed, some interpretations of Hildegard's medicine strongly recall vitalism's anti-industrial, anti-urban, and anti-scientific stance.[32] But why should we assume that Hildegard meant something good by viriditas? In Piotr Sadowski's investigation of the greenness of the Green Knight, for instance, he showed that "greenness" may not always be good even though it is "natural." There are many greens in Nature, he points out, not all of them pleasant—the bright green of plant growth, to be sure, but also the dark green of brackish water and the pale green of death and decay (1999).

Indeed, in what was probably its first appearance in Hildegard's work, in *Causes and Cures*, viriditas simply meant a quality or substance in plants that amplified their medicinal and pharmacologic effects, as we have seen. Thus it seems fair to ask: What *did* Hildegard mean by viriditas? *Did* she mean it to be a life-force or power, a stand-in for God in Nature? Or was it a more material concept? Might its significance for Hildegard have originated in her observations of plants and their effects on bodies rather than in Benedictine chant? If so, then viriditas will add to the evidence of Chapters One and Two that Hildegard was a practicing *pigmentarius*, and will buttress a theme of Chapters Three and Four, that medical/horticultural experience can inspire the construction of philosophic discourse.

Viriditas in Hildegard's Works

Hildegard used viriditas in hundreds of passages, but it is not difficult to parse them into only three distinct but related categories: viriditas simply as the color green; viriditas as a metaphor; and viriditas as a substance or property of some kind. As a color, she usually used viriditas to mean the green of gems—emerald, jasper, or beryl—or the green of vegetation, in fields or meadows. For example:

> I saw the air, clearer than the clearest of waters and brighter than the sun, which breathed, and which contained every greenness of plants and flowers on heaven and earth.[33]

Clearly here viriditas is not a metaphor. It means the actual green color of plants, in field and meadows laid out. This usage, however, was least frequent.

Most often, Hildegard used viriditas as a metaphor. Thus, there was the viriditas of the good man,[34] the viriditas of security,[35] the viriditas of the Word,[36] of holiness,[37] of the living breath,[38] of integrity,[39] of truth,[40] of penance,[41] of good works,[42] and of abstinence,[43] among others. In this usage, the greenness of viriditas was more than its color. It stood for greenness as a metaphor for positive spiritual action—integrity, truth, penance, abstinence.

Given such frequent metaphorical usage, it seems fair to ask, naively: but what was viriditas a metaphor *of*? That is, why, or how, did the color green provide a metaphor for positive action? What was it about green that made its use as metaphor "work"? Previous historiography has assumed that viriditas worked as a metaphor because it stood for the life-force or for spiritual power, but a close look at Hildegard's usage suggests instead that viriditas meant something more substantial, a green plant liquid of some kind. This can be teased out of the strictly metaphorical passages in her work, but it is easiest to see in the non-metaphorical passages of *Causes and Cures* and *Physica*.[44]

Thus in Books One and Two of *Causes and Cures*, viriditas mainly seems to be some kind of plant substance with a physiology of its own, whose particulars can be gleaned with a detailed reading.[45] First, the roots of plants contain viriditas, which turns into flowers in the summer:

> In winter, the roots of plants have viriditas within them, which they put forth in summer as flowers.[46]

This image implies that viriditas was some kind of substance, not an insubstantial power of some kind.

This is supported by Hildegard's remark that in the plant, *swertula*, "all of its effect is in its root, and its viriditas goes up into its leaves."[47] Viriditas, then, does not seem to mean strictly "life-force" because here, at least, it is not the power or effect of a plant but a physical substance that *moves* from root to leaf, and is, therefore, probably a liquid. In another passage, viriditas is said to nourish both flower and fruit from its place in the roots: "Through the root of the tree, which contains viriditas, flowers and fruits are fed."[48]

Viriditas, therefore, is a liquid in the roots of plants; it can travel up their stems and feed their flowers and fruits.

Second, viriditas has some kind of seasonal cycle. In winter it stays in the earth:

> But when the sun pulls its heat to itself, the waters become especially cold and put out their foam, that is, the snow, onto the earth, and this opens the earth and fattens its viriditas.[49]

In spring, viriditas comes out of the earth and appears on the surface. With the warmth of summer, it flourishes as flowers:

> The sun . . . scatters its brightness over the whole earth, and makes it firm, and from the earth emerge flowers and viriditas. . . .[50]

In autumn, viriditas begins to vanish or dry up:

> And so with the sun the two planets move into Libra, where viriditas and dryness are in balance; as viriditas decreases, dryness increases.[51]

In winter, viriditas disappears again:

> In winter the sun bends to the earth, and the earth becomes cold, and all green things dry up.[52]

Viriditas, then, has a seasonal cycle; and it depends on two of the traditional medical qualities—heat and moistness. For instance, it is the *heat* of summer (transmitted by the wind) that draws viriditas out of the earth: "Thus the earth reveals viriditas when it is hot. . . ."[53] And it is *moisture* that gives viriditas to plants: "Water . . . give[es] . . . sap to branches, taste to fruit, viriditas to plants.[54] That viriditas is not simply a property but a substance is suggested by the notion that "plants have viriditas within themselves because of flowing humor."[55]

Rather than being an immaterial "greening power," then, for Hildegard viriditas seems to have been a substance that originated in the earth, entered plants through their roots, and eventually became leaf, flower, and fruit. Its transformation from earth-stuff into flowers and fruits was due to the warmth of wind and sun, and moisture from rain.

That viriditas was not an abstract force but a substance with material characteristics is confirmed by Hildegard's references to bad, as well as to good viriditas, to dry as well as to strong viriditas. For instance, there can be bad viriditas, since "*stutgras* is bad for people to eat because its viriditas is bad."[56] Also there is good viriditas: "*Bertram* is a bit dry but temperate; it has a good viriditas."[57] There is a dry viriditas: "*Berwurtz* is hot and has a dry viriditas."[58] A strong viriditas: "*Ringula* is cold and wet, and has a strong viriditas and works for poisoning."[59] There is a sweet viriditas: "*Asarum* is hot and dry, and has some of the powers of herbs, because its viriditas is sweet and useful."[60] And a harmful viriditas: "Celery gives one a wandering mind, no matter how it is prepared, because its viriditas is harmful."[61]

Finally, viriditas must have something to do with the power or effect of plants, since "dill has to be green, because its strength [that is, its medicinal effect] is in its viriditas."[62]

What these passages imply is that viriditas was not simply the color, green of plants, although it *was* green. Rather, since it could be bad, good, hot, sweet, or dry, it was some *thing* that either indicated or *was* the medicinal effect of a plant. Moreover, just as there turned out to be many humors—qualitative, humoral, and individual—so, too, there were many viriditates. Indeed viriditas seems to have been a kind of humor itself:

> The sun in the fire of its rays sends heat onto herbs [*aromata*] and flowers, while dew and rain give them the humor of viriditas.[63]

Nevertheless, although viriditas may have been a kind of humor, Hildegard does differentiate it from *humor* and juice or sap (*succus*), since some foods can contain all three. For instance, milk has viriditas, humors, *and* juices;[64] *gichtbaum* contains viriditas *and* juice.[65]

Indeed, the earth has its own viriditas, the *viriditas terrae*, which is distinct from the viriditas of plants and different from the *humor terrae*, because "the earth into which he had sunk to his knees, contained *humor* and viriditas and seed."[66] Also, "*crasso* . . . grows more from *viriditas terrae* than from heat,"[67] and so does rue and mustard.[68] In fact, the earth contains a number of substances—*humor, succus,* and viriditas—but it is the viriditas of earth that feeds plants: "Just like lambs suck milk from their mother, so plants suck viriditas from the earth."[69]

To sum up. Hildegard used viriditas occasionally just to signify the color, green; most often she used it as a metaphor for spiritual attributes. Behind its use as metaphor was not a power—greening power or Life Spirit—but a substance, a liquid in earth and in plants that was part of the observed horticultural cycle. In winter, it was in the earth and the roots of plants. In spring, plants drew viriditas up through their roots into their stems, where it was moistened by rain and dew, warmed by sun and wind, and transformed into a variety of "viriditates" that signaled the medicinal effects of the juices of plants.

VIRIDITAS IN OTHER WORKS

Sources of Viriditas

Was this usage idiosyncratic to Hildegard or did it reflect the usual understanding of viriditas? On the face of it, this sounds like the questions of Chapters Three and Four as to how Hildegard understood the elements, qualities, and humors. But it is a very different question. Elements, qualities, and humors were well-known concepts in premodern medicine; viriditas, on the contrary, has been considered to be a concept peculiar to Hildegard. Did Hildegard, then, invent viriditas? If not, where did she find it? In theology, in medicine, in botany? And did she, somehow, transform its meaning?

Viriditas certainly does not occur in the usual natural scientific texts of the twelfth century, including medical, herbal, agricultural, and horticultural texts.[70] In particular, it cannot be found in theoretical medical texts,[71] or in the *capite ad calcem* antidotaries.[72] Herbal material that was available to Hildegard (including Pliny, Dioscorides, Book XVII of Isidore's *Etymologies*, and Macer's *De virtutibus*) were simply lists of plants and their specific medicinal effects; they never mention viriditas.[73] Nor does the ancient agricultural material in Cato, Varro, Columella, and Palladius.[74]

Horticultural material from her era and before is almost nonexistent; and none of it mentions viriditas.[75] Nor is there anything like viriditas in the early *De hortulus* (804–849) of Walafrid Strabo or in the later (thirteenth- and fourteenth-century) texts of Walter of Henley's *On Gardening* and Godfrey of Palladius', *On Husbandry*.[76] As for strictly botanical texts that might have explicitly presented the physiology of plants, and, hence, of viriditas, there were no texts on botany until the translation of *De plantis* by Alfred of Shareshill at the end of the twelfth century.

How then did Hildegard learn of viriditas? There was only one (textual) way. She absorbed her ideas of viriditas from many short passages and allusions in the Church Fathers, not from medical or botanical texts. The reason that she was able to construct a coherent picture of viriditas, however, was that behind their allusions was a botanical concept of viriditas that embodied much of ancient plant physiology, as we shall see.

Viriditas in the Church Fathers

Although Hildegard is said to have learned of viriditas through her reading of the Bible, in the *Vulgate* it occurs only once and then in a negative way:[77]

> The descendants of those who lack reverence will have no branches, and their unclean roots will sound on the pavement. Viriditas will be pulled up before the hay at the lakes and the mouth of the river.[78]

In the Bible, then, viriditas was far from being a common synonym for life-force or God. Rather it meant useless weeds and signified the negative quality of undisciplined and fruitless growth.

But viriditas *was* a common concept in the Church Fathers; it occurs in hundreds of passages in the *Patrologia Latina, Cetedoc,* and the *Acta Sanctorum.* Again all of these can be parsed into three separate but related meanings: viriditas as a color; viriditas as an abstract topos or metaphor; and viriditas as a substance, green sap. It was this substance that was behind the metaphor.

First, viriditas was used to mean simply the color green, either of gems[79] or of vegetation. For instance, Isidore writes of the shining greenness of emeralds[80] and of beryllius, whose greenness resembles that of emerald.[81] Ambrose speaks of the viriditas of fields and seems to mean, simply, their greenness: "The splendor of the earth is the sprouting and greenness of the field."[82] Lactantius, also, speaks of the "greenness and abundance of the woods."[83] There is, also, the greenness of plants: "You have seen the beauty of trees and the greenness of plants."[84] And the greenness of evergreens:

> There is in this land a certain herb called "ever-living" because it never dies when it is in its proper place. Life alone is there, without corruption; and there is no poverty but an eternal greenness obtains. . . .[85]

Sometimes, strangely perhaps, viriditas could mean the color gold and not green; thus both Origen and Augustine allude to the viriditas of gold.[86]

But second, and much more frequently, the Church Fathers used viriditas as a metaphor to signify the fertility or fruitfulness of some virtue. For instance, there was the viriditas of faith:

> Risen is the strength of the reed, that is, the viriditas of preaching, and the strength of the rush, that is, the viriditas of true faith.[87]

There was the viriditas of love:

But because the viriditas of intimate love was lost, the straw is now dry.[88]

The viriditas of hope:

The soul that is filled with the viriditas of hope and good works lives in the gardens of Holy Church.[89]

The viriditas of good works:

Plant in me the roots of true virtues and the seeds of holy contemplation, and with the viriditas of good works make them grow and sprout. . . .[90]

And the viriditas of merits:

Therefore paradise is having many trees, but fruit-bearing trees filled with the sap of virtue, of which it is said: Every tree in the wood is always flowering with the viriditas of merits.[91]

There was the viriditas of the spiritual life[92] and the viriditas of holy conversation.[93] Especially popular was the "fruit of eternal viriditas" (*aeternae uiriditatis fructum*), which seems to have been a stock phrase. For example, Onulfus speaks of him who does not manifest in his actions the fruit of eternal viriditas.[94]

The main characteristic signified by viriditas here was generally that of lush growth. That is, virtues were like viriditas because they were fruitful and caused the growth of like substances. Viriditas usually meant that good engendered good, although it could also signify the growth of more ambiguous things, like pleasure.[95]

But what was behind the metaphor? Again, it seems that what stood behind this metaphor was not an immaterial force but some sort of physical substance found in plants. This unexpectedly came out of an analysis of the hundreds of passages in which the term could be found. That viriditas was often a metaphor or color was clear. But several remarks—for example, that viriditas was in the earth or in the root of a plant; and that it turned into leaves, flowers and fruits—were also common and implied that viriditas was part of a physical system. The following observations summarize these findings.

First, viriditas seems to have had "qualities." Thus it must be "warm" and "wet" because it vanishes in the presence of cold and dryness.[96] Also, it was contained in earth:

> The tree which you see flower, which you see to be fertilized by the best viriditas,[97] is animated underground by a juice of fertility, returning outside what it contains in its root.[98]

This statement must mean that viriditas, a juice of fertility (*succus foecundus*), is in the root of the tree and shows itself externally first as viriditas and then as fruit. Fertile juice is transformed first into viriditas and then into fruit. Although visible as greenness, then, viriditas actually comes from the roots of plants, which are fed by the earth's juice of fertility. In fact, viriditas seems to originate in decaying matter:

> When grain is planted in the ground it lacks shape at first, but as it rots, it acquires viriditas.[99]

From the earth it gets into the roots of plants:

> In such a small seed, where lies hidden the hardness of the trunk, the softness of the middle, the bitterness of bark, the viriditas of the root, the taste of the fruit, the sweetness of the smells, the diversity of colors, and the softness of leaves?[100]

Also viriditas seems to have been a liquid and even a humor, since trees are cut "so that the humor of viriditas can dry out."[101] It participates in the cycle of the seasons[102] since "in summer viriditas is seen in trees; in winter it is as if dried up, although it is still alive, but hidden."[103] From the root viriditas becomes leaf,[104] flower, and fruit.[105]

Apparently, then, viriditas undergoes a series of transformations. The viriditas of the root begins as earth-humor, and eventually becomes the viriditas of leaves:

> But the power in this very small seed is rather amazing; by it the nearby humor, mixed with earth, is turned into wood, branches, the viriditas and shape of leaves, the form of fruit.[106]

The transformation of viriditas does not always stop with its becoming leaves. Sometimes it becomes fruit—though not always, since "although willows have a lot of viriditas they produce no fruit;"[107] and "grass produces viriditas but no solid fruit."[108]

Nor does every plant have its own viriditas. For example, the palm parasitizes the viriditas of the grapevine, by which it becomes green, and from which it produces its flowers:

Palms that grown on vines do not have their own sap, but use the juice of the vine's viriditas from which they become green and produce flowers.[109]

What then, was viriditas? Viriditas was a green stuff that came from rotted, liquefied, organic material in the earth. When the earth warmed up in spring, this green stuff was pulled up into the roots and stems of plants. As this occurred, it was further warmed by sun and wind, and appeared in spring as the viriditas of leaves, and in summer as the viriditas of flowers and fruit. In autumn it dried up and went back down into the roots. Viriditas, that is to say, encapsulated an ancient physiology of plants.

Viriditas in Botany

If so, then there must have been ancient texts on plant physiology that explicitly delineated the process that viriditas implies. The question is: which texts should be used to examine this?[110]

Now ancient botany conformed to the pattern of ancient medicine in that it survived by being translated from Greek into Arabic, and then, toward the end of the twelfth century, into Latin. Unlike ancient medicine, however, summaries of which did pass into the Early Middle Ages, summaries of ancient botany did not. Instead, the first explicit text on plants was *De plantis*, translated into Latin by Alfred Shareshill.[111] It appeared in Europe circa 1180 as *De vegetabilibus*, and was the Latin translation of an Arabic translation of an originally ancient Greek text that has since disappeared (Lulofs, 129).[112] Thus, although appearing thirty years later than Hildegard, *De plantis* does give us some idea of ancient Greek plant physiology in its medieval Latin dress. It is, therefore, the best source for getting at the kind of ancient plant physiology that fed the Latin concept of viriditas.

According to *De plantis*, plants are not very different from animals.[113] They are "alive" because they eat, grow, reproduce, thrive in youth and wither in old age.[114] Like animals, too, they have a finite life span, which is determined by their quantity of inborn *heat* and *moisture*. When these are used up, they weaken, age, and dry up.[115]

Also, like animals, plants have simple parts such as their sap (*humor*) and also complex parts such as their branches, leaves and the rest,[116] which fall each year like the nails and hair of humans.[117] And just as an animal's egg contains both the matter for its own nurture and the power to put forth shoots, so also do the seeds of plants.[118] To thrive and be healthy, plants need the same four things as animals—the right seed, the proper place, the right amount of water, and appropriate air.[119] Above all, they need the proper seasons, the right amount of sun, and appropriate weather.[120]

There are a number of ways to categorize plants. For instance, they can be divided into trees, bushes, or herbs; into wild, garden, or domestic;[121] according to their preferred living conditions, such as hot/dry/wet/salty;[122] according to their shapes;[123] or according to their saps, such as milky, resinous [*pici*] or sticky [*origanalis*].[124] But their food always comes from the earth and their fertility from the sun. So the earth is the mother of plants, and the sun is their father.[125] Because of these two powers of eating and reproducing, some even think that plants are whole and perfect, since in view of their continual leafing and fruiting their own life can be said to continue indefinitely.[126]

Plants produce seeds that are better or worse, and this depends on where they are grown, the work of the gardener and the season of the year.[127] But because they are fixed in place and cannot change position, where they are grown is especially important for their health.[128] For example, plants that grow in the mountains are better for medicine, and those that grow farther away from the sun are weak.[129] (Here we are directly reminded of the importance of place in the Hippocratic *Airs, Waters, and Places* examined in Chapter Three.) Unlike animals, however, the life of plants is hidden, that is, it is difficult to discern.[130] For plants do not have feelings, desire, or will.[131] They do not move but are fixed in the earth,[132] and they cannot be considered to have *spiritus* or *anima*.[133]

All of the various structures in plants—their bark, leaf, etc.—come from various combinations of the four elements,[134] but their cortex, wood, and core specifically come from *humor*;[135] and the root acts as mediator between a plant and its food. Indeed, the Greeks say that the root is the cause of a plant's life.[136] Because of this power to pull *humor* from the earth, plants, therefore, are not entirely inert.

What happens is that this "root power" attracts *humor* out of the earth; and this humor then flows up the cortex of the plant and then out into its branches.[137] This is the way plants eat. Unlike humans, however, who have three digestions, plants have only two. The first takes place inside the earth, and the second inside the cortex.[138] It is during this second digestion (*pepsis*—cooking or ripening) that *humor* changes color. If it gets too hot and dries out, then it turns black; if it is too wet, then it turns yellow. But if it is perfectly cooked, then it turns green. Indeed:

> This is the reason for the *viriditas* of trees and of all things born of the earth. When it is hot outside, the heat draws up what nourishes the tree—*humor* or earth-vapor. If the heat is debilitating and the humor

watery, then its color is green. If the humor is more earthy than watery, then inside it is dried and black. When there is more wateriness than earthiness in the leaves, they do not have such intense *viriditas*, and when they dry up they turn yellow. On the other hand, if it is not too hot and the digestion is better, then the inside of the tree is white or according to the nature of its nutrients.[139]

It is exciting to see viriditas occur here, but at this point it is not entirely clear whether *De plantis* means that viriditas is the *color* of the properly cooked *humor*, or a *name* for the transformed, cooked, green sap itself. What is clear is that the physiology of sap gleaned from the analysis of Hildegard's viriditas and that of the Church Fathers is summarized here. Plants draw *humor* out of the earth into their stems through their roots; from the roots it is pulled up into the branches, leaves, and fruits by the heat of the sun. As this is taking place, *humor* is "digested" by heat from the sun and dries out. At the same time, it is moistened by the water of rain and irrigation.

Thus *humor* is transformed from the initial black of earth-humor (*melancholia*) to the yellow of water; at the midway point of digestion, it becomes green. That is, earth-humor turns into viriditas. If the weather is too hot, then excess viriditas is pulled out into the leaves, and the cortex is left with more of the black earth-humor. When the weather is temperate, then viriditas moves out into the leaves with more water in it and appears yellowish. This physiology makes sense: earth-humor is black and as it flows up a plant, water, which is considered "yellow," is added to it. Heat matures or cooks it until it turns green; that is, *humor becomes viriditas*.

But the digestion or transformation of earth-humor does not stop with viriditas. Earth-humor is cooked or digested until the green of viriditas is transformed into the other colors of leaves, flowers, and fruits:

> And so, as the year goes on, water holds on to color because of the coldness of the air. As it begins to warm up, this heat pulls the moisture out from the cortex and dries it, and so color appears on trees. . . . Thus *humor* [moisture] holds heat and, therefore, the other colors [that is, yellow, red, and black leaves] appear.[140]

What this means is that as digestion proceeds, the greenness of green earth-humor (viriditas) turns into other colors. (This process, of the transformation of viriditas into the other colors of leaves and fruits—yellow, red, black—was explicit in the premodern idea of green, as will be seen below.)

Along with this seasonal maturation of color in leaves and fruits, there is a corresponding transformation of tastes, which, as we saw in Chapter Four, signals the development of the medicinal effects of plant saps:

> At first the fruit will be acidic. As digestion [ripening or maturation] proceeds, this sharp taste dissolves and the fruit becomes sweet. When it is perfectly matured it will be bitter, because of the amount of heat it contains as compared to the amount of humor it then has.[141]

Therefore, green leaves come from the least-digested sap; sap gets digested or transformed, as it goes from root to viriditas to fruit.[142] Moreover, according to their tastes are these transformed saps categorized—into potable, like grape juice and apple juice; oily, like olive oil, nut oil, or pine oil; sweet, like fig; hot and sharp, like oregano and mustard; or bitter, like absinthe and centaury.[143]

This transformation corresponds to what we learned in Chapter Four about the relationship between tastes, medicinal effects, saps, weather, and qualities. Here it finally becomes clear that the characteristic taste of a plant, which represented its specific medicinal effect, was modified by the qualities of weather and climate—like wine.

Viriditas was not simply greenness, not even the greenness of green sap. Rather it *was* green sap, with properties: the qualities of warmth and moistness, the character of balance and transition, the power of "turning into."

To sum up: Ancient Greek plant physiology as transmitted in the circa 1180 *De plantis* did indeed contain the very notion of viriditas that had been passed on in theological texts. The physiology of viriditas implicit in Hildegard's texts and in the Church Fathers was explicit in botany.

The Attributes of Green

It was this very plant physiology—the transformation by heat and moisture of earth-humor into the green sap of viriditas—that was at the root of ideas about the color green itself. Bartholomew Anglicus summed up the twelfth-century conception:[144]

> Green is generated in matter by moderate warmth and moistness, as appears in leaves, fruits, and plants, which are dark but not black. *Viriditas* is generated when this dark is mixed with white and heated but not thoroughly cooked, and so viriditas in plants and fruits is a sign of raw and undigested humor, as Avicenna says. This shows in autumn, when the green color in plants and in fruits turns to yellow as the moistness in

leaves and plants is used up by summer's heat. And so certain trees are
green in spring and in summer, and turn yellow in winter and autumn,
as Aristotle says.[145]

In other words, the first notion of green *did* come from plant life; green
was formed when the raw humor of earth was cooked. As this happens the
originally black humor of earth turns green; when it is finished cooking, in
autumn, the green humor changes to yellow. This was why green was considered to be a balanced, transitional color:

> Green is midway between red and black, which is proved by the trans-
> formation of red choler into black choler, which passes by green choler.
> This is why it especially delights the vision, because of its mixture of
> fire and earth; and so no color is so delightful to the vision as green, as
> appears in emerald which converts and repairs the eyes more than any
> other metal or plant.[146]

Thus the second point about green is that it is a transition stage between
the humors of yellow bile and black choler. (Here Bartholomew hints that
viriditas, the green humor, might also be in the body as something called
green *choler*.)

But green is not only a midway humor, it is a balanced humor, with
just the right amounts of earth and fire, cold and hot, dry and moist. Green
comes from earth-humor, cooked by the sun:

> Green leaves, plants, and seeds and other things born of earth in which
> they take root, have this earth-matter rarefied by fire, and, therefore,
> their smoke is not black or red but green. For blackness tempers red-
> ness, and both are reduced to a middle place and are, therefore, green.
> And although leaves, fruits, and seeds turn green, flowers that are green
> are rare or non-existent because their material is so much more subtle.
> Instead, they are yellow or pale if water prevails; pink or purple if fire is
> more than air; if water predominates over earth, then they are green or
> black, but this is rare.[147]

This, then, is why green was midway between the extremes—of fire and
earth, black and red. In particular, "the green of leaves and plants comes from
its earth parts, which are clarified and colored by the fire which is mixed in
them." Thus just looking at it is good for the eyes:

And thus the color green is between red and black, comforting and repairing the vision, and animals like deer and others pasture in green meadows not only to eat but also to strengthen their sight, which is why hunters dress in green. Since their wild prey take such pleasure in green-ness they are less afraid of the hunter than they would have been if the hunters had dressed in a different color, as Galen says.[148]

But green was not always good; it could also have the negative mean-ing of rawness or immaturity. Green was "a sign of the rawness of humors in fruits and plants . . . as is shown by the fact that as a fruit ripens, it turns from green to other colors, such as white or red or black or yellow."[149]

It seems then, that the characteristics of green, at least in the sum-mary of Bartholomew, derived from the physiology of viriditas. And this was generally true: the accepted characteristics of greenness (as color) did derive from the characteristics of viriditas (as substance).[150] This can be appreciated in many texts, including poetry (Hugh of St. Victor), earlier encyclopedias (Isidore), Church law (Innocent III) and, even technology (Theophilus).

For instance, although in theory green was not a primary color, it was not a mixture of colors, either. While the four primary colors—black, white, red, and yellow—were linked to the four elements, seasons, directions, and humors, green was in a category by itself (Fayet). Its name was related to the Old German "gruoni," the Indo-European root for "grow" (Schipperges, 1992 [1957]) and its character always related to vegetation.[151] In Rome it was the color of earth and spring (Dronke 1972, 66). In twelfth-century Europe, it was linked to seeds and new life.[152]

Green was never unambiguously good or spiritual, either. As a metaphor it could mean temperance and moderation (Pastoureau 1989, 16–18),[153] but also change and disturbance (Pastoureau 1986, 30).[154] As Isidore said, green was green because *viridis* was "vi et suco plenus quasi vi rudis,"—that is, full of power and juice, as if a raw force (Isidore 1856, col. 276).

In sum, greenness—viriditas—*was* related to vegetation but differently from how modern scholars have supposed. Rather than being about "power" (e.g., "greening power") or about an abstract life force (*vis*), its character derived from plant physiology. Greenness was about balance—the balance of qualities (it was moist and warm) and the balance of elements (it had an equal proportion of earth and fire). And it was more than a color; it was a substance, formed in the earth and matured by the sun. When properly warmed (cooked) by the sun, it was transformed into the four humors, as manifested by the color change in leaves and fruits from green to red, yellow, black or white. This was why green signified the power of transformation and of balance.

Viriditas, then, did not take its meaning from green. Green took its character from viriditas, from green sap.

Viriditas in the Body

So Hildegard used viriditas just as it had been used in theological texts, in botanical texts, and in the popular understanding—it encapsulated a set of simple horticultural observations: the transformation of earth, via water and sun, into leaf, flower, and fruit. However, Hildegard did do something unique with viriditas, because she found it not only in plants but also in the body.

For this transfer of viriditas from plant to body she had little precedent. In none of the hundreds of passages that contain the term in *Cetedoc* was viriditas used for something inside the body. On the contrary, viriditas signified the *special* life of plants as opposed to that of humans or animals. This was why Gregory the Great wrote in the quotation at the beginning of the chapter that the life of plants was called viriditas, in contrast to the life of animals, which was called *anima*. As *opposed* to animals, plants and trees live from viriditas—Gregory did not find viriditas in humans.

By contrast, for Hildegard viriditas *was* inside the body, but whether as a metaphor for vigor, health, and, especially, fertility, or as an actual substance, is not entirely clear. For example:

> The menstrual flow of a woman is her generating viriditas, her flowering, because, just like a tree by its viriditas produces flowers and leaves and fruits, so a woman by the viriditas of menstruation brings forth flowers and leaves as the fruit of her womb.[155]

Here Hildegard links menstrual flow to the viriditas of trees. Bodily viriditas seems to be a metaphor for fertility. Also:

> Just as a tree does not bear if its lacks viriditas, so an old woman who does not have the viriditas of her flowering can no longer bear children.[156]

That is, when menstruation stops and a woman is no longer fertile, she no longer is deemed to have viriditas. Again, viriditas seems to be a metaphor taken from the plant world for fertility, procreation, and abundance. In men, too, viriditas serves as a metaphor for fertility and potency:

> Adam was virile because of the viriditas of earth, and strong because of its elements.[157]

Still there is some ambiguity, because when a man is castrated he loses his "virile viriditas":

> If a man lacks his testicles, either naturally or from surgery, then he will also lack the virile viriditas.[158]

Did Hildegard mean this also as a metaphor—a man was impotent without his testicles? Or did she mean that a liquid substance, conferring potency, was stored in the testicles, and related to the green sap of plants? This kind of ambiguous usage would conform to what Hill found for the equally ambiguous *spiritus*, which was "more than breath but less than soul" (1965, 65).

In certain passages, viriditas seems more like an actual substance in the body.[159] It has specific qualities—wetness and heat (and these are the same qualities it has in plants). For instance, viriditas has the power of sending moisture to different parts of the body.[160] Also:

> When there is enough blood, then water is in the body, which thus is moist enough for viriditas to flourish.[161]

Viriditas is warm:

> A person is fertile because of his mixture of cold and heat. . . . Heat is his viriditas, and cold his ariditas and through these he produces everything.[162]

And in a healthy person the veins are full of viriditas:

> And because they have this defect in their body, they are slow, and their veins are not full of viriditas but are fragile like those of reeds.[163]

Still, most of these hints are somewhat ambiguous. However, there are passages where viriditas is certainly a substance. For instance, viriditas is actually in the skin of the scalp and even on the chin:

> Once a man is bald, no medicine can help because both the moisture and the viriditas that were in the skin of the scalp have dried up, and can never come back, whence the hair cannot regrow.[164]

> Women who are a bit masculine on account of their viriditas often have
> a few hairs on their chins.[165]

And viriditas is excreted by the bladder:

> Strong wine dries the bladder, preventing it from properly carrying off
> the viriditas of the marrow.[166]

Indeed, the *anima*:

> Pours itself into all the parts of the body by imparting to the medulla,
> veins and all other parts of the body, viriditas, just as a tree gives juice
> and viriditas to all its branches.[167]

Even animals contain this physical viriditas:

> In the place where moles dig, take the earth that has been mixed with
> its juice and its viriditas, along with the juice and viriditas of the earth,
> which is healthier in that spot than in others.[168]

If we were to describe what these passages imply, then we might say that
viriditas was a warm, moist liquid found in blood and marrow, on the head and
chin as hair, and (perhaps) in the uterus, testicles, and bladder; and that it was
responsible for vigor, sexuality, and fertility. It was, that is to say, a potent liquid,
both substance *and* power. It had many of the characteristics today implied by
the concept of hormone, a special liquid responsible for functional behavior.

But if Hildegard did mean that viriditas was a potent substance in the
body, then what was the relationship between the viriditas of the body and the
viriditas of plants? Well, viriditas gets into the body by the simple process of
ingestion. Thus people are fed by viriditas: " . . . viriditas and those fruits by
which people must be nourished."[169] This suggests that viriditas enters the body
when green things are eaten, such as green plants, leaves, vegetables, and green
fruits. Indeed the earth does feed the body with its juice, viriditas:

> And so pouring heat on him by means of viriditas, the earth is the carnal
> material of man, feeding him with her juice [*succus*] just like a mother
> nurses her child.[170]

Hildegard explicitly summarizes this botano-medical cycle: "Earth nourishes
viriditas; viriditas nourishes fruit, and fruit nourishes animals."[171]

Now, there were other botano-medical overlaps. For instance, there was an ancient connection between plant sap and blood (Pouchelle, 280), but this particular, substantial connection was not precisely made until Hildegard. Earlier literature did hint about a bodily substance that represented or conveyed vitality. For instance, Asaph wrote that in adulthood the vigor of the *blood* reached its peak (Venentianer, 371); and *Ad Maecenatem* compared the waxing and waning of consciousness during the month to the waxing and waning of the medicinal effects of plants.[172] Earlier literature also had certain green (*viridis*) liquids in the body—urine[173] and bile[174]— but, probably, this referred to *viridis* as gold rather than green. In any case, *viridis* urine and bile were markers for illness, not health; and both were dry, not moist like viriditas.

There were other notions that overlapped with Hildegard's human viriditas, although none quite conveyed this substantial connection between the plant and animal world. These were the *humidum radicale*, the *calor innatus*, and the *vis medicatrix naturae*.[175]

The *humidum radicale*, usually rendered as the "radical moisture," is probably best understood as the "root" or basic moisture with which a life begins.[176] At conception semen contributed a fixed, unreplenishable amount of moisture,[177] and as life went on, it was used up like oil in a lamp.[178] When the *humidum radicale* had been totally consumed, the body died. Therefore, one of the tasks of the physician was to minimize this loss with the non-naturals. Hildegard's conception of a bodily viriditas does share some characteristics with this ancient conception of the *humidum radicale*. Both were moist, liquid, and essential to fertility. But Hildegard never suggested that bodily viriditas was finite, any more than it is finite in the earth, or that it diminished with age or activity.

Viriditas also shared some characteristics with the *calor innatus* (the inborn heat).[179] Along with the *humidum radicale*, the *calor innatus* was implanted at birth. Both formed the infant and provided the power for growth and maturation (Durling).[180] Originating in the *pneuma* and the blood, the *calor innatus* dwelt in the heart, traveled in the pulse throughout the body, and distributed the vital spirit. Like the *humidum radicale*, the *calor innatus* was finite.[181] Indeed, the main purpose of respiration was to cool the body and so minimize the loss of the innate heat. Its gradual depletion over the years was what caused illness and explained the coolness of old age and the coldness of death. Misuse of the non-naturals depleted it prematurely.[182] Like the *calor innatus* viriditas, too, was warm, found in the blood, and signified life. But unlike the *calor innatus*, viriditas was not a fixed substance in the body. It did not signify that the life-force was a fixed quantity or that death was inevitable.[183]

A third concept with some overlap with viriditas was the *vis medicatrix naturae*, the healing power of nature.[184] Although the name is relatively recent, the concept was ancient, and meant the body's innate capacity to mend itself. In Greek it was *physis*—Nature—and was the same power that maintained life.[185] It meant the body's innate vigor or strength, the inborn power of the live body to maintain its integrity.[186] Thus viriditas also shared certain characteristics with *physis*. For instance, *physis* was also connected to the plant world; thus Galen compared it to the force that causes the seasonal ripening of fruit.[187] Still viriditas differed from *physis* in important ways. *Physis* was innate and peculiar to a particular body, while viriditas came from outside the body and was generic. Viriditas implied that earth-essence was transformed into plants and signified a continuity and similarity among earth, plants, animals, and humans, while *physis* accented an individual's particularity.

In sum, while viriditas was a common concept in the botanical and theological literature, only Hildegard used it to mean something in the body. Although there are passages in her work that suggest that she sometimes used it simply as a metaphor for fertility, abundance, or life, on the whole she seems to have meant viriditas as an actual substance, a kind of fifth humor. Bodily viriditas came from plants and completed the physical cycle that connected plants with body and gardening with medicine. While Hildegard learned the concept of viriditas from the Church Fathers, its ancient *physical* meaning was revealed to her not in the chapel or in the library but in the garden, and she herself brought it to the infirmary.

Hildegard's *translatio* could "work" because the most important way for understanding the human body *was* as a plant. As Plato had said more than fifteen hundred years before: "We are just like plants only, as their roots are in the earth so ours are in the sky." This was why there could be such a fertile interchange of humors between them—there was not so much difference between plant and body. Both were rooted in the same earth and subject to the same sky, with its winds, weathers, and seasons.

Chapter Six
Conclusion

How, then, *did* the premodern medical system of elements, qualities, and humors work for one particular practitioner? Clearly, the elements, qualities, and humors were multivalent and deeply integrated into Hildegard's thinking and practice. Behind them were horticultural phenomena: the *elements* of gardening—wind, rain, sun, and land; the *qualities* of weather, geography, and season—hot and cold, wet and dry; and the ever-fungible *humors* of plant saps and bodily fluids. But did all of these concepts function together as a system? Was there some kind of overarching model by which all of these concepts were linked?

Two separate figures can summarize what we have learned so far, and when they are placed within the premodern cosmos as Hildegard visualized it, we will find that there *was* a single, unifying image of how the cosmos affected the body. But to see it will require that we change our point of view.

Rooted in the Earth

First: seasons, climates, weather, elements, qualities, and humors were linked by the concepts of premodern plant physiology, as can be seen in a diagram summarizing these relationships (fig. 6.1). Influenced by the *elements*, plants transform *earth-humor* into their own individual humors—saps or essences—through the intermediary of the *qualities*. Each particular plant reflects the sum of all of these effects by its unique flavor. Here wine is our best example and only surviving remnant of this ancient system.

But the concepts of the premodern medical system (elements, qualities, humors) were also linked by a second image, that of the permeable body. Seasons, climate, weather, and geography affected that body both directly—through the physical sensations of hot and cold and wet and dry; and indirectly—through the more abstract qualities associated with the tastes

THE ELEMENTS

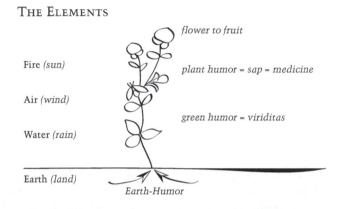

flower to fruit

Fire *(sun)*

plant humor = sap = medicine

Air *(wind)*

green humor = viriditas

Water *(rain)*

Earth *(land)*

Earth-Humor

Fig. 6.1. Plant as Locus for the Effects of Elements, Qualities, and Humors.

of plants. Both kinds of qualitative effects were mediated by the internal circulating wind, as formulated in fig. 6.2.

Thus there were two ways that the body was affected by the elements, qualities, and humors. First, through climate, weather, and geography; and second, through food. The body transformed food into the four humors, although the specific humor or virtue of a plant still, somehow, kept its force. Whatever the body could not absorb, it excreted as *faecium* (dregs), which became earth-humor—that is, food for plants. This neat cycle meant that both body and plant absorbed the elements and qualities and transformed them into humors, indirectly, and directly. Therefore, elements, qualities, and humors were linked, in one way, through their effects on plants and bodies: the humors of plants turned into the humors of bodies, and back into the humors of plants, by the *action* of the elements and *through* the qualities.

But elements, qualities, and humors were also linked by the way the premodern, pre-Copernican universe was supposed to work—as a firmament that revolved around a fixed earth. It was this movement, and the movements of sun, moon, and stars, that created the external conditions of seasons, weather, and climate. And these were what mainly determined the health and sickness of body and plant.

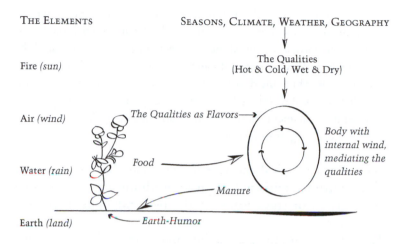

THE ELEMENTS   SEASONS, CLIMATE, WEATHER, GEOGRAPHY

Fire *(sun)*

The Qualities
(Hot & Cold, Wet & Dry)

Air *(wind)*

The Qualities as Flavors →

Water *(rain)*

Food ——

Body with
internal wind,
mediating the
qualities

—— *Manure*

Earth *(land)* —— Earth-Humor

Fig. 6.2. The Body and Plant as Locus for the Effects of Elements, Qualities, and Humors.

Rooted in the Sky

In her last visionary work, the *Liber divinorum operum*, Hildegard explained the relationship among elements, humors, and qualities as resulting from three basic movements over the earth. There was a clockwise movement of sky, propelled by the cardinal *winds* and their collaterals; a counterclockwise movement of sun, moon, and planets through the zodiac (propelled by another wind); and a third movement of the sun from south to north and north to south during the year. These three movements transformed the *qualities* of air. Air then entered the body and changed the *anima* (the internal circulating wind), which, in turn, transformed the internal *humors*.

First:

> This is what I saw: The east wind and the south wind with their collaterals were moving a blast through the firmament, and they were making the firmament revolve over the earth from east to west; and in the same way the west wind and the north wind with their collaterals pushed the firmament back, from west to east, under the earth.[1]

That is, the sky was revolving ("clockwise") around the fixed, spherical earth, propelled by the four cardinal winds. Next:

> From the day on which the days begin to get longer [i.e., the winter sol-
> stice], the south wind with its collaterals also begins to push the firma-
> ment to the north as if it lifted the firmament degree by degree, until the
> day on which days can get no longer [i.e., the summer solstice]. Then,
> on the day in which the days begin to get shorter, the north wind, with
> its collaterals, as if shrinking back from the brightness of the sun, begins
> to repel the firmament from north to south, degree by degree, until the
> length of days begins to increase.[2]

This second cosmological movement then, was that of the sun, which
went from north to south and south to north during the year. Its movement culminates, or stops really, at the two endpoints of summer and
winter solstices.

Last, she explains, there was a third movement—the movement of sun,
moon, and planets backwards through the zodiac during the year:

> And also I saw in the highest fire a circle appear that surrounded the
> firmament from east to west [this is the zodiac]. Out of it came a wind
> that made the sun, moon and planets go in a direction from west to east
> against the rotation of the firmament. And this wind, like the others,
> did not emit its breath onto the earth but only onto the sun, moon and
> planets, and so tempered their motions.[3]

This third movement was the annual, puzzling, but predictable rotation of
sun, moon, and (to some extent) planets backward, in a counterclockwise
direction through the zodiac—the moon in a month, the sun in a year, and
the planets in their confusing and various cycles.

All of these cosmological motions affected the earth, which was, of
course, fixed at the center of the universe. They did so because they changed
the *qualities* of the "air." This changed air modified the body, for better or for
worse, through the *anima*, the internal, circulating wind:

> Then I saw that, as the qualities of winds and air changed, so the humors
> inside the body were correspondingly changed and took up the changed
> qualities. Each of the higher elements contains an air that corresponds
> to its particular quality, and with the help of wind, sun, moon and stars,
> sends this air onto the earth and tempers the earth. . . . And I saw
> that when one of the winds was excited in a certain part of the earth,
> either by the course of the sun [i.e., the seasons] or the moon [i.e., the
> weather] or by the judgment of God, it emits a breath onto the earth

that renders man changeable in his humors, so that he takes it in and lets it out. And by it the *anima* takes up and transmits this air to the inner body, and the humors inside are changed, either for health or illness.[4]

What Hildegard means is that, as the seasons change and the weather changes, the qualities of the external air change. This air enters the body either *directly* as the "qualities" of hot/cold and wet/dry (temperature and humidity) or *indirectly* through the tastes of food, which were also classified, as we have seen, into the qualities of hot/cold and wet/dry. The internal wind is the *anima* and it serves as the final common pathway for all of these effects. It is the internal wind that modifies the internal humors.

Now to understand how this schema provided an overarching and unified model for the premodern body we have to remember what the world looked like before Copernicus. That is, we have to take the point of view of an observer on a fixed earth (fig. 6.3). From this point of view, during the day the sun travels across the sky from east to west. This is what Hildegard meant by the winds which blew the firmament from east to west during the day, and west to east during the night. Of course, as the sun moves, the temperature and humidity of the air change.

From this perspective, the sun also moves during the year, from south to north and north to south. This movement also changes the qualities of hot and cold, and wet and dry because it is what produces the seasons, the

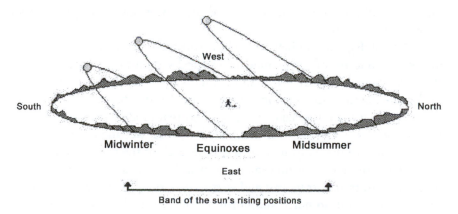

Fig. 6.3. The Course of the Sun During the Year. Taken from *Stones of Wonder* (www.stonesofwonder.com/figure1.gif), by Robert Pollock. Courtesy of Robert Pollock.

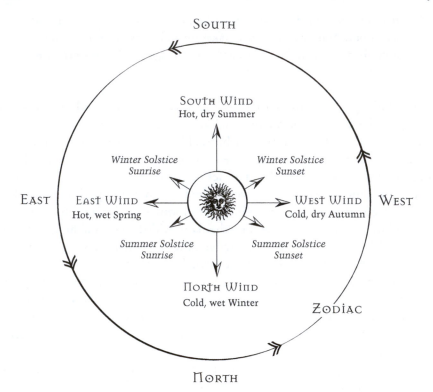

Figure 6.4. Diagram of Hildegard's Wind-based Cosmology.

maturation of earth-humor in plants, and the waxing and waning of the four humors in the body.

A final diagram can help to see how this pre-Copernican point of view unites the concepts of the premodern medical system into a natural holism (fig. 6.4). Since the body (earth and plant) is the focal point for the effects of the moving universe, it belongs at the center of the diagram, on the fixed earth. Inside it is an internal circulating wind that modifies the four internal humors because it tempers their naturally hot, cold, wet and dry qualities.

Outside of the body are four winds, at each of the four cardinal directions. They produce two movements. On a daily basis, they push the sky over the earth from east to west and west to east; on a yearly basis, they push the sky from south to north and north to south. It is this yearly movement that creates the four corners of the universe—the two equinoxes and two solstices—probably the natural basis for the scheme of fours. It is also this yearly movement of the sun over the earth that creates the seasonal change of qualities, from temperate in spring to hot and dry in summer, to cold and dry in autumn, to cold and wet in winter—in northern Europe. (Of course, as we have seen, the specifics of the schema were fungible and related to the actual seasons and weathers of specific localities.)

A third wind also circulates counter-clockwise; it carries sun, moon and planets counter to the clockwise daily movement of the sky. This third wind is what causes the sun to rise a degree earlier in the zodiac every day, and march through the entire zodiac every year; it also causes the moon to rise twelve degrees earlier every day, and so pass through the entire zodiac every month.

These three movements of the firmament change the external qualities of air. As we have discovered, these changed qualities are communicated to the body *indirectly via* their effects on plants (as food); and *directly* through breathing, through the pores and through the orifices. This changed internal air, or circulating wind, is what causes health and disease.

We are at last in a position to see the missing dimension, as it were, of our understanding of premodern medicine. In the introduction, we examined a diagram of how scholars have understood the premodern system of elements, qualities, and humors as primarily a philosophic system based on the number four (fig. 0.1). Erich Schöner's corresponding diagram was put together by a chronology of concepts; for instance, the qualities and elements were at its center because they had been developed first. But the concepts of the premodern system have turned out instead to be linked by a different kind of time—not chronological but cyclical time—the turning of the day, the passing of the seasons, the changing of the year, as they appear to a fixed body on a fixed earth.

This becomes clear if we examine a classic diagram almost contemporary with *Causes and Cures*, the *Concordance of the Elements and Seasons* of the eleventh-century monk, Byrhtferth of Ramsey (see fig. 6.5 following page 124 and fig. 6.6).[5] Like Hildegard, Byrhtferth was a Benedictine, and he created this diagram to gloss Bede's explanation of the cosmos that we have already looked at in Chapter Three. Although Byrhtferth's diagram has often been used as evidence for the schematic nature of the premodern system (Singer 1917), in fact, it provides crucial proof for the experiential and observational basis of the system.

Its orientation is with east on the left, south at the top; the four cardinal winds are at the corners to establish the four directions. Spring and autumn equinoxes, therefore, are at the left and right, just as Aristotle explained; and summer and winter solstices are at top and bottom. Each of the four elements is also located at a corner, one for each season, as Bede explained.

Translation

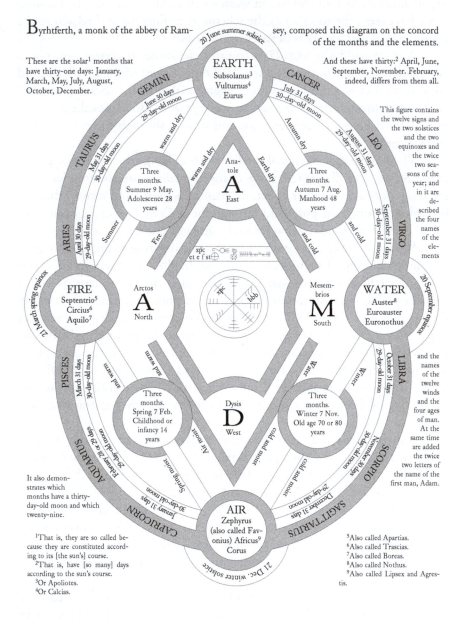

Byrhtferth, a monk of the abbey of Ram-   sey, composed this diagram on the concord
                                              of the months and the elements.

These are the solar[1] months that   And these have thirty:[2] April, June,
have thirty-one days: January,   September, November. February,
March, May, July, August,   indeed, differs from them all.
October, December.

20 June summer solstice

EARTH
Subsolanus[3]
Vulturnus[4]
Eurus

This figure contains
the twelve signs and
the two solstices
and the two
equinoxes and
the twice
two sea-
sons of the
year; and
in it are
de-
scribed
the four
names
of the
ele-
ments

GEMINI June 30 days 29-day-old moon

CANCER July 31 days 30-day-old moon

TAURUS May 31 days 30-day-old moon

LEO August 31 days 29-day-old moon

warm and dry

warm and dry

Autumn dry

Ana-
tole
A
East

Three
months.
Summer 9 May.
Adolescence 28
years

Three
months.
Autumn 7 Aug.
Manhood 48
years

Earth dry

and cold

ARIES April 30 days 29-day-old moon

VIRGO September 31 days 30-day-old moon

Summer

Fire

and cold

21 March spring equinox

20 September equinox

FIRE
Septentrio[5]
Circius[6]
Aquilo[7]

Arctos
A
North

Mesem-
brios
M
South

WATER
Auster[8]
Euroauster
Euronothus

PISCES March 31 days 30-day-old moon

LIBRA October 31 days 29-day-old moon

warm and

Winter

Three
months.
Spring 7 Feb.
Childhood or
infancy 14
years

Dysis
D
West

Three
months.
Winter 7 Nov.
Old age 70 or 80
years

and the
names
of the
twelve
winds
and the
four ages
of man.
At the
same time
are added
the twice
two letters of
the name of the
first man, Adam.

Water

Air moist

cold and moist

cold and moist

SCORPIO November 30 days 30-day-old moon

AQUARIUS February 28 or 29 days 29-day-old moon

Spring moist

cold and moist

SAGITTARIUS December 31 days 29-day-old moon

It also demon-
strates which
months have a thirty-
day-old moon and which
twenty-nine.

CAPRICORN January 31 days 30-day-old moon

AIR
Zephyrus
(also called Favo-
nius) Africus[9]
Corus

21 Dec. winter solstice

[1] That is, they are so called because they are constituted according to its [the sun's] course.
[2] That is, have [so many] days
according to the sun's course.
[3] Or Apoliotes.
[4] Or Calcias.

[5] Also called Apartias.
[6] Also called Trascias.
[7] Also called Boreas.
[8] Also called Nothus.
[9] Also called Lipsex and Agres-
tis.

Fig. 6.6. Byrhtferth's Diagram in Translation. From Peter S. Baker, "Byrhtferth of Ramsey, De concordia mensium atque elementorum" (www.engl.virginia.edu/OE/Editions/Decon.pdf). Courtesy of Peter S. Baker.

These four corners are connected by curved lines, and, reading from outside in, they represent the zodiac, the twelve months of the solar year, the twelve months of the lunar year, the four qualities, and the four ages.

The key to Byrhtferth's diagram is at its center, where a circular figure—body, plant, or earth—is cut into four quadrants by two lines, which are slashed with curious curved strokes, reminiscent of the paddles of windmills and water mills. These odd marks are the key to understanding the diagram because they are meant to indicate that *the diagram is to be read as moving, clockwise.* They indicate, therefore, the same first motion that Hildegard described—the clockwise motion of sky over the fixed central earth.

The diagram, then, is a kind of three-dimensional map of the pre-Copernican universe, its third dimension being time, not space. It shows, therefore, just how cyclic time, which is created by the movement of sun and moon over the earth—daily, monthly and yearly—*engenders* the qualities, elements, and humors. It is to be read from left to right, starting with the spring equinox, as if it rotated like the sky, clockwise over its central figure. It begins at the beginning of the medieval year, with the sign of Aries, when night and day are equal, and, therefore, the qualities of hot and cold, wet and dry, are perfectly in balance.

Now, as the sun begins to move towards the summer solstice, the element fire starts to exert its heating and drying influence; therefore, its qualities—heat and dryness—increase. This change corresponds to the period of transformation from a temperate balance of qualities in youth to the hotness and dryness of the adult. The summer solstice at the top corresponds to a zenith of hotness and dryness.

Right after the summer solstice, the sun begins to descend and move away from the earth towards the autumn equinox; thus the quality of coldness begins to increase. This is indicated by the cold, dry element, earth, which is at the top to signify the beginning of its influence. On the outside of the diagram can be read that this movement takes place in mid-June and lasts through the summer months of July, August, and September. The outside zodiacal circle shows that this is also the time of the passing of the zodiacal constellations of Cancer, Leo, and Virgo; it links the two moments of summer solstice and autumn equinox. The cycle continues in the same way through the autumn equinox and winter solstice, with the rising and falling of the influences of water and air, coldness and wetness, until the cosmos has cycled back to the spring equinox.

All these effects are clearly focused on the central figure, which stands for the body or plant on earth. This figure absorbs the changing qualities, as

Hildegard explained, through its *anima*, the animating principle, or internal circulating wind.

Here, then, all the concepts of the premodern system are linked together, and the links are the equinoxes and solstices, the zodiac, the months, and the seasons of the year—that is, time. But a particular, premodern sort of time: namely, time as created by the movement of the sun over the fixed earth during the year. *The key to the linkages among the concepts of premodern medicine is precisely this*—the generation of the four seasons, the four qualities, and the four elements as the sun moves nearer (that is, more northerly) to the earth, and then farther (that is, more southerly) during the year.

The conceptual system of premodern medicine *did* have an overarching, linking image, then, but it was not the tetrad. Rather it was the fixed earth, around which the moving firmament moved. On that fixed earth, at the center of the universe was—everyone—all people, animals, and plants. Under the moving sky and rooted in the earth, all suffered the powerful, inescapable influence of the winds, blowing from the four corners of the universe, with their qualities of hot and cold, wet and dry; the seasons, the elements, and the zodiac. The sun and, indeed, the entire universe circled the earth, and the qualities, elements, and humors came into being and passed away in consequence.

This same missing element—cyclic time—can be seen in the illumination from Hildegard's *Scivias*, which has stood us in such good stead over the course of these chapters (frontispiece). Although painted in reference to her much earlier visions, which were contemporaneous with *Causes and Cures*, it illustrates very well the cosmological picture she has described. Like Byrhtferth's diagram, it is a roughly diamond-shaped figure, with directions (east at the top) at the four corners; also the winds and the elements. At the center is the fixed earth.

On the outside of this illumination is the element fire. It is clearly meant to be seen as moving clockwise, blown by the east and south winds, and blowing, therefore, the sky across the fixed earth. The elements of air, and then water, come next. The four winds can be seen at the four cardinal points of north and south, east and west, and the four elements are placed centrifugally from most ethereal, fire, to heaviest, earth, at the center. At the east, that is at the top, all seven planets are lined up: Saturn, Jupiter, and Mars, outside the circle of the sun; then Venus, Mercury, and the moon. All the planets are lined up in the east; this is, therefore, the moment of creation, the beginning of time. The blue firmament circles the fixed earth. With its movement, and the counterclockwise movement of planets (just about to begin), and the south to north movement of the sun, the air—that is, the seasons and the weather—will begin to change, and so will change the internal humors, until the last moment of recorded time.

We Never Stopped Being Medieval

What we have learned from Hildegard as our model practitioner is that the premodern system was rooted in an experience of sky, constellations, and seasons, of weather as brought by directional winds, the consequent rise and fall of saps and humors, and the transmission of their qualities to the tastes and medicinal effects of plants. The medical model based on this worldview worked, we have seen, not mainly because of respect for ancient authority, nor mainly because of a fascination with the number four—indeed, many more than four were involved—but because it provided a verifiable picture of the premodern universe. It probably originated as a system that explained the seasons and their observed effects on plants, but it turned out to be a system whose basic concepts—the elements, the qualities, the humors—could be successfully transferred from plant to body.

It flourished in some form in scientific Western culture for millenia, until well into the nineteenth century. Then, in the space of only a few decades, a completely different model for the body emerged, and that machine model replaced the horticultural model almost completely. Within less than a century, however, this machine model began to give way to an entirely novel conception, that of the computer, where the body is best understood as the logical outcome of DNA code.

What, then, does this study contribute to the contemporary discourse of medicine and the body? Perhaps its most salutary conclusion is that medical models "work" when they reflect the world they are in. In a world of machines, bodies will be understood as machines; in a world of computers, bodies will be understood as computers; and in an agrarian world, bodies will be understood as plants.

To visualize the premodern world, therefore, we have to take its point of view. This is not quite as difficult as it sounds, however, because, in many ways, we never stopped being medieval. In a very real sense, the sun rises for us every morning and sets every evening. True, as moderns we live at the edge of a nondescript galaxy, in a not-very important solar system, on a small, spinning earth. But as premoderns, we do not only live there. We also stand, like Hildegard, within a dark blue, orderly cosmos, surrounded by the cycling heavens, on a perfectly spherical earth, which is the center of the universe and completely still.

Appendix

The Only Surviving Witness

In trying to learn as much as possible about the materiality of Hildegard's life, I examined the only surviving manuscript of Disibodenberg's library, MS Bern Burger-bibliothek 226. What can be inferred from this manuscript alone about Hildegard's textual and scriptoral surroundings at Disibodenberg?

Description: The manuscript is a well-preserved parchment codex of average quality, without holes, erasures, palimpsests, or extraneous marks. There are three distinct hands, and pricking but no lining. Rubrication is generous. Illuminations appear in the *Calendarium* and precede each month. They are highly-stylized figures that look like a K or an R, and vary only slightly from month to month. Some seem expertly done, with additional lines that turn the figures into the head or body of an animal, while others (particularly the illumination for June) seem cruder and less mature, with a blurring of lines and a rather tentative form.

In the catalogue the MS is described as a twelfth-century parchment manu-script containing three texts, a *Calendarium*, a *Tabula calendarium*, and a *Regula monachorum*.[1] These texts have previously been identified as Usuard's *Martyrol-ogy*, a *Tabulum computi*, and a *Rule of St. Benedict*, respectively.[2] Preceding the *Tabulum* is a short introduction by "Henry the Monk" never previously com-mented on.

The manuscript is not a collection of manuscripts bound together; and so it must have been copied at Disibodenberg as we have it, since the first two texts, the *Martyrology* and *Computus*, share fol. 47b, and the *Computus* and *Rule* share fol. 50. On the basis of internal evidence, it has been placed and dated (see below.)

Calendarium romanorum ecclesiasticum: fols. 1a–47b
MS Bern Burgerbibliothek 226 begins with three lines of what appears to be classical Latin:

    Procedunt duplices in martia tempora pisces (AL 640)[3]
    Dando diem pruinam dabit horam martius ipsam
    Quarta nec est munda cuius nocet hora secunda.

An illuminated figure comes next, following which is a fourth metric line on the days of the month of March. The chapter from Usuard's *Martyrology* for March then occurs.[4] This pattern—a short calendrical poem, an illuminated figure, a second verse on the days of the month, and the relevant chapter from Usuard—repeats through December. But January and February (which may have been either the first, or the last months of the text, since the medieval year began either with January or with spring, in March) are missing. The rest of the calendrical poem is transcribed here:

    Respices apriles aries frixee kalenda
    Horis in primis decimis suffocat aprilis
    Undecimis nonas ferit inde diesque per hora

    Maius agenorei miratur cornua tauru
    Integrunt terra madii horaque sexta
    Inq. mali moris in denis septimus hori

    Iunius equatos celo videt ire laconas
    Sexta nocet die vini satis hora diei
    Horis quaterni quindenis mordet et arguus

    Solsticio ardentis cancri fert iulius astrum
    Dampnat tredennia iuliu verat hora secunda
    Huius et in nonis decimus quo faciat horis

    Augustum mensem leo fervidus igno perurit
    Horas dat prunas auguste datque kalenda
    Inde secunda dies septem turbinat horas

    Sydere virgo tuum bachum september opimat
    Horis septembris perunit lux tertia ternis
    Eiusdem mensis necat horis dena quaternis

    Equat october sementis tempore libram
    Sauciat octobris in quivis tertius horis
    Inqu. die dena cuius ferit hora novena

Scorpius hibernum preceps iubet ire novembris
Pungit in octavis horis lux quinta novembris
Cuius terna migram facit horam cum fino quintam

Terminat arcitenens medio sua signa decembri
Vulnerat in pruinis horis septenis decembris
In denis horis decimus sit causa doloris.

Thus this piece provides hard evidence that classical Latin was indeed available at Disibodenberg, as has been inferred from the analysis of Hildegard's own texts.

Although these strange lines about the months and their climatologic effects do support the notion that Hildegard had considerably more sophisticated Latin available than she claimed, the story behind the lines is not straightforward. Poems on the zodiacal constellations often did precede the martyrologies in manuscript, but have rarely been included in the printed editions (Hennig 1958). It turns out that there were many versions of this poem, which were constructed just as this one was, by taking lines from a variety of authors—Ausonius, Cicero, Bede, and even Wandalbert of Prüm—although some lines have no certain author. John Hennig has suggested that because of this variability it should be possible to use the particular versions found in individual manuscripts to establish connections among monasteries, since a martyrology copied from one monastery would, presumably, carry its poems with it.[5]

Unfortunately, the verses in MS Bern Burgerbibliothek 226 are unique. Although the first line of each stanza comes from Ausonius, the sources for the other lines are unknown. This may mean that an author at Disibodenberg had composed them, implying a high standard of Latin in the monastery, or, more likely, that the lines had come from a martyrology that has disappeared. In any case, we cannot use it to establish a monastery from which Disibodenberg may have borrowed its copy.

As noted above, each stanza is followed by the relevant chapter for the month from Usuard's *Martyrology*. According to Hennig, the combination of these ancient verses on the zodiacal constellations with the martyrologies had to do with the necessity of combining the old, circular, zodiacal time with the new, linear, Christian time. The Christian saints, he suggested, performed a similar function in the Middle Ages as did the zodiacal constellations in the ancient calendar; the saints came and they went with the year. Theirs was not a linear, once in a lifetime existence; their stories were not, that is to say, history. Rather the saints provided a recurrent, cyclical theme, a way of demarcating the year (77). And in this way, it is, perhaps, not surprising that of all things, Disibodenberg's *Martyrology* should have survived.

The first martyrology was a (lost) Syrian collection of *Saints' Lives*, but the earliest surviving text is by Pseudo-Jerome; dates and other details were added by Bede and Hrbanus Maurus.[6] The most common texts were those by Usuard and Adon, although they remained locally variable and, like the calendric verses, can sometimes be used to localize manuscripts.

Indeed, it was the *Martyrology* of MS Bern Burgerbibliothek 226 that allowed it to be placed and even dated, because Disibodenberg had made three idiosyncratic additions to the known text of Usuard. It listed July 8 as "natale sancti patris nostris Dysibodi"; September 8 as "item ipso die, translatio Sci. Dysibodi episcopi et confessoris";[7] and September 29 as the date on which its church had been dedicated, and these all agree with the dates in Disibodenberg's annals. The tone is typical of martyrologies, as the entry for May 15 (the day I happened to examine the manuscript) makes clear:

> This is the birthday of the saint-confessors, Torquatos, Ctesiphons, Secondus, Indalecius, Esitius, and Euphrasius. They were ordained bishops by the holy apostles and went into Spain to preach the word of God to the Spaniards, who at that time were still pagan. When they arrived in Accitans, seeking rest from their journey, they sent some of the disciples out to buy food. Quite soon a large number of pagans, who at that very moment were celebrating the festivals of their gods, chased them down to the river over which was a very sturdy bridge. The saints were able to escape, by God's will, but when the crowd tried to follow, the bridge collapsed. At which miracle, the rest, terrified, followed the example of their great senetrix Ludanias, who, divinely inspired, had converted, and, leaving their idols they believed in Christ. After this, the saints preached in numerous cities, convincing many. To this day, there stands a marvel for the commemoration of their precious death. Near the city of that same town, Accitans, by the grave of that same Torquatos, stands a flowering olive tree which, by God's grace, bears mature fruit at the same time. In addition, on the island of Chius, the birthplace of St. Isidore, is his basilica, where the well stands in which he had been thrown. From it the insane, those with fevers, and other ill people, by drinking its waters, are often made well.[8]

This style of Disibodenberg's *Martyrology* certainly recalls the style in Hildegard's *Lives* of St. Disibod (which is notably missing from this *Martyrology*) and St. Rupert; and there are similarities with her own *Vita* as well. For instance, there is the miraculous framing of otherwise explicable events (here the bridge falling in, in Hildegard's *Vita* the birth of a spotted calf); there is God's active participation in the world through a miracle of Nature, here evidenced by a horticultural miracle, and in Hildegard's *Vita*, by her miraculous recoveries from illness. There is also the theme of preaching to the heathen (or the heathenized Church.)

Tabula calendaria Henrici monachi: fols. 47b–50a

The martyrology is followed by a short passage written in a coarse hand on 47b:

Henry the monk, although uneducated, has properly put together this table for the purpose of helping whoever must hold to it; that is, all those of Christian name who must fast for the seven weeks of Lent—the people, and also the bishops, abbots, canons, deacons and subdeacons, cantors, and all priors, keepers and masters, untaught children, nuns and monks, presbyters, enclosed ladies, abbesses, prelates, soldiers, even the ignorant . . . and students, youths, adults, and the aged . . . princes, and kings and the Count Palatine . . . farmers and wives. Come, learn something wonderful, and listen to Henry, the poor monk (fol. 47b).[9]

The coarse penmanship certainly suggests that Henry the monk scribed this passage himself, although he did not copy the text to which this passage was the introduction. Perhaps we can assume, then, that he was the author or composer of the text that follows, and that he used a secretary or scribe. Hildegard's use of Volmar as scribe or secretary, then, would not be peculiar to her gender, at least at Disibodenberg.

Henry's list also provides us with a snapshot of the social ordering of Disibodenberg's universe; and it has similarities to those gleaned from Hildegard's own social ordering in the *Lingua ignota*. For instance, once again we find that "enclosed ladies" have been clearly distinguished from "nuns"; that members of the church precede secular rulers, and that secular rulers precede workers and farmers. Like Hildegard, too, Henry includes a modesty trope, bringing into question the usual ascription of Hildegard's modesty to her awareness of her gender.

Regula monachorum: fols. 50a–97a

The third text has been identified as a copy of the Benedictine Rule, so dear to Hildegard.[10] Its script is larger and clearer than the other two texts, perhaps because it was read aloud at meals. A few passages are underlined lightly by a second hand, suggesting that it had also been read privately, however.

Observations

MS Bern Burgerbibliothek 226 proves that Disibodenberg did have a scriptorium, as provided for in the Benedictine Rule. The scriptorium had at least two copyists, and probably more, since it took only three years for the manuscript to be copied (from the dedication of the church in 1141 to the beginning of the *Tabulum* in 1144). The scriptorium probably also had illuminators, some of whom were inexpert, and, therefore, probably students. The manuscript consists of three basic monastic texts, the two main texts being those meant to be read aloud at meals, the *Martyrology* and the *Rule*. Thus the only surviving witness to Disibodenberg's library does support the notion that Disibodenberg conformed to the main dictates of the Benedictine Rule. Referring back from the Rule to life in Hildegard's first monastery, particularly with regards to infirmary practice, is, therefore, warranted.

Notes

NOTES TO THE INTRODUCTION

1. Identifying modern plants using their medieval names is tricky; on this, see Irmgard Müller, "Zur Verfasserfrage der Medizinisch-Naturkundlichen Schriften. Hildegards Von Bingen," in *Tiefe des Gotteswissens: Schönheit der Sprachgestalt bei Hildegard Von Bingen*, ed. Margot Schmidt (Stuttgart-Bad Cannstatt: Frommann-Holzboog, 1995), 1–17. Throughout the book when rendering Latin or German plant names, I will mainly use the modern cognate, although the plant signified may *not* be what Hildegard meant. If the plant is only identified in German I use the English equivalent as given in various sources, including Kaiser's "Index verborum germanicorum," in Hildegard of Bingen, *Hildegardis Causae et curae*, ed. P. Kaiser (Leipzig: Teubner, 1903), 252–254, as well as in a variety of German dictionaries. If none is secure, then I retain the original in a transliterated form.

2. The full translation and quotation is: "A person who wants to make and use potions should take ginger, and one half as much licorice, and one third zedoary [a spice like turmeric still used in Eastern health preparations], and reduce them to a powder and strain them. Add equal sugar. Then let him take up of this mixture about as much as the weight of thirty coins. Then let him take the purest of flour, about as much as a half-shell of walnut will hold, and as much springroot juice as the cut quill of a scribe can hold; that is, as much as a scribe takes up when he dips his quill. Then let him make a sticky mass and a flat cake, and divide it into four. Dry these little cakes in the sun in March or April, since the sun in those months is mild. For the heat of the ginger and the cold of the zedoary will unite the humors, and the heat and moistness of the sugar will retain and humidify them. But, at the same time, the heat and strength hold the humors in check so that they will not disappear. The juice of the springroot by its coldness makes the humors well-tempered, because ginger and zedoary and sugar retain good humors in the body, and springroot draws the bad humors out. This medicine should

be prepared in the above months, because the sun and the wind are of good tempers at that time." ("Homo, qui potiones facere et sumere uult, zinziber accipiat et dimidiam partem tanti de lequiricio et tertiam partem de zituar, quantum de zinzibero, et hec in puluerem redigat et colet, et deinde eundem puluerem simul penset. Postea suma tantum de zuccaro, quantum puluis iste pensat. Quod cum fecerit, omnia simul penset ad pondus triginta nummorum. Deinde accipiat de purissima farinula simile, quantum dimidia testa nucis capit, et tantum lactis de citocacia quantum inscisa penna scriptoris in inscisione tenere potest, id est sicut scriptor pennam in incausto intingit, et sic de predicto puluere et de farinula et de lacte citocaciae tenuissimam massam uelut tortellum faciat et hanc massam in quatuor partes diuidat et ad solem in Martio uel in Aprili siccet, quoniam in mensibus istis radius solis ita temperatus est. . . . Nam calor zinziberi et frigiditas zituaris humores congregant, et calor et humiditas zucchari eos retinet et humectat, calor uero et fortitudo simile eos continent, ne iniuste effluant, succus autem citocacie frigiditate sua eos suauiter et conpetenter educit, cum sic temperantur, ut predictum est. Zinziberum autem et zituar et zuccharum et similia bonos humores in homine retinent, et citocacia malos humores educit. . . . In hoc predictis autem mensibus hec potio parabitur, quia et sol et aura boni temperamenti sunt.") CC, IV, 236:1–26.

3. Scholars do not, however, agree completely on the timetable for this evolution. For instance, Elizabeth Sears writes that it was Ptolemy in the second century A.D. who first introduced the analogy between the Ages of Man and the seasons (15–16). Erich Schöner gives this to Galen; and for him it was Antiochus who defined a system that contained a whole set, not only of the zodiac, the seasons, and the ages, but also of the winds, qualities, conditions, humors, temperaments, and color.

4. For this argument as applied to later texts, notably those of Hildegard and Honorius Augustodunensis, see Barbara Maurmann, *Die Himmelsrichtungen im Weltbild des Mittelalters. Hildegard von Bingen, Honorius Augustodunensis und andere Autoren* (Munich: Wilhelm Fink, 1976).

5. I have simplified things here, but this is the general scholarly explanation for the development of premodern Western medicine. Sears applies Schöner's argument to the Middle Ages: "The quadripartite life was defined with a cosmological system developed in antiquity and subsequently transmitted to the Middle Ages. A system for understanding the natural order, it found the roots of nature to lie in the tetrad" (9). It was, she writes, an academic, literary system, transmitted by scholars in school texts as philosophical theory: "The theme of the ages of man was a bookish one . . . which do[es] not mirror life so much as . . . reflect medieval meditations"(6).

6. The examples can be multiplied; for instance, see the description of the system in Lois Ayoub, "Old English Waeta and the Medical Theory of the Humours," in *Journal of English and Germanic Philology*, no. 3, (July 1995):

332–346; in Sabina Flanagan, "Hildegard and the Humors: Medieval Theories of Illness and Personality," in *Madness, Melancholy, and the Limits of the Self*, eds. Andrew Weiner and Leonard Kaplan (Madison, Wisconsin, 1996), 14–23; and in Margret Berger, *Hildegard of Bingen: On Natural Philosophy and Medicine* (Cambridge: D. S. Brewer, 1999), 16–18.

7. In his work on nineteenth-century medicine, Charles Rosenberg gets at the more existential roots of the theory's popularity—its ability not only to explain but to prove, *via* the known and expected effects of its therapy, its own truth; see Charles E. Rosenberg, "The Therapeutic Revolution: Medicine, Man, and Social Change in Nineteenth-Century America," in *Essays in the Social History of American Medicine*, eds. Morris Vogel and Charles E. Rosenberg (Philadelphia: University of Pennsylvania Press, 1979), 3–23.

8. See the many variations of the diagram of the fours illustrated in Elizabeth Sears, *The Ages of Man: Medieval Interpretations of the Life-Cycle* (Princeton: Princeton University Press, 1986), plates 1–13; and in Barbara Obrist, "Le diagramme isidorien de l'année et des saisons: son contenu physique et les représentations figuratives," *Mélanges de l'Ecole Française de Rome: Moyen Age*, 108, no. 1 (1996): 95–164.

9. For discussion, see Wesley D. Smith, *The Hippocratic Tradition* (Ithaca: Cornell University Press, 1979), especially "Galen's Hippocratism," 62–178, and "From Hippocrates to Galen," 179–246; and Owsei Temkin, *Galenism: The Rise and Decline of a Medical Philosophy* (Ithaca and London: Cornell University Press, 1973).

10. There are other medical authors for whom practice was probably central: Henry de Mondeville, Arnauld of Villanova, and Jean de St. Amand, for instance. But the medical text of Mondeville, while based on practice, is "bookish" in tone, although Marie-Christine Pouchelle has been able to extract much "mentality" from it; see Marie-Christine Pouchelle, *The Body and Surgery in the Middle Ages*, trans. Rosemary Morris (New Brunswick: Rutgers University Press, 1990). The same can be said for the writings of Jean de St. Amand and Arnauld of Villanova; see Walton Orvyl Schalick, III, "'Add One Part Pharmacy to One Part Surgery and One Part Medicine': Jean de Saint-Amand and the Development of Medical Pharmacology in Thirteenth-Century Paris" (Dissertation, Johns Hopkins University, 1997); and Michael R. McVaugh, ed., *Arnaldi de Villanova opera medica omnia: aphorismi de gradibus*, vol. 2 (Granada-Barcelona: Seminarium Historiae Medicae Granatensis, 1975). Although these medical authors were experienced practitioners, they were writing for an academic audience; hence the relationship between their practice, their actual use of the ancient theory to understand their practice, and what they say in their books, is unclear.

11. On how to get at practice through, or in spite of, texts, see the essays in Peregrine Hordren, ed., *The Year 1000: Medical Practice at the End of the First*

Millennium, vol. 13, (Oxford: Society for the Social History of Medicine, 2000).

NOTES TO CHAPTER ONE

1. Hildegard apparently kept better track of her birthdate than most medieval people did; in the prologue to her first book, *Scivias*, she noted that in 1141 she was forty-two years and seven months old: "Actum est in millesimo centismo quadragesimo primo Filii Dei Jesu Christi incarnationis anno cum quadraginta duorum annorum septemque mensium essem. . . ." (*Scivias*, in PL197, col. 383A). For biographies in English, see Sabina Flanagan, *Hildegard of Bingen, 1098–1179: A Visionary Life*, 2nd ed. (London: Routledge, 1998); Fiona Maddocks, *Hildegard of Bingen: The Woman of Her Age* (New York: Doubleday, 2001); and Barbara Newman, *Sister of Wisdom: St. Hildegard's Theology of the Feminine* (Berkeley: University of California Press, 1997[1987]), among many others. For primary sources gathered and translated into English, see Anna Silvas, *Jutta and Hildegard: The Biographical Sources* (University Park: The Pennsylvania State University Press, 1998).

2. Trithemius called her "Hildegard of Boeckelheim" while Marianna Schrader concluded from a study of land documents that she came from Bermersheim (Silvas, *Jutta and Hildegard: The Biographical Sources*, 40). At stake in this seemingly arcane dispute are different interpretations of Hildegard's early life. Bermersheim does lay claim to her to this day; its beautiful small church advertises itself as Hildegard's baptismal chapel. It houses some (but not all) of her bones, and they are removed daily for Protestant services by the sacristan, so as not to give offense to the non-Papists. (She puts them in a box in her kitchen.) But Bermersheim is a dark and gloomy place, and far from Sponheim and Disibodenberg. Boeckelheim makes more sense given that it was within walking distance of Hildegard's first monastery, Disibodenberg, and of Sponheim, the home of her first teacher, Jutta. But Boeckelheim also happens to be the village where the excommunicated Holy Roman Emperor, Henry IV, was imprisoned when Hildegard was a young girl. If, as Miriam Marsolais has proposed, her father was a noble servant (*ministeriale*) of Henry's, then she would have seen at close hand the negative results of uncareful politics. Such lineage would also mean that Hildegard was from a lower class than previously assumed, throwing a different light on the class issues in her well-known disagreement with the nun, Tenxwind. For an account of the issues, see Sabina Flanagan, "'For God Distinguishes the People of Earth as in Heaven': Hildegard of Bingen's Social Ideas," *Journal of Religious History* 22, no. 1 (1998): 14–34; and Miriam Marsolais, "'God's Land is My Land': The Territorial-Political Context of Hildegard of Bingen's Rupertsberg Calling" (Dissertation, University of California, Berkeley,

2002). All we do know is that her upbringing was rural, since the only story she tells about her childhood places her with a nurse, in a field, looking at a pregnant calf (*Vita* [1998], 264:24–266:5).

3. For useful genealogies of both Hildegard and Jutta, see Silvas, *Jutta and Hildegard: The Biographical Sources*, 278–281.

4. In the Rupertsberg charter of 1158, five (and a half!) houses in Bermersheim were donated to the monastery by Drutwin and Hugo (Silvas, *Jutta and Hildegard: The Biographical Sources*, 241).

5. Since the discovery of a *Life of Jutta* in 1992 (thought possibly to have been written by Hildegard), we know considerably more about Jutta, although her *Life* is certainly hagiography, not biography. Jutta was very well-connected. The Emperor, Lothar, was a relative (Silvas, *Jutta and Hildegard: The Biographical Sources*, 279) as was Richardis von Stade, Hildegard's future protectress (Staab, "Aus Kindheit," 66). Her mother, the Lady Sophia, was not only devout but also wealthy, as an 1128 land contract makes clear, referring to Lady Sophia as a very devout woman ("mater domna Sophia mulier religiosa") in Heinrich Beyer, Leopold Eltester, and Adam Goerz, *Urkundenbuch zur Geschichte der mittelrheinischen Territorien 2 Band*, vol. 1. (Koblenz, repr. Hildesheim: Scientia Verlag Aalen, repr. Georg Olms, 1860 [1974]), Doc. 521. On Jutta, see Staab, "Aus Kindheit, and Staab, "Reform und Reformgruppen." See also Miriam Schmitt, "Blessed Jutta of Disibodenberg. Hildegard of Bingen's Magistra and Abbess," *American Benedictine Review* 40, no. 2 (1989): 170–189.

6. Silvas suggests that Uda might have lived with Jutta's mother, Sophia, so that Hildegard would have spent the years from eight to fifteen at Sponheim (52). Van Engen thinks that such a timeline, where Hildegard lived outside in the world for several formative years, might account for Hildegard's surprising fund of worldly knowledge; see John Van Engen, "Abbess, Mother, and Teacher," in *Voice of the Living Light: Hildegard of Bingen and Her World*, ed. Barbara Newman (Berkeley: University of California Press, 1998) 32–33.

7. Jutta spent three years with Uda, from the age of twelve to fifteen, that is, from 1104–1107. Since Hildegard arrived when she was eight, this would have been in 1106. It is unclear whether there were two or three other girls with them because the *Annales Sancti Disibodi* refers to three ("aliae tres cum ea, scilicet Hyldegardis et suimet vocabuli duae." See *Annales Sancti Disibodi 841–1200*, ed. G. W. Waitz, in vol. 17 of *Monumenta Germaniae Historica SS* (Hannover, 1861), 25. The *Vita Juttae* refers to only two girls (Staab, "Reform und Reformgruppen," 176).

8. A history of Disibodenberg is provided by Wolfgang Seibrich, "Geschichte des Klosters Disibodenberg," in *Hildegard von Bingen 1179–1979: Festschrift zum 800. Todestag der Heiligen*, ed. Anton Ph. Brück (Mainz: Selbstverlag der Gesellschaft für mittelrheinishen Kirchengeschichte, 1979), 55–75.

9. "Hoc anno, scilicet 1108, inceptum est in monte sancti Dysibodi novum monasterium construi. . . . 2. Kalend Iulii primum fundamenti lapidem . . ." (Waitz, *Annales Sancti Disibodi 841–1200*, 20).

10. Despite archaeological work, not much is known for certain about the campus where Hildegard spent half of her life, but a hypothetical plan is reproduced in Silvas, *Jutta and Hildegard: The Biographical Sources*, 276.

11. The scanty remains of Disibodenberg's twelfth-century church are today displayed in a little museum at its base, and include mosaic floors, painted frescoes, and stained glass.

12. Perhaps this location is what accounts for the extraordinary knowledge of Rhenish fish that Hildegard reveals in *Physica*, as well as for her use of mills (which were a new twelfth-century technology) as metaphors. Even today, the Nahe is lined with watermills.

13. Disibodenberg's monastic isolation was more rhetorical than real. Its annals document a surprising familiarity with politics and even gossip, as well as frequent, often long-staying, visitors. For instance, the Archbishop of Mainz often visited. For the dedication of the altar (1138) the Abbot of St. Maxim in Trier and the Abbot of Sponheim (along with his entire congregation) were present (Waitz, *Annales Sancti Disibodi 841–1200*, 25). Siward, the Bishop of Uppsala (from whom it is thought Hildegard may have learned some of her medicine) stayed for many months during 1136–1138.

14. The story is not entirely clear on this point. The *Vita Hildegardis* relates that Hildegard entered Disibodenberg when she was eight years old, which is not very likely since that would have been in 1106, before it was refounded. The *Vita Juttae* suggests the more plausible scenario that she entered Disibodenberg with Jutta in 1112, when she was fourteen. However, it is unclear whether that meant that Hildegard formally became a nun at that time, or did so later in a separate ceremony, as suggested by Klaes in the *Vita* (1998), 13–14. Jutta's ceremony was conducted by the famous Otto of Bamberg and not by the new Archbishop of Mainz, Adalbert I (1111–1137), because he had been imprisoned by Henry V (Böhmer, *Regesten zur Geschichte der mainzer Erzbischof von Bonifatus bis Heinrich*, vol. 2 [Innsbruck: Neudruck der Ausgabe, 1877], 246–247). According to Theodorich, Otto of Bamberg also professed Hildegard, but this may have been on a separate occasion (Silvas, *Jutta and Hildegard: The Biographical Sources*, 211).

15. The record of Jutta's lands (1128), though a legal document, conveys a sense of Jutta's devotion, as well as of her wealth. To Disibodenberg she gave her "rights in the town of Neuvenkirchen with all things pertaining—that is, the church, with its right to a tenth of the serfs, and the fields, woods, meadows, pastures, cultivated and wild areas, and plowed, pathed and pathless, waters, and mills. . . . Of her free will, she offered to God and to St.

Disibod these possessions which she had from her mother, the Lady Sophia, ardently desiring to leave the pomp of the secular world and be enclosed at Disibodenberg. Which her mother, the Lady Sophia, a religious woman, had obtained from the venerable Rutthard, Archbishop of Mainz, gratefully, on account of her devotion," in Beyer, *Urkundenbuch zur Geschichte der mittelrheinischen Territorien*, vol. 1., Doc. 521. The record of Hildegard's gift, by contrast, has not survived.

16. Constable argues that Hildegard's interpretation of monastic practice was lenient, perhaps in contrast to Jutta's asceticism, and implied a certain disapproval of her; see Giles Constable, "Hildegard's Explanation of the Rule of St. Benedict," in *Hildegard von Bingen in ihrem historischen Umfeld, Internationaler wissenschaftlicher Kongress zum 900 jährigen Jubiläum, 13–19 September 1998, Bingen am Rhein*, ed. Alfred Haverkamp and Alexander Reverchon (Mainz: Philipp von Zabern, 2000), 163–187.

17. Later at Rupertsberg manual labor would include "the writing of books, the weaving of robes, and other work with the hands" ("scribendis libris, vel texendis stolis, vel aliis manuum operibus intendunt") in *Guiberti Gemblacensis epistolae quae in codice B. R. Brux. 5527–5534 inveniuntur*, ed. Albert Derolez in vol. 66A of CCCM (Turnhout: Brepols, 1988), 368:59–60.

18. What we know about Jutta's ascetic practices from the *Vita Juttae* are reminiscent of those described in the *Martyrology*, which was read at meals, and which happens to be part of the only surviving manuscript from Disibodenberg's library. (For an account of this manuscript, see the Appendix.) Jutta's healing practices were not only spiritual but also physical; that is, she touched patients as well as prayed for them.

19. For instance, Van Engen demonstrates that the papal letter authorizing her to write was forged by Volmar; see "Letters and the Public Persona of Hildegard," in *Hildegard von Bingen in ihrem historischen Umfeld, Internationaler wissenschaftlicher Kongress zum 900 jährigen Jubiläum, 13–19 September 1998, Bingen am Rhein*, ed. Alfred Haverkamp and Alexander Reverchon (Mainz: Philipp von Zabern, 2000), 379–390.

20. Her letter to Bernard of Clairvaux took the naive and helpless stance of the "poor little woman" that she called herself, but it is unlikely that Hildegard was unaware of Bernard's ruthlessness towards those who ignored him. It was, after all, Bernard who had engineered Abelard's imprisonment for heresy in 1141 and it was Bernard's protégé, Eugenius, who was Pope in 1145. Hildegard probably met Bernard in person a few years later; possibly when he was traveling in the Rhineland preaching the Second Crusade (Van Engen, "Letters and the Public Persona of Hildegard," 381), or perhaps when he visited Rupertsberg, as Trithemius recounts (*Johannis Trithemii opera historica*, vol. 1, 250).

21. It was Hildegard herself who wrote to the Pope in 1148 (Van Engen, "Letters and the Public Persona of Hildegard," 383).

22. Although the hagiographers write that "she was shown the place by the Spirit" ("demonstratur illi per Spiritum locus, ubi Naha fluvius Rheno confluit, videlicet collis a priscis diebus sancto Ruperto confessori ex nomine attitulatus," Godfrey, *Vita Hildegardis,* PL197, col. 95), like many of the other events in Hildegard's life, things were more complicated than the hagiography or her own accounts implied. Bingen was certainly not unknown to Hildegard, since the monastery possessed vineyards there (Beyer, *Urkundenbuch zur Geschichte der mittelrheinischen Territorien,* vol. 1., Doc. 552) and had had legal problems with the canons of Bingen (Böhmer, *Regesten zur Geschichte der mainzer Erzbischof,* 339). Indeed, Haverkamp argues that the move was a move for Hildegard from "periphery to center," rural to urban, remote to involved (Haverkamp, "Hildegard von Disibodenberg-Bingen: Von der Peripherie zum Zentrum," 15–69).

23. Many reasons for Cuno's reluctance have been suggested, including finances, since the nuns would be taking their previously endowed properties with them; reputation, since Hildegard had become as famous as Jutta (Newman, *Sister of Wisdom,* 9); and even bureaucratic, since Disibodenberg would then lose Volmar, probably their most scholarly monk (Silvas, *Jutta and Hildegard: The Biographical Sources,* 4).

24. Exactly when she left is not entirely clear. According to the 1210 inventory of Rupertsberg's possessions, the move was made in 1147, but this would severely compress the above events (Silvas, *Jutta and Hildegard: The Biographical Sources,* 250–251).

25. Hildegard describes the scene in her *Life of St. Rupert:* "A loco illo ubi reliquiae ipsius conditae sunt; videlicet, ubi Naha fluvius Rheno influit, sursum per ripam Rheni usque ad Selsam fluvium se extendebat; et deinde ad alia duo flumina quorum primum Wiza, secundum Apsa dicitur, transibat. . . . Habitatio autem . . . propter suavitatem defluentium aquarum, in ipso erat . . . civitas vero ipsorum, ibidem sita et fortissimis aedificiis munita, per totam adjacentem planitiem usque a radicem vicini montis et usque ad ripam Rheni tendebatur. Sed ex altera parte Nahae fluvii vicus erat, in qua habitacula famulorum et piscatorum eorum et stabula equorum ipsorum ac horrea ubi frumentum eorum condebatur, et torcularia, ubi vinum ipsorum exprimebatur, fuerunt. In ipsis quoque locis major celebritas et major copia divitiarum et omnium saecularium dignitatum illo tempore pollebat, quam in aliis civitatibus ejusdem regionis vigeret; quoniam ibi concursus et transitus multorum hominum diversarum provinciarum assidue frequentabatur." *Vita Sanctae Rupertis,* PL197, col. 1089.

26. "Nullamque habitacionem seu habitatorem hic invenimus excepto veterano quodam et uxore eius ac filiis." *Vita* (1993), V, 28:46–47.

27. Rupertsberg's properties, which mainly came from the dowries of the nuns, were in more than twenty-two locations. They were substantial, however, most consisting of several houses and 120–150 acres; see Maria Laetitia

Brede, "Die Klöster der heiligen Hildegard Rupertsberg und Eibingen," in *Hildegard von Bingen 1179–1979: Festschrift zum 800. Todestag der Heiligen*, ed. Anton Ph. Brück, (Mainz: Selbstverlag der Gesellschaft für mittelrhein-ische Kirchengeschichte, 1979), fn. 13. For a list of Rupertsberg's holdings, see Silvas, *Jutta and Hildegard: The Biographical Sources*, 240–242.

28. Constable emphasizes Hildegard's practical bias, with her focus on the kinds of foods the monks could be allowed to enjoy. He also notes her telling reference to what she called "the unmitigated boredom of silence" (184).

29. Brother Hugo was present for the dedication and performance (Silvas, *Jutta and Hildegard: The Biographical Sources*, 239, fn. 3).

30. It seems fair to note that Hildegard's attitude towards Cuno was somewhat truculent. After an unpleasant trip by horseback back to Disibodenberg, in order to arrange transfer of the nuns' properties, she counseled Cuno to take more care for his spiritual affairs and less for his temporal affairs, "since time in this world is short for you!" ("Nunc tempus deficiendi tibi adest. . . .") (Van Acker, *Epistolarium*, Letter 74r, 161:14). Cuno did die shortly thereafter, on July 2, 1155, leaving open the question of whether Hildegard's admonition had more to do with her diagnostic acumen or her frustration.

31. In her letters, Hildegard was surprisingly outspoken with Frederick, whom she had met at Ingelheim in 1154. She had made some successful predictions (whose content is still unknown) and ever after Frederick was an impressed supporter (Newman, *Sister of Wisdom*, 11, fn. 31).

32. Much of the middle Rhineland was sacked at this time, on Frederick's orders. Although Rupertsberg itself was not touched, Bingen had its walls pulled down and Hildegard may have experienced the results of war. "Predicti milites cum valida manu Pinguiam et adjacentiam omnia devastant . . . Rudesheim et Geisenheim . . . furibunda strage multorum," in Böhmer, *Regesten zur Geschichte der mainzer Erzbischof von Bonifatus bis Heinrich*, vol. 2, p. 8.

33. "We have also decided that no one of the imperial court, neither great nor small, no judge, count, lawyer, villager, and no tax collector may demand a collection from the possessions of the monastery against the will of the abbess or of the nuns" (Beyer, *Urkundenbuch zur Geschichte der mittelrhein-ischen Territorien*, vol. 1, Doc. 636).

34. This strife surfaces even in the legal agreement between Hildegard and Disibodenberg. For instance: "The sisters shall hold their property free from the brothers. . . . For when Lady Hildegard of Disibod went to Ruperts-berg with her nuns, she bought the place with its vineyards from a number of people, and the brothers, by common consent, gave back eight of the houses which came from the sisters' oblations. . . . But, so that there will be no doubt, We declare that whoever in the future is abbot [at Disiboden-berg] shall manage the care of the souls so that the nuns may have whom-ever they choose. . . . And We hold that the sisters, after the death of

their spiritual mother, are entitled to choose their own new abbess" (Beyer, *Urkundenbuch zur Geschichte der mittelrheinischen Territorien*, vol. 1, Doc. 615). Among the witnesses was Hildegard's brother, Hugo, now cantor of Mainz Cathedral.

35. "What wonder is it that God can inspire you, when He was able . . . to allow even the ass to speak with human words?" ("Nam quid mirum est, si ille inspiratione sua te docet . . . et asinam humana verba proferre fecit.") (Van Acker, *Epistolarium*, v. 91, Letter 20, p. 56).

36. "Anima mea valde tristis est quoniam quidam horribilis homo . . . filia nostra Richarde, abstrahens eam de claustro nostro. . . ." (Van Acker, *Epistolarium*, v. 91, 27:13–16).

37. "Nunc audi me, non abiciens verba mea sicut mater tua et soror tua et comes Hermannus ea abiecerunt. . . ." (Van Acker, *Epistolarium*, v. 91, 28:36–37).

38. "Heu, me, mater, heu me, filia quare me dereliquisti sicut orphanam? Amavi nobilitatem morum tuorum et sapientem et castitatem et tuam animam et omnem vitam tuam, ita quod multi dixerunt, quid facis?" (Van Acker, *Epistolarium*, v. 91, 147:14–17).

39. For a fuller investigation of this relationship, see Susan Schibanoff, "Hildegard of Bingen and Richardis of Stade: The Discourse of Desire," in *Same Sex Love and Desire among Women in the Middle Ages*, ed. Francesca Canade and Pamela Sheingorn (New York: Palgrave, 2001), 49–83.

40. According to the *Acta Inquisitionis*, the record of the papal commission, also known as the *Canonizatio*, there were at this time at Rupertsberg fifty nuns, two priests, and eight charity inmates ("In quo monasterio quinquaginta dominarum prebendas instituit et duas sacerdotum, preterea pauperum matronarum septem in honore sancti Spiritus et unam in honore beate Marie. Preterea trans flumen Reni ad dimidiam leucam aliud monasterium fundavit ubi triginta prebendas instituit.") *Vita* (1998), 266:20–25. The actual population of the houses would have been considerably larger, since each monastic required almost two workers to support him or her, according to calculations by Walter Horn and Ernest Born in *The Plan of St. Gall: A Study of the Architecture and Economy of Life in a Paradigmatic Carolingian Monastery*, 3 vols. (Berkeley: University of California Press, 1979), v. 1, 344f.

41. Her visits included trips to Cologne, Trier, Metz, Wurzburg, and Bamberg; Disibodenberg, Siegberg, Eberbach, Hirsau, Zwiefalten, Maulbronn, Rothenkirchen, Kitzinger, Krauftal, Herde, Honningen, Werden, Andernach, Marienberg, Else, and Winkel (*Vita* [1993], III, 17:7–15). Monasteries visited included Cistercian, Augustinian, Benedictine, and Premonstratensian (Silvas, *Jutta and Hildegard: The Biographical Sources*, 191). There may have been only four distinct journeys: along the Main to Bamberg (1158–1161); along the Rhine into Lotharingia to Metz (1160); along the Rhine

(1161–1163); and into Swabia (1170–1171). See Änne Bäumer, *Wisse Die Wege: Leben und Werk Hildegards von Bingen. Eine Monographie zu Ihrem 900. Geburtstag* (Frankfurt am Main: Peter Lang, 1998), 34.

42. Mews gives 1176 as the year of Volmar's death, which would compress the events quite a bit (Constant Mews, "Hildegard and the Schools," in *Hildegard of Bingen: The Context of Her Thought and Art*, eds. Charles Burnett and Peter Dronke (London: Warburg Institute, 1998), 95.

43. Tellingly, although Hildegard and the Abbot of Disibodenberg were reconciled in 1170, and she agreed to compose a *Life of St. Disibod*, and even paid the monks a visit, still "some monks had greeted her there like some terrible creature." (Van Engen, "Letters and the Public Persona of Hildegard," 390).

44. The records of the thirteenth-century papal investigation of the question of her sainthood (the *Acta Inquisitionis* or *Canonizatio*) present the disappearance of the tomb as a miracle. "Ipsa (beata H.) tumulum eius baculo suo signum crucis (faciens) signavit, et sic sepulcrum eiusdem adhuc non poterat inveniri." (*Vita* [1998], 260:25–262:2).

45. The *Vita* has been shown to be a complex, many-layered composition made up of Hildegard's own memoir, a second version constructed out of this by Godfrey, a third by Theodorich, and a fourth, contained in a "Letter to Bovo" by Guibert. In her edition of the *Vita*, Monica Klaes lays this out in her introduction; see Monika Klaes, *Vita* (1993), 17–156. See also Barbara Newman, "Three-part Invention: The 'Vita S. Hildegardis' and Mystical Hagiography," in *Hildegard of Bingen: The Context of Her Art and Thought*, eds. Charles Burnett and Peter Dronke (London: Warburg Institute, 1998), 189–210.

46. Theodorich wrote that she had died on the 15 Kalends of October in the eighty-second year of her life ("Aliquamdiu itaque infirmitate laborans octogesimo secundo etatis sue anno quinto decimo Kal. octobris ad sponsum celestem felici transitu migravit.") *Vita* (1998), 230:18–21; and that in the evening of that Sunday lights appeared over her grave. ("Nam supra habitaculum, in quo sancta virgo primo crepusculo noctis dominice diei felicem animam Deo reddidit, duo lucidissimi et diversi coloris arcus in firmamento apparuerunt, qui ad magnitudinem magne platee se dilataverunt in quatuor partes terre se extendentes, quorum alter ab aquilone ad austrum, alter ab oriente ad occidentem procedebant.") (*Vita* [1998], 232:3–8). But computerized programs of sidereal astrology, which calculate the days of the week (and take into account the change in the Gregorian calendar), show that September 17 was not a Sunday but a Monday. However, it *was* a full moon. Thus if Hildegard died as Theodorich wrote, in the evening, then the sun must have been setting in the constellation of Virgo, while a full moon was rising. Perhaps that is what Theodorich meant by the lights playing on her grave.

47. For more on this, see Barbara Newman, "Hildegard and Her Hagiographers: The Remaking of Female Sainthood," in *Gendered Voices: Medieval Saints and Their Interpreters*, ed. Catherine M. Mooney (Philadelphia: University of Pennsylvania Press, 1999), 16–34.

48. The literature on hagiography is enormous; for a readable review of current literature, see Cynthia Hahn, *Portrayed on the Heart: Narrative Effect in Pictorial Lives of Saints from the Tenth through the Thirteenth Century* (Berkeley: University of California Press, 2001). For examples as well as a useful survey, see Thomas Head, ed., *Medieval Hagiography: An Anthology* (New York: Garland Publishing, 2000). An important issue is how to clear away the hagiographical tropes to uncover biographical truth; for a discussion, see Dieter R. Bauer and Klaus Herbers, eds., *Hagiographie im Kontext: Wirkungsweisen und Möglichkeiten historischer Auswertung* (Stuttgart: Franz Steiner, 2000).

49. The literature on women saints is daunting; I found helpful Jane Tibbetts Schulenberg, *Forgetful of Their Sex: Female Sanctity and Society, ca. 500–1100* (Chicago: University of Chicago Press, 1998). See also Jo Ann McNamara and John E. Halborg, eds. *Sainted Women of the Dark Ages* (Durham: Duke University Press, 1992); Sabina Flanagan, "Hildegard and the Gendering of Sanctity," in *Hildegard of Bingen and Gendered Theology in Judaeo-Christian Tradition*, eds. Julie Barton and Constant Mews (Clayton, Victoria, Australia: Centre for Studies in Religion and Theology, Monash University, 1995), 81–92; and John Kitchen, *Saints' Lives and the Rhetoric of Gender: Male and Female in Merovingian Hagiography* (New York and Oxford: Oxford University Press, 1998).

50. Twelfth-century women made up 11.8% of the twelfth-century saints (Weinstein 1998, 220); oddly, the exact percentage of tenured female faculty in today's universities.

51. Except for Hildegard, who, as we have seen, *did* preach. Newman, therefore, puts her into the more ancient category of "virile saint," although this non-congruence of Hildegard's life with the usual *Life* probably provides an example of biography showing through hagiography.

52. The editors of Hildegard's work were not against modifying the facts in order to increase her holiness, as they did regarding the day of her death, and as is evident in her letters; see for instance, Van Engen, "Letters and the Public Persona of Hildegard," and Mews, "Speculum," 238.

53. "Cum vix noticiam litterarum haberem, sicut indocta mulier me docuerat." *Vita* (1993), 24:90–91.

54. "She was ignorant of the art of grammar, of its cases, tenses, and genders." ("Qui ad evidentiam grammatice artis, quam ipsa nesciebat, casus, tempora et genera quidem disponere. . . .") *Vita* (1998), 120:8–10.

55. "Still, although she had no earthly teacher, in her forty-second year she began to write more than a few books, revealed to her by the Holy Spirit."

("Preterea cum magistrum terrenum non habuerit quadragesimo secundo etatis sue anno libros non paucos scribere incepit Spiritus sancti revelatione.)" *Vita* (1998), 268:14–16.

56. "Truly, although she was ignorant of Latin, and besides the Psalms had been taught nothing by Man, she was learned by the internal teacher of the Holy Spirit." ("Verum cum esset Latini sermonis ignava et praeter simplicem psalmodiam nihil ab homine didicisset interno Spiritus sancti magisterio edocta.") Trithemius, *Johannis Trithemii opera historica*, vol. 1, p. 257).

57. This is Dronke's summary of Liebeschütz in "The Allegorical World-Picture of Hildegard of Bingen: Revaluations and New Problems," in *Hildegard of Bingen: The Context of Her Thought and Art*, eds. Charles Burnett and Peter Dronke (London: Warburg Institute, 1998), 1–16.

58. See Dronke's introduction in Derolez, *Liber divinorum operum*, viii–lxxxvi.

59. "Et repente intellectum expositionis librorum, videlicet psalterii, evangelii, et aliorum catholicorum tam veteris quam novi Testamenti voluminum sapiebam. . . ." (*Vita* [1998], 88:11–14). Also, "In eadem visione scripta prophetarum, evangeliorum et aliorum sanctorum et quorundam philosophorum sine ulla humana doctrina intellexi. . . ." (*Vita* [1998], 128:10–12). Dronke also quotes the problematic "Berlin Fragment" where Hildegard alludes more specifically to her reading of "pagani philosophi ut Donatus, Lucanus . . . erant precurrens sucus et precurrens vox philosophorum ecclesia." ("Problemata Hildegardiana," 108).

60. "Et eadem per Spiritus sancti revelationem, que preter psalterium litteras non didicerat, multos libros composuerit. . . ." (*Vita* [1998], 244:19–246:1).

61. "Recluditur in monte sancti Disibodi cum pia Deoque dicata femina Iuttha. . . ." (*Vita* [1998], 86:17–19).

62. "And besides the rather small window, through which visitors could speak and provisions could be passed, access was closed off, not with wood, but with cement." ("Et preter fenestram admodum paruam, per quam aduentantibus certis horis colloquerentur et uictui necessaria inferentur non lignis sed cementatis solide lapidibus omni aditu obturato.") Derolez, "Letter 38," in *Guiberti Gemblacensis epistola*, 373: 227–229. Later, after more nuns had arrived, "the entrance to her tomb was opened up, she brought inside with her the girls who were to be nurtured under the guidance of her disciplined guardianship . . . [and] what was formerly a sepulcher became a kind of monastery, but in such a way that she did not give up the enclosure of the sepulcher, even as she obtained the concourse of a monastery." From Guibert's "Letter to Bovo," in Silvas, *Jutta and Hildegard: The Biographical Sources*, 109–110 in Silvas' translation.

63. "Reclusa seorsim in arcta et bene murata custodia." (Trithemius, *Johannis Trithemii opera historica*, vol. 1, 239).

64. Gronau makes exactly this point: "Hildegard ist unabhängig von der medizinischen und naturkundlichen Literatur ihrer Zeit . . . weilen steht fest, daß
sie keine eigen medizinische Praxis ausgeübt hat" in Eduard Gronau, *Hildegard
Von Bingen 1098–1179: Prophetische Lehrerin der Kirche an der Schwelle und
am Ende der Neuzeit* (Stein-am-Rhein: Christiana, 1985), 216. On the other
hand, Van Engen is correct to point out that if Hildegard had stayed at Sponheim with Jutta until she was fourteen, then she would have had ample experience of court and village life ("Abbess, Mother, and Teacher," 33).

65. Indeed, Van Engen changes the whole frame of Hildegard's life when he
summarizes that "by the later 1140s Hildegard was fifty years old, had lived
in religious life at the Disibodenberg for thirty-five years, and led the wom-
en's community there for more than a decade. She had already introduced
some of the usages, together with the chant and lyrics that would set her
community apart" ("Letters and the Public Persona of Hildegard," 380).

66. For example, "In octavo autem anno meo in spiritualem conversationem
Deo oblata sum. . . ." (*Vita* [1998], 124:20–21); or "Quod de loco in
quo Deo oblata fueram," (*Vita* [1993], 27:18); "A loco in quo Deo oblata
fueram. . . ." in *Vita* (1993), II, 31:3–4; and "Locum in quo oblata Deo
eram," in *Vita* (1993), II, 34: 2.

67. "Nos, qui vos fere a cunabilis novimus, et apud quos per plurimos annos
fuistis. . . ." Adalbert, prior of Disibodenberg (1150–1155) in Van Acker,
Epistolarium, v. 91, Letter 78, 175:7–8.

68. For more on Hildegard's enclosure, see Julie Hotchkin, "Enclosure and
Containment: Jutta and Hildegard at the Abbey of Disibodenberg," *Mag-
istra: A Journal of Women's Spirituality in History* 2, no. 2 (1996): 103–23;
and Sabina Flanagan, "Oblation or Enclosure: Reflections on Hildegard of
Bingen's Entry into Religion," in *Wisdom which Encircles Circles, Papers on
Hildegard of Bingen*, ed. Audrey Ekdahl Davidson (Kalamazoo: Medieval
Institute Publications, 1996), 1–14.

69. "Non solum enim omnes isto in loco degentes eius salutaribus monitis atque
consiliis devote obtemperabant rerum" (Staab, "Aus Kindheit," 178).

70. "His quae sub magisterio eius educatae sunt. . . ." (Staab, "Aus Kindheit,"
187).

71. Klaes takes it as a given that at some point during these twenty years the
enclosure became a monastery, and indeed, Guibert wrote that, "sepulchrum
illus prius factum est quasi monasterium." See Derolez, *Guiberti Gemblacensis epistolae*, 374:265–266. Atherton in his 2001 sketch also assumes that
"the anchorage soon became a convent at what was in effect a double monastery"; see Mark Atherton, ed. *Hildegard of Bingen: Selected Writings* (Suffolk: Penguin Books, 2001), xiii.

72. Based on his archaeological research, Nikitsch suggested that the nuns lived
in a number of different dwellings, beginning with the eleventh-century
cemetery chapel; see Eberhard J. Nikitsch, "Wo lebte die heilige Hildegard

wirklich? Neue Uberlegungen zum ehemaligen Standort der Frauenklause auf dem Disobodenberg" in *"Im Angesicht Gottes suche der Mensch sich selbst,"* ed. Rainer Berndt (Berlin: Akademie Verlag, 2001), 147–156.

73. In her childhood she is described by herself and by her hagiographers as weak and frail: see *Vita* (1993), 14:23–4, and Book II, Section 5, p. 29; and generally "in poor health," (Mews, "Hildegard and the Schools," 102, quoting Van Acker, "Letter 40r," 104–5).

74. "Ad scribendum propter crebras infirmitates et humidum caput penitus indisposita . . . " (Trithemius, *Johannis Trithemii opera historica,* vol. 1, 257).

75. For instance, "Hildegard was fascinated by these metaphors of health and well-being, perhaps because she herself was so much aware of suffering and illness in her own body" (Mews, "Hildegard and the Schools," 104). Or, "Her interest in medicine came not only from ill visitors, who came from near and far for consults, but also from her own experience of illness . . . " (Führkötter, *Hildegard Von Bingen,* 32).

76. Approaching the *Vitae* as palimpsests of accounts (Hildegard's, Godfrey's, Theodorich's, and Guibert's), Newman has analyzed the ever-changing descriptions of Hildegard's illnesses as an ever-new relationship to her symptoms; see "Three-Part Invention," 198.

77. "In octavo autem anno meo . . . usque in quintum decimum annum . . . de frequenti egritudine quam a lacte matris mee hucusque passa sum que carnem meam maceravit et ex qua vires mee defecerunt" (*Vita* [1998], 124:20–126:4). Tellingly, Silvas translates this as "a recurring ailment" (159).

78. "Tunc in eadem visione magna pressura dolorum coacta sum palam manifestare. . . . Vene autem et medulle mee tunc plene virium erant in quibus ab infantia et iuventute mea defectum habebam" (*Vita* [1998], 126:25–128: 2).

79. "Quodam inquit tempore ex caligine oculorum nullum lumen videbam tantoque pondere corporis deprimebar quod sublevari non valens in doloribus maximis occupata iacebam. . . . Me quoque quadam vanitate deceptam esse dicebant. Cumque hec audissem cor meum contritum est et caro mea et vene aruerent, et per dies plurimos lecto decumbens vocem magnam audivi me prohibentem, ne quicquam amplius in loco illo de visione hac proferrem vel scriberem" (*Vita* [1993]), 27–28). Godfrey adds that Hildegard was paralyzed and lay like a pile of heavy stones. No one could budge her from the bed on which she had collapsed (Silvas, *Jutta and Hildegard: The Biographical Sources,* 145).

80. Newman argues that what was going on was not many illnesses but a group of symptoms whose interpretation changed over time (Newman, "Three-Part Invention," 199).

81. "In lectum egritudinis quodam tempore me Deus stravit et aeriis penis totum corpus meum infudit, ita quod vene cum sanguine, caro cum livore, medulle cum ossibus in me aruerunt. . . . In isto strepitu triginta dies

fui, ita quod ex calore aerii ignis venter meus fervebat. . . . At ego in vera
visione aciem magnam angelorum humano intellectui innumerabilem per
hos dies interdum vidi qui de exercitu illo erant, qui cum Mychael contra
drachonem pugnabant et hi sustinebant. . . . Tantum spiritus meus de die
in diem plus quam prius in me confortabatur. . . . In his autem languori-
bus triennio finito vidi. . . . Mox spiritus meus in me pleniter revixit . . .
sicque tota convalui" (*Vita* [1993], 33).

82. Here I summarize a long passage; for the Latin, see *Vita* (1993), II, Sections
    9–10, pp. 33–35.

83. Again, Newman interprets this as a recurrence of some kind of recurring ill-
    ness; see Newman, "Three-Part Invention," 201.

84. "Posteaquam . . . tanto pondere egritudinis gravata sum, quod in lecto
    egritudinis iacens ullo modo me levare valebam. Et eadem egritudo de flatu
    australis venti in me afflata est, ita quod corpus meum de die in diem egro-
    tando tantis doloribus et ardoribus conterebatur, quod etiam anima mea in
    me vix sustinebat. Sic autem dimidio anno finito flaus predicti venti totum
    corpus meum perforaverat quod in tanto agone semper fui. . . . In afflic-
    tione autem ista per integrum annum fui. . . ." (*Vita* [1993], Book III,
    Section XI, Vita III, 20:5–65).

85. "All or nearly all, present certain characters. . . . A prominent feature is a point
    or a group of points of light, which shimmer and move, usually in a wave-like
    manner, and are most often interpreted as stars or flaming eyes. . . . Some are
    accompanied by concentric circular figures of wavering form, and often are
    definite fortification figures." From Charles Singer, "The Scientific Views and
    Visions of Saint Hildegard (1098–1179)," in *Studies in the History and Method
    of Science*, ed. Charles Singer (Oxford: Clarendon Press, 1955), 53.

86. www.observer.co.uk/review/story/0,6903,449641,00.html.

87. Hildegard gives numerous treatments in *Causes and Cures* for headaches;
    there is even a passage on *emigranea*, although whether she meant what we
    mean by migraines is unclear. See, for example, CC, III, 208:6–13.

88. Translated and quoted by Charles Singer, "Scientific Visions," 55, from
    PL197, col. 18A. ("Lumen enim quod video locale non est, sed nube quae
    solem portat multo lucidius; nec longitudinem, nec latitudinem in eo con-
    siderare valeo. Illudque umbra viventis luminis mihi nominatur: atque sicut
    sol, luna et stellae in aqua apparent, ita scripturae et sermones, et virtutes, et
    quaedam opera hominum formata in illo mihi resplendent. . . . Attamen
    aspicio interdum in eodem lumine aliam lucem, quae mihi lux vivens nomi-
    natur; sed hanc non video frequenter . . . Dumque istam lucem intueor,
    omnis mihi tristitia omnisque dolor aufertur de memoria, ita ut tunc mores
    simplicis puellae, et non vetulae mulieris habeam.")

89. "Et tunc in eadem visione sensi venas meas et medullas plena viribus resti-
    tutas, quibus per multas infirmitates a juventute mea defeceram" (PL197,
    col. 18C).

90. Guibert gives "Tunc in eadem visione magna pressura dolorum coacta sum palam manifestare, que videram et audieram. . . ." (*Vita* [1993] Book II, p. 24).

91. For instance, ". . . ut significet se habitum corporis quem induit, non sine *pressura dolorum*, sed longa crucis tribulatione deponere." (Bede, in *S. Joannis Evangelium expositio*, PL92, col. 802B); or Marbod, "Quoque dies partes cruciatos dissipat artus/Es sortita torum, quem nec *pressura dolorum*/Nec timor infestat, nec sordida culpa molestat" (Marbod, *Marbodi, Redonensis episcopi carmina varia*, PL171, col. 1654A).

92. "In eadem visione scripta prophetarum, evangeliorum et aliorum sanctorum et quorundam phylosoforum sine ulla humana doctrina intellexi . . ." (*Vita* [1993], 24:88–91).

NOTES TO CHAPTER TWO

1. What follows is my description of the contents of Ny kgl. saml. 90b. Jessen describes the contents similarly but, probably due to the loss of some of his notes, more sketchily.

2. For an example of such a text, see Klaus-Dietrich Fischer, "The Isagoge of Pseudo-Soranus: An Analysis of the Contents of a Medieval Introduction to the Art of Medicine," *Medizinhistorisches Journal* 35, no. 1 (2000): 3–30.

3. For examples, see Faith Wallis, "Signs and Senses: Diagnosis and Prognosis in Early Medieval Pulse and Urine Texts," *Social History of Medicine* 13, no. 2 (2000): 265–278; Pedro Gil-Sotres, "Derivation and Revulsion: The Theory and Practice of Medieval Phlebotomy," in *Practical Medicine from Salerno to the Black Death*, eds. Luis Garcia-Ballester, Roger French, Jon Arrizabalaga, and Andrew Cunningham (Cambridge: Cambridge University Press, 1974), 110–155.

4. A sixth Book, a lunary that uses the moment of conception to predict character, is separate in the manuscript and in the new edition of Moulinier, but was incorporated into the fifth Book in Kaiser's edition. Many scholars follow Heinrich Schipperges in asserting that this sixth Book was spurious; see Heinrich Schipperges, ed., *Hildegard von Bingen, Heilkunde: Das Buch von dem Grund und Wesen und der Heilung der Krankheiten*, 6 ed. (Salzburg: Otto Müller, 1992 [1957]), 41–42; and Laurence Moulinier, "Hildegarde ou Pseudo-Hildegarde? Réflexions sur l'authenticité du traité 'Cause et Cure,'" in *"Im Angesicht Gottes Suche der Mensch sich Selbst": Hildegard von Bingen (1098–1179)*, ed. Rainer Berndt (Berlin: Akademie Verlag, 2001), 142–143. Jessen describes only five Books (104). Dronke, however, sees the lunary as compatible with her general work ("Hildegard of Bingen," 177–179). On lunaries, see Laurel Means, *Medieval Lunar Astrology: A Collection of Representative Middle English Texts* (Lewiston: E. Mellen, 1993).

5. Actually, Jessen relates the title as *Hildegardis curae et causae* (104), doubtless a mistake due to his lost notes.

6. For thoughtful essays on what criteria should be used to determine whether a text was "practical," see Peregrine Horden, ed., *The Year 1000: Medical Practice at the End of the First Millennium*, vol. 13 (Oxford: Society for the Social History of Medicine, 2000); Linda E. Voigts, "Anglo-Saxon Plant Remedies and the Anglo-Saxons," *Isis* 70 (1979): 250–68; and Faith Wallis, "The Experience of the Book: Manuscripts, Texts and the Role of Epistemology in Early Medieval Medicine," in *Knowledge and the Scholarly Medical Traditions*, ed. Don Bates (Cambridge: Cambridge University Press, 1995) 101–126. A practical text is one that reflects actual therapy (Alvarez-Millan, "Practice Versus Theory: Tenth-Century Case Histories from the Islamic Middle East," 215–218), with methods and recipes that were what the practitioner *did* "employ" (Fischer, "Dr. Monk's Medical Digest," 244), as shown by their recommending *real plants grown in the area* and which are also *effective* for the given indication (Meany, "The Practice of Medicine in England About the Year 1000"). Hints that Hildegard was providing her own *methodus medendi* can be found throughout the text and will be pointed out explicitly in Chapters Three, Four, and Five.

7. For an outstanding new edition, see Moulinier, *Hildegardis Bingensis causae et curae*, 2003. No completely reliable English translation is yet available. Versions range from the New Age, bowdlerized *Hildegard's Medicine* of Strehlow to excerpts with glossed commentary by Margret Berger. All have been translated from Kaiser's 1903 edition, which, fortunately, has very few differences with Moulinier's new edition. I use both in conjunction with a microfilm of Ny kgl. saml. 90b.

8. In the same article, Charles Singer also rejected the *Physica*, because it was "clearly a compilation, and numerous passages in it can be traced to such sources as Pliny, Walafrid Strabus, Marbod, Macer" (13). For that matter, he also rejected the ascription of medical writings to Trotula (13, n. 1).

9. The peculiar historiography of Hildegard's medicine has interested several historians. For a helpful analysis of the way the trends of nationalism, feminism, and New Ageism have influenced Hildegard studies over the last century, see Sabina Flanagan, "Zwischen New Age und wissenschaftlicher Forschung: die Rezeption Hildegards von Bingen in der englisch-sprachigen Welt," in *Prophetin durch die Zeiten: zum 900. Geburtstag*, ed. Edeltraud Forster (Freiburg: Herder, 1997), 476–484; Sue Spencer Cannon, "The Medicine of Hildegard of Bingen. Her Twelfth-Century Theories and Their Twentieth-Century Appeal as a Form of Alternative Medicine" (Ph.D. diss., UCLA, 1993); and Monica H. Green, "In Search of an 'Authentic' Women's Medicine: The Strange Fates of Trota of Salerno and Hildegard of Bingen," *Dynamis* 19 (1999): 25–54. Laurence Moulinier expressed her frustration with the incomplete and partisan readings of the text that are

due to "Catholic fervor, faith in naturopathy, or feminist conviction," in *Le Manuscrit Perdu à Strasbourg: Enquête sur l'oeuvre scientifique de Hildegarde* (Paris: Publications de la Sorbonne, 1995), 10.

10. Thus, "Hildegard's medicine has a special place, coming not only from ancient and monk medicine, but also from folk medicine. . . . Her sources were, first, her Benedictine life, whose daily hours gave her many of her concepts, such as 'viriditas'" (Schipperges, *Heilkunde*, 1992 edition, 44); see also Heinrich Schipperges, "Menschenkunde und Heilkunst bei Hildegard von Bingen," in *Festschrift zum 800. Todestag der Heiligen*, ed. Anton Ph. Bruck (Mainz: Selbstverlag der Gesellschaft für mittelrheinische Kirchengeschichte, 1979), 295–310; Heinrich Schipperges, *Die Welt der Hildegard von Bingen* (Freiburg: Herder, 1997); Heinrich Schipperges, *Hildegard of Bingen: Healing and the Nature of the Cosmos*, trans. John Broadwin (Princeton: Princeton University Press, 1997); Heinrich Schipperges, *Hildegarde de Bingen (1098–1179)*, trans. Pierre Kemner (Paris: Brepow, 1996). Schipperges' mainly religious approach has been taken by many; see, e.g., Klaus-Dietrich Fischer, "Mensch und Heilkunde bei Hildegard von Bingen," *Arzteblatt Rheinland-Pfalz* 51, no. 5 (1998): 165–168.

11. Gottfried Hertzka has created quite a Hildegardian industry. His approach is illustrated by the title page of one of his texts, where Hildegard is shown vanquishing the devils of disease with sword and cross. His titles include such suggestive notions as *The Wonder of Hildegard-Medicine*; *So God Heals*, *The Medicine of Saint Hildegard*, etc. For an example, see Gottfried Hertzka and Wighard Strehlow, *Handbuch der Hildegard-Medizin* (Freiburg im Bresau: Hermann Bauer, 1987).

12. Even what purport to be translations of Hildegard's work are often infused with this movement's peculiar tone; see, for example, Hildegard of Bingen, *Hildegard von Bingen's Physica: The Complete English Translation of her Classic Work on Health and Healing* (Rochester, Vt.: Healing Arts Press, 1998). Irmgard Müller argues against the practice of "Hildegard-Medizin" in many articles; see, for example, Irmgard Müller, "Wie 'authentisch' ist die Hildegardmedizin? Zur Rezeption des 'Liber simplicis medicinae' Hildegards von Bingen im Codex Bernensis 525," in *Hildegard von Bingen. Prophetin durch die Zeiten. Zum 900. Geburtstag*, ed. Edeltraud Forster (Freiburg: Herder, 1997), 420–430.

13. Engbring's 1940 article connects the first with the third wave of feminism; see Gertrude Engbring, "Saint Hildegard, Twelfth-Century Physician," *Bulletin of the History of Medicine* 8 (1940): 770–784.

14. Thus a recent translation of the *Physica* leaves out unpalatable recipes as unworthy of the saint, according to Glaze in "Medical Writer" 42, n. 82. Carl Jessen in his 1862 article even comments on this: "I cannot let it go unremarked that the oft-made objection that it is indecent for a nun to speak about the kind of diseases and treatments that Hildegard does, is precisely an objection of *our* generation, where to say what one thinks is quite

improper. . . . On such topics the author speaks as succinctly as possible and, of course, we should not forget that when she was writing this text she was almost sixty years old" (99).

15. "Subtilitates diversarum naturarum creaturarum," *LVM*, 8.

16. " . . . ubi tunc expositio naturarum diversarum creaturarum . . ." Van Acker, 91A, 443:23–444:1.

17. " . . . quasdam diversarum rerum naturales virtutes," *LDO*, 381:51–52.

18. "Quedam de natura hominis ac elementorum diversarumque creaturarum, et quomodo homini ex his succurrendum sit. . . ." *Vita* (1998), 118:5–6.

19. "Librum simplicis medicine, secundum rerum creationem octos libros continentem, librumque eius medicine composite de egritudinum causis, signis atque curis." Quoted by Moulinier, "Hildegarde ou Pseudo-Hildegarde," 115–116.

20. "Scripta etiam eius, que conventus iuratus confessus est sua esse, scilicet librum Scivias, librum Vite meritorum, librum Divinorum operum, Parisius per theologie magistros examinatos; librum Expositionis quorundam evangeliorum, librum Epistolarum, librum Simplicis medicine, librum Composite medicine. . . ." *Vita* (1998), 272:27–32.

21. As Moulinier notes, "son temoinage confirme l'identification du *Liber medicinae compositae* avec le *Cause et cure* que nous connaissons" ("Réflexions," 119).

22. Thus Moulinier simply says that "this formulation [of two, not one, medical texts] is found for the first time in a list of Hildegard's works submitted to the Inquisition, a material division of her naturalist writings nowhere attested while she was alive" (*Manuscrit*, 31).

23. For example, although acknowledging that the idea of a fixed text is modern, Fischer repeats that Books Three and Four were taken from *Physica*, and that its disorder leads to the conclusion that the whole work must simply be a collection of excerpts from a larger work (Fischer, "Mensch und Heilkunde bei Hildegard von Bingen," 102–104).

24. Dronke notes similar repetitions between *Scivias* and the *Ordo Virtutem* and argues that the more unusual reading was more likely to be by Hildegard. Her secretaries, who tended to "conventionalize her expressions" were probably responsible for the later, more conventional reading ("Problemata," 103).

25. The earlier thirteenth-century manuscript of *Physica*, the Florentine manuscript, is not yet edited and published; it provides many different recipes and may lead to a different conclusion; see Benedikt Konrad Vollmann, "Auf dem Weg zur authentischen Hildegard. Bemerkungen zu den nur in der Florentiner 'Physica' Handschrift uberlieferten Texten," *Sudhoffs Archiv* (2003), 87 (2):159–172.

26. Many of these overlapping passages were pointed out in 1903 by *Causes and Cures'* editor, Paul Kaiser, although he assumed that the copying took place

in the opposite direction. Eliza Glaze notes that, if the texts were written by the same author, such resemblances would be expected. In any case, repetitions were "an authentic feature of Hildegard's habitual patterns of activity as a writer" ("Medical Writer," 147).

27. Kaiser noted this correspondence but interpreted it in the opposite sense—that *Physica* derived from *Causes and Cures*.

28. Again this correspondence was noted by Kaiser but interpreted in the opposite sense.

29. This is my analysis of the microform; it agrees with Glaze, "Medical Writer," 148, who dates it to 1250–1260, and with Moulinier (*Manuscrit Perdu*, 48). Müller gives it a much later date, from the 1300s. Moulinier now takes it for certain that it was copied at St. Maxim's of Trier ("Hildegarde ou Pseudo-Hildegarde?" 123).

30. Moulinier posits that there must have been three manuscripts ("Hildegarde ou Pseudo-Hildegarde?" 117–123).

31. As Madeline Caviness noted with regard to the similar argument regarding the illustrations in *Scivias*: "Yet if it is to be argued that Hildegard had nothing to do with the pictures in the Rupertsberg *Scivias*, some tangible alternative has to be suggested, with a workplace, intellectual context, and other extant productions and not some phantom atelier" (Madeline H. Caviness, "Hildegard of Bingen: Some Recent Books—a Review Essay" *Speculum* 77, no. 1 (2002), p. 120).

32. Eliza Glaze also points out how difficult it is to imagine what possible kind of text could have been the source for such different kinds of medical texts as *Causes and Cures* and *Physica* ("Medical Writer," 146).

33. Glaze also argues that the core of the problem is not only the "rather vague statement in the *Liber vitae meritorum* which has been interpreted as signifying her authorship of one united medical book, already completed by 1158" ("Medical Writer," 145), but also the lateness of manuscripts and the absence of inclusion in the Reisenkodex. But most important, "these doubts have arisen more from our modern assumptions about female medical authorship, as well as general ignorance of early medieval medical literature" (146).

34. Of course, medieval "authenticity" is problematic. As Baird noted about Hildegard's letters, "the public nature of letters could also have disquieting consequences for what we like to think of as authenticity. The texts of letters . . . were subject to various kinds of tampering, by recipient as well as original writer after the letter had been sent out" (*Letters*, v. 1, p. 9). Likewise Monica Green points out: "Recent understandings of medieval practices of composition raise further questions about the unity of authorship. The fact that Hildegard dictated to a series of four different male scribes has long been known; what is still not known is how much each scribe put his own imprint on her words as he pressed them into writing" ("In Search of an Authentic Women's Medicine," 53).

35. "Hodie aperuit nobis/clausa porta/quod serpens in muliere/suffocavit/unde lucet in aurora/flos de virgine Maria" (Peter Dronke, "Hildegard's Inventions. Aspects of her Language and Imagery," in *Hildegard von Bingen in ihrem historischen Umfeld*, eds. Alfred Haverkamp and Alexander Reverchon [Mainz: P. von Zabern, 2000], 299–320).

36. In *Causes and Cures*, Hildegard does not refer to *suffocatio*, but it was a well-known syndrome. For example, "De curatione suffocationis" is discussed in Berlin MS Phillips 1790, which Fischer has called a "typical representative of a collection of assorted medical texts" ("Isagoge," 240).

37. This is a common image. For example, "Clausa enim janua thalamum uteri introivit, humanam naturam sibi conjunxit, et *clausa porta*, ut verus sponsus de thalamo processit." Honorius, *Elucidarium*, PL172, Liber Primus, Ch. 19, col. 1123D.

38. Although Judy Chicago did include Hildegard in her "Dinner Party," along with a vaginal image, she did not use this illustration but one that she apparently created herself. Nor has this concept yet entered academic discourse, which still invariably interprets this image as an "egg." Hildegard simply describes it as an "egg-shaped or oval thing"—an oval *instrumentum*. However, see Jennifer Borland, "Subverting Tradition: The Transformed Female in Hildegard of Bingen's 'Scivias'" (paper presented at the Seeing Gender: Perspectives on Medieval Gender Conference, King's College, London, 2002), www.medieval gender.org.uk/King's_2002/ abstracts. htm#Borland.

39. See Alvarez-Millan, "Practice versus Theory: Tenth-Century Case Histories from the Islamic Middle East."

40. "Most of the more than 145 extant medical books produced during these centuries [of early medieval medical literature, e.g., ninth through eleventh] stand as anthologies of short treatises either entirely anonymous, or attributed to late ancient authors about whom little independent evidence survives, or falsely attributed to the twin giants of antiquity, Hippocrates and Galen." See Florence Eliza Glaze, "The Perforated Wall: The Ownership and Circulation of Medical Books in Medieval Europe Ca. 800–1200" (Dissertation, Duke University, 2000), 12–13.

41. See Monica H. Green, "The Development of the Trotula," *Revue d'Histoire des Textes* 26 (1996): 119–203, especially 154; and Monica H. Green, *The Trotula: A Medieval Compendium of Women's Medicine* (Philadelphia: University of Pennsylvania Press, 2001).

42. See Paul Oskar Kristeller, "Bartholomaeus, Musandinus, and Maurus of Salerno and Other Early Commentators of the Articella, with a Tentative List of Texts and Manuscripts," *Italia Medioevale e Humanistica* 19 (1976): 57–87; and Charles H. Talbot, "A Letter from Bartholomew of Salerno to King Louis of France," *Bulletin of the History of Medicine* 30 (1956): 321–328.

43. "1142: Dedicata est capella in hospitali 5 Kal. Iunii. . . . Proxima die anni eiusdem dedicata est capella in infirmaria . . . 4 Kalends. Iunii" (Waitz, *Annales Sancti Disibodi*," 26:23–26).

44. "Locus autem aegrotantium remotus erit a Basilica vel cellulis fratrum ut nulla inquietudine, vel clamoribus impediantur." S. Isidori Hispalensis Episcopi in *Regulam Monachorum*, Lucas Holstenius, ed., *Codex regularum monasticarum et canonicarum* (Graz: Akademische Druck-U.Verlagsanstalt, 1957(1759), 138.

45. "Infirmi quolibet morbo defessi in una jaceant domo et una qui aptus est delegentur." S. Fructuosi Brasanensis Episcopi, in Holstenius, *Codex regularum*, 211t.

46. "Cellulae infirmorum infirmitatis tempore sub sigillo minime habeantur, quatenus a fratribus visitari possint," in the *Regula Solitariorum Grimlaici*, in Holstenius, *Codex regularum*, 330.

47. "Separatim cellam habeant cum omnibus opportunitatibus," in the *Regula cujusdam patris*, Holstenius, *Codex regularum*, 401.

48. I calculate six to eight beds based on the fact that Disibodenberg had about twenty nuns and thirty monks; each would have been bled on a quarterly basis and spent three days in the infirmary. Therefore, at any time, two people would have been in the infirmary simply due to the requirements of bleeding. Second, although it was a young monastery, by 1142 when the infirmary chapel was dedicated, many of its monks and nuns would have been in their forties and fifties. It does not seem unlikely that one or two would have been incapacitated by some kind of chronic problem—a stroke or heart attack for example—and one or two, on average, hospitalized for some kind of acute illness.

49. Bird, however, makes the point that female "recluses continued to attach themselves to male religious houses and scarce resources often meant that it was difficult to maintain segregated living spaces." Jessalynn Bird, "Texts on Hospitals: Translation of Jacques de Vitry, Historia Occidentalis 229 and Edition of Jacques de Vitry's Sermons to Hospitallers," in *Religion and Medicine in the Middle Ages*, eds. Peter Biller and Joseph Ziegler (Oxford: York Medieval Press, 2001), 98.

50. In Hildegard's *Lingua ignota* she provides a specific word for the medicinal sauna—*stoinz*—suggesting that there was a medical sauna at Rupertsberg.

51. A twelfth-century letter from the Abbot of Tegernsee documents the existence of a medicinal herb garden at Benedictbeuren, the oldest Benedictine monastery in Bavaria (Baader, "Mittelalterliche Medizin," 286).

52. Given that it provides a picture of the 1100 most important objects in Hildegard's world—in hierarchical order—the *Lingua ignota* has received too little scholarly attention. It begins with words for spiritual beings (God, angel, patriarch, saint); then human beings (person, man, and woman) and familial relations. Positions in the church (bishop, etc.), follow; then

the buildings of a monastery and the contents of an altar. Positions in the secular world come next, from king to peasant, to several types of workers. There are also words for the days of the week, the times of life, and the names of the months, and for pieces of clothing, for the tools of gardening, and the tools of writing. There are words for many types of food and drink. It was first edited in part by Wilhelm Grimm, who argued that it was not practical, because it contained words for trees like the fig and pepper, and birds like the peacock and parrot, which Hildegard could not have known, although, perhaps, she did. Grimm did not know of the weather change during this period, a century-long warming of Europe (the mini-climactic optimum) that did allow for the growing of grapes in England. See Wilhelm Grimm, "Wiesbader Glossen," *Zeitschrift für deutsches Alterthum* 6 (1848): 321–340. For editions see Marie-Louise Portmann and Alois Odermatt, *Wörterbuch der Unbekannten Sprache (Lingua Ignota) in der Reihenfolge der Manuskripte* (Basle: Basler Hildegard-Gesellschaft, 1986); also, F. W. E. Roth, "Glossae Hildegardis (Das Wörterverzeichnis der *Lingua ignota*)," in *Die althochdeutschen Glossen*, eds. Emil Elias Steinmeyer and Eduard Sievers (Berlin: 1895), 390–404. Scholars have used it rarely. But Dronke pointed out that it documents "the sensory world of a keenly observant naturalist" ("Hildegard's Inventions," 306). See also W. J. A. Manders, "Lingua ignota per simplicem hominem Hildegardem prolata," in *Sciencaj Studoj* (Copenhagen: Internacia Scienca Asocio Esperantista, 1958), 57–60; and Jeffrey T. Schnapp, "Virgin Words: Hildegard of Bingen's 'Lingua Ignota' and the Development of Imaginary Languages Ancient to Modern," *Exemplaria, A Journal of Theory in Medieval and Renaissance Studies* 3, no. 2 (1991): 267–298.

53. Because it is arranged hierarchically, the *Lingua ignota* also reveals the structure of Hildegard's world. For example, while man (*vir*) comes before woman (*mulier*), human (*homo*) comes before man. Spiritual people are more important than family, but family is more important than worldly position, either within or without the monastery. Moreover, it provides a record of the people in Hildegard's world, who are not only monks and servants but also princes and prelates, workers and farmers, gardeners and artisans. A remarkably similar list was put together by Disibodenberg's only known scribe, Henry the Monk; see the Appendix.

54. For examples of such medical "consultations," see Charles H. Talbot, "A Letter from Bartholomew of Salerno to King Louis of France," *Bulletin of the History of Medicine* 30 (1956): 321–328; and Urban T. Holmes, Jr. and Frederick R. Weedon, "Peter of Blois as a Physician," *Speculum* 37, no. 2 (1962): 252–256.

55. On the use of hagiography for documenting disease, see Clare Pilsworth, "Medicine and Hagiography in Italy (c. 800–c. 1000)," *Social History of Medicine* 13, no. 2 (2000): 253–264. For an example, see Jerome Kroll and Bernard

Bachrach, "Sin and the Etiology of Disease in Pre-Crusade Europe," *Journal of the History of Medicine and Allied Sciences* 41 (1986): 395–414.

56. "Fever" provides an instructive example. In modern terms it means a measurement taken with a mercury or electronic thermometer that is over one standard deviation from a mean of 98.6 F. But this is certainly not what the numerous medieval tracts on fevers were referring to, although they may have meant that the skin was hot to the touch. See the discussion in Peregrine Horden, "The Millennium Bug: Health and Medicine around the Year 1000," *Social History of Medicine* 13, no. 2 (2000): 201–220.

57. "Nullus fere egrotus ad eam accesserit, quin non continuo sanitatem receperit." *Vita* (1998), 186:3.

58. On the lack of evidence for medical practice by nuns in convents, see Katrinette Bodarwé, "Pflege und Medizin in mittelalterlichen Frauenkonventen," *Medizinhistorisches Journal* 37, no. 2 (2002): 231–63. There is also, of course, a lack of evidence for practice by men, as well.

59. Since Benedictine monasteries had about two servants for every monk (Horn, *Plan of St. Gall*, v. 1, 344f), and ten nuns gathered around Jutta's bedside in 1136, while in 1150, twenty nuns (but not all) left Disibodenberg for Rupertsberg, Hildegard must have had a patient base of at least sixty people within the convent itself.

60. There were many ill visitors for Jutta, according to the *Vita Juttae*. They included nobles and commoners, rich and poor, pilgrims and guests, ("nobiles, ignobiles, divites ac pauperes, peregrini et hospites" (Staab, "Aus Kindheit, 178:2–4); and Jutta often "put her hands on those ill in body or in spirit" ("super afflictos quoque corpore et anima pia semper gestans viscera") Staab "Aus Kindhert," 178:4.

61. On the training of English monks in Salerno; see E. A. Hammond, "Physicians in Medieval English Religious Houses," *Bulletin of the History of Medicine* 32 (1958): 105–20.

62. "Cura infirmorum sano sanctaeque conversationis viro committenda est." S. Isidori Hispalensis Episcopi in *Regula Monachorum*, in Holstenius, *Codex regularum*, 196.

63. "Infirmi quolibet morbo defessi in una jaceant domo et uni qui aptus est delegentur." S. Fructuosi Brasanensis Episcopi in Holstenius, *Codex regularum*, 211t.

64. "Deputetur autem frater timens Deum, et diligens atque sollicitus, qui eis ea, quibus opus habuerint, misericorditer administret et balnearum usus quoties expedit offerat," *Regula solitarium Grimlaici* in Holstenius, *Codex regularum*, 330.

65. "Ergo cura maxima sit Matri, ne aliqua negligentia patiantur . . . et si hoc necessitas infirmarum exegerit, et Matri Monasterii justum visum fuerit, cellariolum et coquinam suam infirmae in commune habeant." S. Donati Vesont, in Holstenius, *Codex regularum*, 381.

66. "Et ideo discite quidem naturas herbarum commixtionesque specierum
     sollicita mente tractate . . ." R. A. B. Mynors, ed., *Cassiodori Senatoris
     Institutiones* (Oxford: Clarendon Press, 1937), 78. Glaze argues that his
     instructions were directed to the "monastic reader" not to the infirmarian,
     because the texts were to be in the library not the infirmary ("Perforated
     Wall," 59–67). But the title of the section is "On doctors" ("De medicis")
     and the instructions, which are not theoretical but practical, seemed to be
     aimed at doctors as practitioners, not as readers.

67. "Medicinis autem, excepto cauterio et sanguinis minutione, peraro utimur."
     Statuta Ordinis Carthusiensis, in Holstenius, *Codex regularum*, 325.

68. "Habeat igitur unumquodque Monasterium Medicum Physicum stipen-
     dio conductum qui sit boni nomini in arte sua ac bonae famae in moribus,
     itemque Chirurgum, quos moneant Praesules . . . Nullus item Monachus
     consulat Medicum etiam ordinarium sine licentia Praelati, neque sumat
     medicinam aliquam quae custodiam vitae requirat sine speciali licentia
     Medici." *Constitutiones Camaldulensis*, in Holstenius, *Codex regularum*, 246.

69. "Magister Infirmarius in Coquina sua cum omnibus solatiis loquatur. . . ."
     Ordinis Cisterciensis, in Holstenius, *Codex regularum*, 427.

70. "Nullum officium deputetur Infirmariae, vel ejus solatiae extra Infirmito-
     rium, dum aliqua graviter aegrotat." *Regula ordinis sempringensis sive Gilber-
     tinorum chronicorum*, in Holstenius, *Codex regularum*, 535.

71. For example, the Abbot of St. Alban's (1183–1195), had been, formerly, a
     trained physician.

72. See Darrel W. Amundsen, "Medieval Canon Law on Medical and Surgical
     Practice by the Clergy," *Bulletin of the History of Medicine* 52 (1978): 22–44;
     and Vern. L. Bullough, "Training of the Nonuniversity-Educated Medical
     Practitioners in the Later Middle Ages," *Journal of the History of Medicine* 15
     (1959): 446–58; and E. A. Hammond, "Physicians in Medieval English Reli-
     gious Houses," 120. Actually the relevant portions of Mansi, *Sacrorum Con-
     ciliorum*, XXI, are equivocal. Thus in 1130 the Council of Claremont ruled
     that "it was a bad and detestable custom that monks and regular canons, after
     taking up the habit and making their profession, spurn the Rule, and learn
     law and medicine for the sake of gain." (Mansi, cols. 432–458). This exact
     wording was simply repeated at the Second Lateran Council in 1139, with
     the addition that "monks and regular canons are not to learn temporal law
     or medicine" (528). Then in 1162, at the Council of Montpellier they added
     that "it is also prohibited under the greatest severity, that a monk or regular
     canon, or any other religious, go to learn secular law or physic." The intent,
     therefore, seems to be less a prohibition on practicing medicine than of prac-
     ticing medicine for the sake of gain or leaving the monastery to learn it.

73. " . . . in prima habetis Herbarium Dioscoridiis qui herbas agrorum mira-
     bili proprietate disservit atque depinxit. Post haec legite Hippocratem
     atque Galienum latina lingua conversos, id est Therapeutica Galieni ad

philosophum Glauconem destinata et anonyma quendam, qui ex diversis auctoribus probatur esse collectus, deinde Caeli Aureli de medicina et Hippocratis de Herbis et Curis diversosque alios medendi arte compositos quos vos in bibliothecae nostrae sinibus reconditos Deo auxiliante dereliqui." In Mynors, 78–79.

74. For a catalogue of early medieval monastic medical holdings (through the twelfth century), see Glaze, "Perforated Wall," 268–292. For information on the frequency of surviving manuscripts, see Pearl Kibre, *Hippocrates Latinus: Repertorium of Hippocratic Writings in the Latin Middle Ages* (New York: Fordham University Press, 1985); and Augusto Beccaria, *I Codici di Medicina del Periodo Presalernitano (Secoli IX, X, e XI)*, vol. 53 (Rome: Storia e Letteratura, 1956).

75. Indeed, most of the four dozen medical miscellanies I examined were collections of just such texts. For a discussion of typical medical manuscripts of this period, see Fischer, "The Isagoge of Pseudo-Soranus"; and Klaus-Dietrich Fischer, "Dr. Monk's Medical Digest," *Social History of Medicine* 13, no. 2 (2000): 239–252.

76. The infirmary had medical books separate from the library; on what these were, see Nebbiai-Dalla Guarda, "Les livres de l'infirmerie."

77. For a discussion of contemporary German medical manuscripts, which generally contain comparable collections of texts, see Baader, "Mittelalterliche Medizin," 284–285. For instance, the thirteenth-century MS CLM 22056 contains an anonymous *Practica*, a *Hippocratic Aphorisms*, and a recipe book; Baader argues that it was indeed meant for practicing medicine in the infirmary.

78. A partial list of her correspondents includes: the Archbishop of Mainz; the Bishops of Spire, Worms, Constance; the Abbots of Hirsau, St. Emmerman, and Springerbach; and more than twenty abbesses. Also, because her brother, Hugo, was magister at the cathedral school in Mainz, she presumably had access to its large library.

79. This is in MS CLM 13002, according to Baader, "Mittelalterliche Medizin," 282.

80. The will mentions an "herbarium, lapidarium in uno volumine, gemmam anime, elucidarium . . . medicinales sex" (MGH SS XXV, 502–503).

81. I have used the words (and their meaning) from the German glossary to Kaiser's edition, *Hildegardis causae et curae*, 252–254. Many of the German words, are, however, adjectives—for example, *volmudich, wortselich, otmudich, stolz*—and are only used in the lunary. Since these are words for which Hildegard often used the Latin, this finding would support the idea that the lunary was written by someone else.

82. Not much has been written on medieval German folk medicine, but see M. Höfler, "Altgermanische Heilkunde," in *Handbuch der Geschichte der Medizin*, eds. Max Neuburger and Julius Pagel (Jena: Fisher, 1902),

453–477; and Jerry Stannard, "Greco-Roman Materia Medica in Medieval Germany," *Bulletin of the History of Medicine* 46 (1972): 455–468. German Jewish folk medicine of the same period provides a different but complementary source; see Joseph Shatzmiller, "Doctors and Medical Practice in Germany around the Year 1200: The Evidence of 'Sefer Hasidim,'" *Journal of Jewish Studies* 33 (1982): 583–593; and Joseph Shatzmiller, "Doctors and Medical Practice in Germany around the Year 1200: The Evidence of 'Sefer Asaph,'" *Proceedings of the American Academy for Jewish Research* 50 (1983): 149–164.

83. See also William C. Crossgrove, "The Vernacularization of Medieval Science, Technology and Medicine, Introduction," *Early Science and Medicine* 3, no. 2 (1998): 81–87.

84. For further analysis and commentary, see Bernhard Schnell, "Das Prüller Arzneibuch: Zum ersten Herbar in deutscher Sprache," *Zeitschrift für deutsches Altertum und deutsche Literatur* 120 (1991): 184–202; Bernhard Schnell, "Vorüberlegungen zu einer 'Geschichte der deutschen Medizinliteratur des Mittelalters' am Beispiel des 12. Jahrhunderts," *Sudhoffs Archiv* 78, no. 1 (1994): 90–97; and Gundolf Keil's articles on the "Innsbrucker (Prüler) Kräuterbuch" in Keil, *Verfasserlexicon*, vol. 4, 396–398; on the "Innsbrucker Arzneibuch," in Keil, *Verfasserlexicon*, vol. 4, 395–396; on the "Arzenîbuoch Ipocratis," in Keil, *Verfasserlexicon*, vol. 1, 505; on the "Prüler Steinbuch," in Keil, *Verfasserlexicon*, vol. 7, 875–876; all in Kurt Ruh, et al., eds., *Die deutsche Literatur des Mittelalters. Verfasserlexikon*, vols. 1–8 (Berlin: Walter de Gruyter, 1978).

85. According to Schnell, the text (Munich's MS CLM 536, fols. 82v–83v) is sandwiched between a short passage on the "seven wonders of the world" and a (Latin) text on "ophthalmology"; this is followed by a tantalizing allusion to a "visio cuiusdam pauperculae muliebrae" ("Das Prüller Arzneibuch").

86. The text is found on fols. 86r–87v of the MS CLM 536 of Munich and in MS Innsbruck 652, where it is known as the Innsbruck Kräuterbuch. Keil hazards that it was written by a "medical monk" for the infirmary or the pharmacy, around 1100 in Bayern; see Gundolf Keil, "Innsbrucker (Prüler) Kräuterbuch," in Keil, *Verfasserlexicon*, 396–98.

87. See Gundolf Keil, "Arzenîbuôch Ipocratis," in *Die deutsche Literatur des Mittelalters. Verfasserlexikon*, vol. 1, Kurt Ruh, et al., eds. (Berlin: Walter de Gruyter, 1978), 505–506.

88. The following is based on Zurich's Zentralbibliothek MS C58, fols. 44v–47r; and Friedrich Wilhelm, *Denkmäler deutscher Prosa des 11. und 12. Jahrhunderts*, 2 vols. (Munich: G. D. W. Callwey, 1914–1918), vol. 1, 53–64. I examined the manuscript myself but here I use Wilhelm's edition.

89. "Ad capitis dolorem . . . ad oculos dolentes . . . ad dolorem dentium . . . qui non potest urinam continere . . . contra ydropicam passionem . . ." Wilhelm, *Denkmäler deutscher Prosa*, vol. 1, 53–54.

90. "Ad cappillos cadentes, Brenne den linsamen u. mische in mit ole u. salbe das hâr." Wilhelm, *Denkmäler deutscher Prosa*, vol. 1, 54.

91. "Ad oculos dolentes . . . Diz collirium ist wundirliche gôt ze der finster-nisse der ôgen. Nim daz gôte cinimin, u daz caferan, beider geliche u mil-wez u nim des ephes wrcunso u honec u mischez allez zesamine vil harte, u sich ez durch ein tôc u gehalt ez. So du disses bedurfist. so trôfe mit einir federe einin trôfin indaz ôge." Wilhelm, *Denkmäler deutscher Prosa*, vol. 1, 55:76–82.

92. "In dirre stete ist gescribin .v. geordonot. wie man ineineme iegelichen manote sol luturtranc machon. vzzer crvteren .vn picmentis. Diz lvtertranc ist vil got .v heilit .v. gehaltet. .v gedobit die vberfluzzigin humores. die dir sint in den menneschin. Zi dirre wis sol man ez machon. In martio. sol man ez machon uzir einem teile faluiun .v sol lman da zo nen. XII corn piperis. pertheram. gingiber. spic. wol gesotinhonec. vn. XXX mez wines. Disv alliv suln wol gemilwet sin. dar nach gestan daz sie gelvteren .v daz div clara potio svze si zitrinchinne. . . . In aprile, sol man zo diseme tranche ton die wormate .v allez daz da vor gescribin ist . . . In iulio gam andream . . . In novembere millefolium . . . etc." Wilhelm, *Denkmäler deutscher Prosa*, vol. 1, 59:231–60:258. My abridged translation.

93. "Ad singultum. Acetum acrum olfactum bibat. mox desinet. Ad tussicos. dictammum dabis bibere cum vino prodest, Ade eos qui cibum continere non possunt. Millefolium tritum cum vino tepido bibat, etc." Wilhelm, *Denkmäler deutscher Prosa*, vol. 1, 63:361–372.

94. "Adiuro te mala malanna. per patrem et filium. et spm. ssm. ut non crescas. sed evanescas in nomine p. et f. et s. sci. kx.k Pater nostri ter et pater nos-ter." Wilhelm, *Denkmäler deutscher Prosa*, vol. 1, 64: 397–403.

95. I am thinking here of the "uberfluzzigin humores die dir sint in den menn-eschin" of the ever-changing, monthly "lutertranc" quoted above. Wilhelm, *Denkmäler deutscher Prosa*, vol. 1, 59: 234–238.

96. A similar conclusion is reached when another vernacular medical text avail-able in the twelfth-century Rhineland, the Hebrew *Sefer Asaph*, is analyzed. Mentioned by both Rashi and Eliezer ben Nathan of Mainz, it makes an inter-esting comparison with *Causes and Cures*. Its oldest manuscript is probably MS Bodley 2138, written in the Rhineland c. 1150. Its first chapter reviews natural science according to the four elements; its second chapter goes over the medicinal uses of water, minerals, and plants. The third chapter covers monthly dietary recommendations, and the fourth explains the pharmaceutical uses of trees—the walnut, almond, and hazelnut among others—and of herbs, such as opium and belladonna. MS Bodley 2138 reportedly also contains a set of 150 prescriptions whose ingredients are identified by their German, not Hebrew, names, suggesting that it, too, was a written version of information learned in the field and structured by ancient theory. On the text, see A. Bar-Sela and Hebbel E. Hoff, "Asaf on Anatomy and Physiology," *Journal of the History of*

Medicine 20 (1965): 358–389; Elinor Lieber, "Asaf's 'Book of Medicines': A Hebrew Encyclopedia of Greek and Jewish Medicine, Possibly Compiled in Byzantium on an Indian Model," in *Symposium on Byzantine Medicine*, ed. John Scarborough (Washington, D.C.: Dumbarton Oaks Research Library and Collection, 1984), 233–249; and Ludwig Venentianer, *Asaf Judaeus, der alteste medizinische Schriftsteller in hebraeischer Sprache*, vols. 1–3 (Budapest: Teubner, 1915–1917).

97. For comparison, *infirmarius* occurs 76 times in the *PL* and 23 times in the *CLCT4*; *physicus* occurs, respectively, 144 and 177 times; *herbarius* 12 times; and *apothecarius*, 6 times and once.

98. *Medicus* occurs 17 times in the *PL*197 and 12 times in the works of Hildegard in the *CLCT4* (several of these are the same passages), and 0 times in *Causes and Cures*. By way of comparison, *medicus* occurs 4091 times in the *PL* and 748 in the *CLCT4*. Hildegard uses the term *hortulanus* only twice (used 334 times in the *PL* and 53 in the *CLCT4*). Hildegard uses the term *pigmentarius* 3 times in the PL197 and 12 times in *CLCT4* (it is used 54 times in the PL and 235 in the *CLCT4*).

99. For example, "Disce quoque *vulnera* peccatorum cum penitentia misericordia curare, quemadmodum summus *medicus* salutare exemplum uobis ad saluandum populum reliquit." *Epistulae Hildegardis*. CLM, up. : 27 R, line : 33; or "Sed si *vulnera* accepteris, *medicum* quare ne moriaris." *Scivias*, PL197, col. 431B.

100. "Hominem in se *vulnera* habetis et ea cum oleo aspergentem, sed tamen infusum uinum in *uulneribus* suis sustinere non ualentem *medicus* frequenter cum misericordia *ungat*, nec fetentem liuorem in illo esse sinat, quia lepra per summum *medicum* abstergitur, ubi se homo sacerdoti ostendit." *Epistulae Hildegardis*, CCCM 91A, ep. : 189, linea: 1.

101. "Quod si dolorem suum inspexerit et *medicum* quaesierit, ille inuentus ostendit ei *amarum sucum pigmenti* per quod saluari potest, quod sunt amara uerba per quae probandus est utrum." *Sciuias*, part : 1, vision: 4, cap.: 30 (commentarii), line: 967.

102. "Ego enim sum magnus *medicus* omnium languorum, faciens uelut medicus qui languidum uidet qui *medelam* ardenter desiderat." *Sciuias*, Part: 1, Vision: 3, cap.: 30 (commentarii), line : 600.

103. "Sed si sacerdos habuerit crimen in putredine carnis aut in petulantia gule et lasciuie, ilico surgat et *medicum* requirat, ac illud *euomat* quasi uenenum biberit, nec in se diutius uelut secure teneat, scilicet ut domesticum suum." *Epistulae Hildegardis*, CCCM 91A, ep.: 170 R, linea: 24.

104. "Magnus enim *medicus* uigilantes suscitat et dormientes corripit et in malis suis perseuerantes occidit." *Epistulae Hildegardis*, CCCM 91A, ep.: 189, linea: 20.

105. " . . . quia lepra per summum medicum abstergitur . . ." *Epistulae Hildegardis*, CCCM 91A, ep. : 189, linea: 1–5.

106. "Fidelis autem homo dolorem suum consideret et *medicum* quaerat, antequam in mortem cadat." *Scivias*, CCCM 43, Part: 1, vision: 4, cap.: 30 (commentarii), line 966.

107. According to the OLD, the post-classical meaning for *pigmentum* is "the juice of plants," OLD, s.v. "pigmentum," 1375.

108. This is according to the *Lexicon Latinatis medii aevi*, s.v. "pigmentarius." In the Oxford Latin Dictionary, the term has a more limited scope; the *pigmentarius* is a dealer in paints or cosmetics; see OLD, s.v. "pigmentarius."

109. By *pigmenta*, Hildegard did not mean paints or dyes but the "juice of medicinal plants. For instance, ""I will sow roses and lilies and others of the best *pigmenta* in that wonderful garden." ("Ego autem rosas et lilia ac alia optima virtutum *pigmenta* in agrum illum seminabo . . .") *Scivias*, Part 3, Vision 10, PL197, col. 697C. For another example, she writes of the "bitter sap of medicine through which one can be saved" (*amarum sucum pigmenti* per quod saluari potest " in *Scivias*, CLCT4, part 1, vision 4.: cap. 30, line 970).

110. "Just as when at the dawn of justice the *pigmentarius* runs outside with humility and wisdom" ("ita omnes *pigmentarii* ad primam auroram justitiae cum timore verecundia et sapientia currant" in "Epistula 26," PL197, col. 186A).

111. "Just like the good *pigmentarius* who plants medicinal herbs and spices in his garden, and always takes care that his garden stays green. ("Ut bonus *pigmentarius* qui in horto suo quaeque pigmenta et aromata plantat, semper studens ut hortus suus viridis et non aridus sit.") *Vita Sancti Disibodi*, PL197, col.1098A.

112. "The *pigmentarius*, who has a well-irrigated and good-smelling garden " ("*Pigmentarius* qui irriguum et bene olentem hortum habet." "Epistula 33," PL197, col. 198D.

113. "Velut bonus pigmentarius hortum suum ab inutilibus herbis purgat." *Epistulae Hildegardis*, CCCM 91, ep.: 84 R, line: 208.

114. "Quia *pigmentarius* qui irriguum et bene olentem hortum habet, uideat ut hortus ipsius utilitatem fructuum afferat, ut non deficiat." *Epistulae Hildegardis*, CCCM 91A, ep.: 144 R, line: 53.

115. "Tunc etiam *pigmentarios* suos consituit, qui eundem hortum rigare sciant et qui fructum eius colligant et exinde diversa *pigmenta* faciant." *Scivias*, CCCM 43, part 1, vision: 2, line 751.

116. "Sed qui ex his in clausura subiectionis amore Filii mei se continent et in moribus suis institutionem maiorum suorum quam me inspirante protulerunt observant, sollicitudinem curationis *pigmentariorum* non habentes." *Scivias*, CCCM 43, part: 1, vision: 5, line 263.

117. "Deinde domus infirmorum habens fumariam sive focariam capellulam lobiam cameram dispensatoriam cameram privatam aliam que privatiorem ortum ante eam postibus tabulis spinis munitum ut esset ex aere et viriditate infirmorum aspectibus refrigerium." Gislebertus Trudonensis. *Gest. abbatum Trudonsium*, lib. 10: par 13, pag. (MGH); 295, linea: 26.

118. "Each of the abbey's gardens was under the control of a monk . . . and in the case of the Infirmary Garden, this was the Infirmarer." John H. Harvey, "Westminster Abbey: The Infirmarer's Garden," *Garden History* 20, no. 2 (1992), 98.

119. In the Rule of the Carthusians, there was a specific office of the *hortulanus.* "Hortus et quae ad eum pertinent, uni deputatur ex Fratribus" in the *Ordinis Carthusiensis* in Holstenius, *Codex regularum,* 329.

120. "In omnibus his *hortulanus* inutilia de horto suo proiciet et utilia ad se colliget." *Epistulae,* CM 91A, ep: 223R. linea 172.

121. "Filius namque Dei, sicut bonus et sapiens *hortulanus,* bonas et ad cujusque utilitatem perfectas colligit herbas," in "Epistula 47," PL197, col. 243B.

122. As Laurence Moulinier observed, "the cloister garden combined with the care of the sick, among other things, contributed to forging her pharmacobotanical culture" (1995, 258).

NOTES TO CHAPTER THREE

1. "Humans are made from the four elements—from fire, water, winds, and earth." ("Constat homo ex quattuor elementis—de igne, aqua, ventis et terra." MS BN Lat. 7028, fol. 4r).

2. "Deus ante creationem mundi absque initio fuit et est, et ipse lux et splendor fuit et est et uita fuit." CC, I, 19:1.

3. For comparison, see Richard C. Dales, ed., *Marius: On the Elements* (Berkeley, Los Angeles, and London: University of California Press, 1976).

4. For instance, in two relatively recent works, scholars examined the elements in philosophy, music, hagiography, alchemy, and literature, but not in medicine: see Danielle Buschinger and André Crepin, eds., *Les Quatre Elements dans la Culture Médiévale* (Goppingen: Kummerle, 1983); and Francesca Rigotti and Pierangelo Schiera, eds., *Aria, terra, acqua, fuoco: I quattro elementi el le loro metafore* (Bologna: Società editrice il Mulinon, 1996). McKeon posed the question abstractly; see Richard McKeon, "Medicine and Philosophy in the Eleventh and Twelfth Centuries: The Problem of the Elements." *Thomist* 24 (1961): 211–256. Finally, see Susan Warrener Smith, "Bernard of Clairvaux and the Natural Realm: Images Related to the Four Elements," *Cistercian Studies Quarterly* 31, no. 1 (1996): 3–19.

5. For instance, Barbara Maurmann used Hildegard's *LDO* and *Scivias* in conjunction with Honorius against a background of Greek philosophy to look at the traditional relationships between the four elements and the four directions; see Maurmann, *Die Himmelsrichtungen.* Schipperges does look at the elements in *Causes and Cures* but glosses the material in *Heilkunde* with Hildegard's much later, visionary *LDO.* For a discussion of Hildegard's elements, qualities, and humors, see Liebeschütz, *Das allegorische Weltbild,* 99–106.

6. "Et elementa mundi deus fecit, et ipsa in homine sunt, et homo cum illis operatur. Nam ignis, aer, aqua, terra sunt, et hec quatuor elementa sibi ita intricata et coniuncta sunt, ut nullum ab alio separari possit. . . ." CC, I, 22:6–8.

7. "Nam deus mundum quatuor elementis colligauit, ita quod nullum ab alio separari potest, quoniam mundus subsistere nequiret, si aliud ab alio separari ualeret. Sed indissolubiliter sibi concatenata sunt." CC, II, 68:1–3.

8. "Ignis enim superat ac domat et incendit aerem et fortior illo est. Aer uero proximus igni ut fullo facit eum flagrare et temperat eum, quoniam ignis quasi corpus aeris est atque aer quasi uiscera et uelud ale et penne ignis." CC, II, 68:3–6.

9. "Et ut mundus in prosperitate est, cum elementa bene et ordinate officia sua exercent, ita quod calor, ros et pluuia singillatim et moderate in tempore suo se diuidunt ac descendunt ad temperiem terrae et fructuum et multum fructum et sanitatem afferunt, quoniam, si simul et repente ac non in tempore suo super terram caderent, terra discinderetur ac fructus eius et sanitas interiret; sic etiam cum elementa ordinate in homine operantur, eum conseruant et sanum reddunt; sed cum in eo discordant, eum infirmum faciunt et occidunt." CC, II, 83:16–23.

10. The customary rendering of *anima* as "soul" is misleading in a medical context. It means, rather, that which "animates," that is, moves, the body. *Anima* is that invisible presence that causes the lungs to breathe, the pulse to beat, and the limbs to move, and in whose absence the body dis-integrates.

11. "Aer autem cum quatuor uiribus suis, ut predictum est, in spiramine et rationalitate hominis est. Ipse enim uiuenti spiramine, quod anima est, in homine ministrat, que eum portat, ac ala uolatus eius est, ubi homo spiramen in se inducit et educit, ut uiuere possit. . . . Aer etiam rorem in emissione, uiriditatem in excitatione, flatum in motione, calorem in dilatatione hominis ostendit." CC, II, 73:10–74:1, 74:2–4.

12. "Adam enim cum terra fuit, ignis eum excitauit, et aer eum suscitauit, et aqua eum perfudit, quod totus mouebatur." CC, II, 75:24–25.

13. "Vnde etiam diuersos febres a diuersis qualitatibus aeris et ceterorum elementorum in se habet, scilicet a calore, a frigore et ab humore. . . ." CC, II, 201:22–24.

14. "Ignis enim, aer, aqua et terra in homine sunt, et ex hiis constat. Nam ex igne habet calorem, ex aere alitum, ex aqua sanguinem et ex terra carnem. . . ." CC, II, 83:12–16.

15. "Tunc et stellae de eodem calore aerem calefaciunt et roborant, et aer calefactus sudorem suum, scilicet rorem suum, super terram mittit et eam fecundat." CC, I, 29:1–3.

16. "Uolucres autem et animalia et bestie, que in usus hominum uenire possunt et sunt, uitam de aere secundum dispositionem dei sumunt, unde super terram uersantur." CC, II, 62:1–3.

17. This is from Pliny as quoted by Barbara Obrist, "Wind Diagrams and Medieval Cosmology," *Speculum* 72 (1) (1997), 75.

18. "Sed et aer est uentus. . . ." CC, II, 69:5. This was, apparently, an ancient conception; for instance, Honorius gives "air is the ground substance of the wind" (quoted in Maurmann, *Die Himmelsrichtungen*, 27).

19. "In comparison with other domains of the sublunary world—above all, the four elements and seasons, which might be expected to have provoked equal diverse treatment—the winds received the most developed attention" (Obrist, "Winds Diagrams and Medieval Cosmology," 34).

20. Other scholars have examined the winds in Hildegard but mainly in the context of her visionary material. For instance, Dronke has a fascinating discussion of the winds in the *LDO*; and Maurmann looks at them in *Scivias* and the *LDO*, focusing on the idiosyncratic link Hildegard made between animals, planets, and winds. Schipperges also discusses the winds in *Heilkunde*, 71–72; and Hildegard's winds are presented in Obrist, "Wind Diagrams and Medieval Cosmology," 76–78; and Thomas Raft, "Die Ikonographie der mittelalterlichen Windpersonifikationen," *Aachener Kunstblätter* 48 (1978–79), 157–159.

21. "Quatuor etiam cardinales uenti sub sole et super solem firmamento assunt et illud continent ac totum orbem scilicet de inferiori parte usque ad superiorem partem firmamenti sicut cum pallio circumdant. Orientalis enim uentus aerem amplectitur et suauissimum rorem super arida mittit. Uentus enim occidentis fluentibus nubibus se admiscet, ut aquas sustineat, ne erumpant. Australis uero uentus ignem in magistratione sua retinet et eum prohibet, ne omnia conburat. Uentus quoque septentrionalis exteriores tenebras retinet, ne modum suum excedant. Isti quatuor uenti ale potentie dei sunt. Qui cum simul mouebuntur, omnia elementa conplicabunt et se diuident ac mare concutient et omnes aquas exsiccabunt." CC, I, 24:11–20.

22. Since all of the medieval planets—sun, moon, Mercury, Venus, Mars, Jupiter, and Saturn—are conjunct, that is, clustered together, in the east at the moment illustrated in *Scivias*, what is probably portrayed here is the moment of (astrological) creation, when all of the planets were created at once.

23. Liebeschütz was the first to emphasize the originality of Hildegard's winds in *Das allegorische Weltbild*, 61–186. See also Maurmann, *Die Himmelsrichtungen*, 38–73 and Dronke, *LDO*, xxxiii. On the traditional iconography that represented the winds as three-headed male figures, see Raff, *Die Ikonographie der mittelalterlichen Windpersonifikationen.*

24. "Et unusquisque uentus istorum principalium duos alios debiliores sibi astantes uelud duo brachia habet, in quos interdum quasdam uires suas expirat. Isti enim debiliores uenti eandem naturam habent, quam et principales uenti eorum tenent, ita ut unusquisque inferiorum uentorum principalem uentum uelud caput suum imitetur, quanto multo minores

uires habeant et quod unam uiam cum principali uento suo habent, ut due aures unam uiam auditus in capite tenent. Et cum per diuinam iussionem ultorem excitantur, flatum et fortitudinem a principalibus uentis suis accipiunt, et tunc in tali inquietudine sunt et tales ac tantos fragores et collisiones periculorum faciunt, qualiter etiam pericula inquietudinum mali humores in hominibus faciunt, cum eos in egritudinem deiciunt." CC, I, 25:16–26.

25. "Nam ut anima totum corpus hominis tenet, ita etiam uenti totum firmamentum continent, ne corrumpatur, et inuisibiles sunt, uelud etiam anima inuisibilis est de secreto dei ueniens. Et sicut domus absque angularibus lapidibus non constat, sic nec firmamentum nec terra nec abyssus nec totus mundus cum omnibus compositionibus suis absque hiis uentis esset, quia hec omnia cum ipsis conposita et retenta sunt. Nam terra tota scinderetur et rumperetur, si isti uenti non essent, quemadmodum etiam homo totus scinderetur, si ossa non haberet. Principalis enim uentus orientalis totam orientalem plagam continet, et principalis uentus occidentalis totam plagam occidentalem, ac principalis uentus australis totam plagam australem, atque principalis uentus septentrionalis totam plagam septentrionalem." CC, I, 25:5–16.

26. Dronke uses this passage to argue for Hildegard's acquaintance with the obscure writer, Nimrod (or Nemroth), but in view of her practical acquaintance with the body and with the constant building on Disibodenberg and Rupertsberg, one wonders whether that literary connection is necessary. But see Peter Dronke, "Liber Nemroth," in *Dante e le tradizioni latine medievali* (Bologna: Il Mulino, 1990), 179–187.

27. "Sed illi principales uenti ab initio mundi numquam in uiribus suis pleniter moti sunt nec mouebuntur usque in nouissimum diem. Et cum tunc fortitudinem suam ostenderint et flatus suos tunc pleniter emiserint, pre fortitudine et collisione eorum nubes scindentur et superiora firmamenti conplicabuntur et rumpentur, uelud corpus hominis scinditur et omnia membra eius deiciuntur, cum anima ipsius de corpore suo exeundo soluitur." CC, I, 26:6–12.

28. Hildegard's contemporary, William of Conches, saw the winds in very much the same way, as keepers of order at the four corners of the universe. For him they represented the powers of universal cohesion; the world survived its constant change only because of the winds, perfectly-compensated balance. The same notion appears in Chinese medical cosmology; see Shigehisa Kuriyama, "The Imagination of Winds and the Development of the Chinese Concept of the Body," in *Body, Subject, and Power in China*, edited by Angelo Zito (Chicago: University of Chicago Press, 1995), 35–36.

29. The idea that on the Last Day all the winds would blow at once was conventional; see Raff, *Die Ikonographie der mittelalterlichen Windpersonifikationen*, 97.

30. "Uentus enim orientalis duas alas habet, per quas omnem orbem ad se trahit, ita ut ala una a superioribus ad inferiora solis cursum contineat et ala altera soli occurrat, ita ut ei obstaculum sit, ne ibi ulterius procedat, quo ultra non tendit. Et uentus iste omnem humiditatem humectat et omnia germina germinare facit." CC, I, 26:12–17.

31. This inference is confirmed in Vision Three of the *LDO*, in which Hildegard sees the east and south winds moving the earth from east to west, and the west and north winds moving it back from west to east. "From the winter to the summer solstice, the south winds lift the firmament northward, and then the north winds push it south again until the winter solstice." Quoted by Dronke, in *LDO*, xlvi. This key passage will be examined in depth in Chapter Six.

32. "Ventus autem occidentalis uelud os ad omnes aquas dissipandum et dispergendum habet, ita ut omnes aquas in recta itinera diuidat et dispergat, ne ulla aqua supra aliam ascendat, sed ut recte incedat, quia potestatem habet super aerem illum, qui aquas portat. Et uentus iste omnia uirida arefacit et ea, que iuxta eum sunt." CC, I, 26:17–21.

33. "Sed uentus australis uelud fereum baculum tenet, qui superius tres ramos habet et qui inferius acutus est. Nam quasi calibineam fortitudinem habet, que firmamentum et abyssum continet. Sicut enim calips omnia eramenta superat et domat, et ut cor hominem confortat, ita et fortitudo ueni huius firmamentum et abyssum plage illius continet, ne dilabatur. Et superius tres uires quasi tres ramos habet, quarum una calorem solis in oriente temperat et una ardorem eius in meridie deprimit et una calorem illius in occidente refrigerat, ne modos suos in plagis illis excedat. Sed inferius acutus est, quod fortitudo eius etiam in abissum fixa est, ne humiditas et frigus de abysso supra modum sursum ascendant. Et idem uentus omnia matura facit, ita ut folia siluarum et gramina, segetes, poma, uinum et omnes reliquos terrae fructus ad maturitatem producat." CC, I, 26:21–27:7.

34. "Uentus uero septentrionalis quatuor columpnas habet, per quas totum firmamentum et totum abyssum continet. Sed cum eas sursum traxerit, tunc firmamentum ad abyssum conplicabitur. Hee autem quatuor columpnae quatuor elementa tenent, que ibi in eadem septentrionali plaga simul conglutinantur et finiuntur et etiam uelud columpnis innixa sunt, ne cadant. Cum uero idem uentus fortitudine sua quatuor columpnas suas in nouissimo die commouerit, tunc etiam firmamentum complicabitur, uelud tabulae conplicari solent. Et uentus iste frigidus est et frigus affert et omnia frigore constringit et simul tenet, ne dilabantur." CC, I, 27:7–16.

35. Maurmann argued, instead, that east and south were connected with warmth and light, and west and north with darkness and cold because of the daily experience of sunrise (in the east) and sunset (in the west). See Maurmann, *Die Himmelsrichtungen*, 202.

36. In Ayurvedic medicine, the internal wind and the external winds play a remarkably similar role. Just as the winds temper the two predominant climate regions of *jangala* and *anupa*, so the internal wind mediates between the hot-dry humor of bile and the cold-wet humor of phlegm; see Francis Zimmermann, *The Jungle and the Aroma of Meats: An Ecological Theme in Hindu Medicine*, trans. Janet Lloyd (Berkeley: University of California Press, 1987) 162.

37. For material on the history of wind, see Jean DeBlieu, *Wind: How the Flow of Air has Shaped Life, Myth and the Land* (Boston: Houghton Mifflin, 1998); Joseph Needham, "Winds," in vol. 3 of *Science and Civilization in China*, eds. Joseph Needham (Cambridge: Cambridge University Press, 1959), 477–479; Lyell Watson, *Heaven's Breath: a Natural History of the Wind* (Great Britain: Hodden and Stoughton, 1984). On the cross-cultural symbolism of winds, see Jean Chevalier and Alain Gheerbrant, *Dictionnaire des symboles* (Paris: Seghers, 1978), 997–998. The connection between air as an element and the winds has also been noted in the Koran by Toëlle, where "air is most often represented by the winds." See Heidi Toëlle, "Vents, pluies liquides et pluies solides dans le Coran," in *Le temps qu'il fait au Moyen Age: Phénomènes atmospheriques dans la littérature, la pensée scientifique et religieuse*, eds. Claude Thomasset and Joëlle Ducos (Paris: Presses de l'Université de Paris-Sorbonne, 1998), 194. Toëlle's article is also important for the clear way in which she parses the influence of regional and local winds on a wind schema and reviews the observational role of the winds in navigation and agriculture. For a revised dissertation that exhaustively covers the iconography of the winds in myth, as natural phenomena, as symbols, in cosmology, in ancient and medieval periods; in the Bible, in encyclopedias, cosmologies, and as wind personifications, angels, and devils, see Thomas Raff, *Die Ikonographie der mittelalterlichen Windpersonifikationen*.

38. On the medieval tradition of the winds, see Obrist, "Wind Diagrams and Medieval Cosmology," and Obrist, "Le diagramme isidorien de l'année et des saisons: son contenu physique et les représentations figuratives," *Mélanges de l'Ecole Française de Rome: Moyen Age* 108, no. 1 (1996): 95–164. For a brief review of wind diagrams, see John E. Murdoch, *Album of Science: Antiquity and the Middle Ages* (New York: Charles Scribner's Sons, 1984), 343–345. According to Obrist, the Greco-Roman tradition entered the Middle Ages "by way of what appear to be variants of the lost *De natura rerum* of Suetonius, and based on Varro, two of which variants were poetic versions and the third, Isidore's chapter 37 of the *DNR*" (Obrist, "Wind Diagrams and Medieval Cosmology," 39). Before the twelfth century, this chapter, along with Isidore's *Etymologies* 13.11 and the two wind poems, formed the core of meteorological knowledge along with a series of wind diagrams (Obrist, "Wind Diagrams and Medieval Cosmology," 41). Hildegard may also have had access either to the *Cosmology* of Nimrod,

which had a small section on winds, or to some lost common intermediary. On Nimrod, see Dronke, "Liber Nemroth," 179–187; see also S. J. Livesey and R. H. Rouse, "Nemrot the Astronomer," *Traditio* 37 (1981): 203–266.

39. For the poems, see Franciscus Buecheler and Alexander Riese, eds., *Anthologia Latina: sive poesis latinae supplementum* (Leipzig: Teubner, 1894–1926) #484, 6–8; and Obrist, "Wind Diagrams and Medieval Cosmology," fn. 39.

40. Neither of the Greek poems were available c. 1150 (Ducos, "Météorologie," 14). For a review of the early Greek tradition, see the introduction to Theophrastus, *De ventis*, trans. Victor Courant and Val L. Eichenlaub (Notre Dame: Notre Dame University Press, 1975). Galen did incorporate material on the winds into his commentary on Hippocrates (Kühn, v.16:394, ff.), but his text was not available until after the time period we are concerned with here; see G. Kaibel, "Antike Windrosen," *Hermes* 20 (1885): 579–624. On the winds in the ancient world, see Kora Neuser, *Anemoi: Studien zur Darstellung der Winde und Windgottheiten in der Antike* (Rome: G. Bretschneider, 1982); in Rome, see Henry S. Robinson, "The Tower of the Winds and the Roman Marketplace," *American Journal of Archaeology* 47 (1943): 291–305; on the relationship between the winds and animals, see C. Zirkle, "Animals Impregnated by the Wind," *Isis* 25 (1936): 95–130.

41. Two authors contemporary with Hildegard, Honorius Augustodunensis and William of Conches, also included versions of an abbreviated wind schema that followed Isidore. For example, Honorius defined the air as that empty stuff which can be seen between the earth and the moon, from which the vital spirit is taken, and in which birds fly like fish swim in the water. ("Aer est omne quod inani simile, a terra usque ad lunam conspicitur, de quo vitalis spiritus hauritur. Et quia est humidus, ideo in eo volant aves, ut in aqua natant pisces.") As for the wind, it is simply moving air. There are four cardinal winds, situated at the four cardinal directions, each with two helpers. Each is associated with two qualities, and with a certain weather that each brings. Thus "the third cardinal wind is Auster, also called Notus, and it brings moist heat and storms. On its right is Euroaster, which brings heat." ("Tertius cardinalis Auster, qui et Notus, humorem calorem atque fulmina gignens. Hujus dexter Euroauster, calidus.") See Honorius Augustodunensis, "Cap. 53, De aere," "Cap. 54, De ventis," and "Cap. 55, De cardinalibus ventis," in *De imagine mundi*, PL172, col. 136.

42. "Ventus est aer commotus et agitatus. . . . Quod etiam in loco tranquillisimo et ab omnibus ventis quieto brevi flabello adprobari potest, quo etiam muscas abigentes aerem commovemus flatumque sentimus." Jacques Fontaine, ed., *Isidore de Seville, Traité de la Nature* (Bordeaux: Feret et Fils, 1960), 293:1–6.

43. " . . . occultiore quodam motu caelestium vel terrenorum corporum. . . ." Fontaine, *Traité de la Nature*, 293:7–8.

44. "Fructus germen concipiat aestiusque ardor terre sumat." Fontaine, *Traité de la Nature*, 293:14.

45. My summary of Fontaine, *Traité de la Nature*, 295–299.

46. "Ventorum primus cardinalis Septentrio, frigus et nivalis; flat rectus ab axe et facit arida frigora et siccas nubes . . . quartus cardinalis Zephyrus qui et Favonius, ab occidente interiore flat; iste hiemis rigorem gratissima vice relaxat, flores producit." Fontaine, *Traité de la Nature*, Chapter 37, and sections 1 and e.

47. Again, east is at the upper left corner. It is notable that Fontaine has mislabeled the diagram with north at the top; see *Traité de la Nature*, 297.

48. Exactly the same derivation is given for the eight Chinese winds; see "The Concepts of Wind and Ch'i" in Paul Unschuld, *Medicine in China: A History of Ideas* (Berkeley: University of California Press, 1985), 67–73.

49. The twelve-wind scheme only gradually took over, and never completely. For example, there were only eight winds in the thirteenth-century rose window of Lausanne and the earliest weathervanes had only eight directions (Robinson, "The Tower of the Winds").

50. "Et dum aspiceret ad celum, vidit illud volvens et inclinatum in sua rotunditate, exponens tunc illi duos cardines, unum a septentrione super terram, et alterum a parte meridiana sumptus terra. . . ." Quoted in Dronke, "Liber Nemroth," 181.

51. This is my interpretation. Thompson follows Aristotle in stating that north and south were mathematically generated by a perpendicular, but support for my interpretation can be found in Hildegard's own later explanation in the *LDO* (Third Vision, Part I). "From the winter to the summer solstice the south wind and its allies lift the firmament northward; then the north wind and its allies push the firmament southward again, till the next winter solstice." Dronke's translation in Dronke, *LDO*, xlvi.

52. In the following I use the earliest Latin translation in conjunction with the Greek text and English translation of Jones, and Jouanna's new edition. For the Latin, see H. Kühlewein, "Die Schrift 'Peri Aeron Udatron Topon' in der lateinischen Ubersetzung des Cod. Paris 7027," *Hermes Zeitschrift für klassische Philologie* 40 (1905): 248–74. For Greek and English, see W. H. S. Jones, ed., *Hippocrates' Airs, Waters, and Places*, vol. 1 (Cambridge: Harvard University Press, 1984(1923). For a new Greek edition, see Jacques Jouanna, ed., *Hippocrate Tome II, Airs, Eaux, Lieux* (Paris: Belles Lettres, 1996).

53. This is Jones' translation; see Jones, *Airs, Waters, Places*, 71. The medieval Latin gives, "Flatus ventorum calidos et frigidos, maxime *communes* omnibus; deinde et *singulis* provinciis regionales qui sint. . . ." Kühlewein, "Die Schrift 'Peri Aeron Udatron Topon,'" 254.

54. The Latin states, "Quae in terra gignuntur omnia consequentia terra." Kühlewein, "Die Schrift 'Peri Aeron Udatron Topon,'" 274.

55. The Latin is, "Regio in qu(s) nutriuntur et aquae invenies regiones conse-
quentes, hominum et mores." Kühlewein, "Die Schrift 'Peri Aeron Udatron
Topon,'" 274.

56. My summary of the material in Kühlewein, "Die Schrift 'Peri Aeron Udatron Topon'"; for an English translation, see Jones, *Airs, Waters, Places*, 72–
83.

57. Jones, *Airs, Waters, Places*, 77 and Kühlewein, "Die Schrift 'Peri Aeron Udatron Topon,'" 256. In Latin the term "bilious" is absent.

58. The Latin is "homines vero capite humida habere et fleumatica." Kühlewein, "Die Schrift 'Peri Aeron Udatron Topon,'" 255.

59. Four ancient agricultural treatises survived into the Middle Ages, the latest
and most popular (as defined by numbers of surviving manuscripts) was
the fourth-century Palladius. He was followed in popularity by Columella,
Varro, and Cato. For editions, see Raoul Goujarol, ed., *Caton de l'agriculture*
(Paris: Belles Lettres, 1975); Jacques Heurgon, ed., *Varron Economie Rurale*,
vol. 1 (Paris: Belles Lettres, 1978); Lucius Junius Moderatus Columella,
On Agriculture, with a Recension of the Text and an English Translation, ed.
Harrison Boyd Ash, vol. 3, *Loeb Classics* (Cambridge, Mass.: Harvard Uni-
versity Press, 1941–1955); Robert H. Rodgers, ed., *Palladis Rutilis Tauri
Ameiliani viri inlustrus opus agriculturae* (Leipzig: Teubner, 1975). For
later authors with notions about wind, see Roderich König, ed., *C. Plinius
Secundus Naturkunde* (Munich: Artemis and Winkler, 1973). For the late
medieval authors, Pietro da Crescenzi and Godfrey, see Will Richter, ed.,
*Petrus de Crescentiis (Pier de' Crescenzi). Ruralia Commoda. Das Wissen des
volkommenen Landwirts um 1300*, 3 vols. (Heidelberg: Universitätsverlag
C. Winter, 1995); and William C. Crossgrove, "Medicine in the Twelve
Books on Rural Practices of Petrus de Crescentiis," in *Manuscript Sources for
Medieval Medicine*, ed. Margaret R. Schleissner (New York: Garland Publishing Inc., 1995), 81–103; W. L. Braekman, "Bollard's Middle English
Book of Planting and Grafting and its Background," *Studia Neophilologica*
57, no. 1 (1985): 19–39. For general material on the practice of agricul-
ture, see René Martin, *Recherches sur les agronoms latins et leurs conceptions
économiques et sociales* (Paris: Belles Lettres, 1971); René Martin, ed., *Palladius Traité d'Agriculture* (Paris: Belles Lettres, 1976); Bertha Tilly, *Varro the
Farmer. A Selection from the Res rusticae* (London: University Tutorial Press,
1973); Mauro Ambrosoli, *The Wild and the Sown: Botany and Agriculture in
Western Europe, 1350–1850*, trans. Mary McCann Salvatorelli (Cambridge:
Cambridge University Press, 1997).

60. That there were local winds "peculiar to each particular region, just like its
waters and its earth," which also affected the body was also well understood;
see Jones, *Airs, Waters, Places*, 71.

61. The Koran, too, takes note of many winds and their importance for farming, as well as navigation; see Toëlle, "Vents, pluits liquides," 195–197.

62. " . . . Sed hoc flante ne arato frugem ne serito, semen ni iacito, praestringit enim atque praegelat hic radices arborum. . . ." König, *C. Plinius Secundus naturkunde*, Book 18, p. 200: Section 334.

63. "Now the third [wind] from the northern line [that is, the east wind] is the equinoctial wind; it brings gentle rain, drier than Favonius. Cato says that olive trees should then be planted. The vines should be pruned, fruits cared for, trees tapped, fruit trees planted." ("Tertia a septentrione linia, quam per latitudinem umbrae duximus et decumanam vocavimus, exortum habebit aequinoctialem ventumque subsolanum, Graecis aphelioten dictum. In hunc salubribus locis villae vineaeque spectent . . . et siccior favonius. . . . In hunc spectare oliveta Cato iussit. Hic ver inchoat aperitque terras tenui frigore saluber, hic vites putandi frugesque curandi arbores serendi poma inserendi oleas tractandi ius dabit adflatuque nutricium exercebit.") König, *C. Plinius Secundus Naturkunde*, Book 18, p. 202: Section 337.

64. "And they don othir provys there-to whyle the wynde is in the northe, ffore thenne wynys wul be freyishe and fayre. And thenne thei wul kepe lengest there colowre. But the lyas of wyn, looke and taste them and preve hem whyle the wynde is in the southe. For thenne most lyghtlly and sunnest they shew yf they have febulnesse." Braekman, "Bollard's Middle English Book of Planting and Grafting and its Background," 323:511–518. This is not because of any specific characteristics of wind or winds, but simply because wind transmitted the four qualities: "Suche harmys þat fallyn to wyn thorw wyndys and sodeyn chaungeings of wedirs must be shewd . . . thee wynde chaungithe hote and cold. . . ." Braekman, "Bollard's Middle English Book of Planting and Grafting and its Background," 324:539–541 and 544–545.

65. William Crossgrove teases these connections out of the later work of Petrus de Crescentiis; see William C. Crossgrove, "Das landwirtschaftliche Handbuch von Petrus de Crescentiis in der deutschen Fassung des Bruder Franciscus," *Sudhoffs Archiv* 78, no. 1 (1994): 98–106; and Crossgrove, "Medicine in the Twelve Books on Rural Practices of Petrus de Crescentiis," 81–103.

66. On Asaph, see Simon, *Asaph Ha-Jehudi*; Shatzmiller, "Doctors and Medical Practice in Germany around the Year 1200: The Evidence of 'Sefer Asaph'"; Newmeyer, "Asaph's Book of Remedies"; and Lieber, "Asaf's 'Book of Medicines.'" Since it is still not translated in full, I have depended on the two volume summary of Venentianer, and the partial English version of Bar-Sela; see Venentianer, *Asaf Judaeus*; and Bar-Sela, "Asaf on Anatomy and Physiology."

67. "Wenn der Charakter der Jahreszeiten und der herrschenden Winde mit dem Elementen-Charakter des Körpers nicht harmoniert. . . ." Venantianer, *Asaf Judaeus*, 91. "Unter Rücksicht auf die herrschenden atmosphärischen

Einflüsse, zur Wahl der Heilmittel bestimmen. . . ." Venantianer, *Asaf Judaeus*, 117.

68. "Denn der Ostwind ist warm und trocken, der Westwind kalt und feucht; der Südwind warm und feucht, der Nordwind kalt und trocken." Venantianer, *Asaf Judaeus*, 78.

69. "A very strong, hot wind." ("Uehemens calidus uentus." CC, II, 97:10). Also, "a fiery and airy wind in the loins mixed with the smoke of black bile." ("Et uentus, qui in lumbis eorum est, in tribus modis exsistit, ita quod igneus est et uentosus ac fumo melancolie permixtus. . . ." CC, II, 110:10–12).

70. "A wind in the loins more fiery than windy." ("Uentus quoque, qui in lumbis eorum est, magis igneus quam uentosus est. . . ." CC, II, 106:21–22.)

71. "A burning wind that surges out of the marrow and moves a person to plea-sure." ("Sed prefatus ardens uentus, qui ad delectationem carnis ex medulla hominis surgit. . . ." CC, II, 182:6–7.)

72. "A squalid, dry wind that dries the dew." ("Quemadmodum etiam squalidus et aridus uentus uires roris inminuit, ita quod aptam humiditatem calori solis subministrare non ualet. . . ." CC, II, 181:12–14.)

73. "A cold winter wind that dries up the leaves and branches of trees." ("Et ut frigidus uentus et gelu et hyemps folia et ramos arborum arefacit." CC, II, 147:12–13.)

74. "A great wind in the loins that tempers heat" [and is, therefore, cool]. ("Plurimus uentus, qui in femoribus eorum est, ignem in eis compescit et temperat." CC, II, 109:2–3.)

75. "A sweet and gentle wind." ("Nam medulla uigiliis attenuata et debilitata mox uires anime suauissimum et dulcissimum uentum ex medulla produ-cunt." CC, II, 120:19–20.)

76. "A wind that brings forth seeds." ("Ut uentus et ros gramina educunt." CC, II, 153:6–7.) Because it brings forth seeds, I infer that it must be a spring or summer wind.

77. "A wind that makes people happy, even laugh out loud." ("Ventus . . . uenas eiusdem splenis inplet et se ad cor extendit et iecur replet ac ita homi-nem ridere facit atque uocem eius similem uoci pecorum in caccynnis edu-cit." CC, II, 189: 10–13.)

78. "Quod si calorem solis non habuerit, in lenem uentum aut ad lenem auram ponat, ut suauiter exsicccetur." CC, IV, 234:9–10.

79. "A strong and cooling wind." ("Sed ualidus uentus de stomacho uelud aer exiens ignem medulle aliquantum infrigidat. . . . CC, II, 180:22–23.) Also, a "cooling, balancing wind" ("Quia uentus et ignis in femoribus parentum eorum recte ipsos temperauit; quoniam ignis excedendo uentum non super-auit, sed uentus ignem temperauit." CC, II, 109: 21–23.)

80. "A wind that blows the dust away." ("Uelud uentus, qui puluerem disper-git." CC, VI, 293:17.)

81. "A dangerous wind that clashes and brings storms," ("Uelud periculosus uentus efflauerint, quasi in conmotionem uentorum uertuntur atque peri-culosum sonum faciunt uelud sonum tonitrui." CC, II, 87:10–12.)

82. "A powerful wind that brings storms." ("Sed quemadmodum ualidus uentus tempestatem in flumine mouet. . . . CC, II, 143:13.) And "a strong wind that brings storms." ("Et auditus eorum uelud ualidissimus uentus cum eam audierint, et cogitationes eorum quasi procella tempestatum. . . ." CC II, 107:13–15.) "A very powerful wind [of hearing]." ("Et ubi auditus aliorum quasi ualidissimus uentus ad ipsas sunt, ibi auditus istorum uelud sonum cythare habent, et ubi cogitationes aliorum quasi procella sunt, ibi isti pru-dentes amatores in honorificentia uocantur." CC, II, 109:13–16.)

83. "Like a strong wind that injures plants and fruit." (". . . ut ualidus uentus, qui omnibus herbis et fructibus inutilis est." CC, II, 66:13–14.)

84. "Sunt praeterea quidam innumerabiles ex fluminibus aut stagnis aut finibus nominati. . . ." Fontaine, *Traité de la Nature*, 299.

85. For instance, Swiss newspapers supply wind proverbs along with satellite maps; and even Los Angeles has its "Santa Ana," a hot, dry wind from the east that makes people agitated and restless.

86. Thus, when interviewed, a sailor on Lake Geneva was able to identify five distinct winds, each of which he spontaneously identified with a direction, a weather, a quality and a disease-state. (The ever-uncertain relationship between experience and textuality is highlighted by the fact that the sailor's winds number only five, while the official, written version regarding Lake Geneva's winds display the customary set of twelve.) He told me that there was the south wind, called the Föhn; and it was hot and dry and caused headaches. There was the north wind, called the Bise, which was cold and dry and brought good weather. The northeast wind brought gales, and there was a "white wind" from the southwest which was tepid, humid, and tem-pestuous. Finally, there was the Vaudère, which was strong and unpredict-able and agitated the lake. He also explained that the mechanism of their effect was that the winds brought the weather from the localities through which they blew. Thus, the north wind brought cold, dry weather from England, and the Föhn brought hot Italian weather.

87. On the place of wind and winds in contemporary German folk culture, see E. Hoffmann-Krayer and Hanns Bächtold-Stäubli, *Handwörterbuch des deutschen Aberglaubens* (Berlin: Walter De Gruyter and Co., 1938/1941), s.v. "wind."

88. Toëlle notes this same phenomenon for Arabic culture and the winds. "Le Coran en fait ne constitue pas un calque pur et simple de la réalité cli-matique observable. Cette description s'accompagne, au contraire, d'une

schématisation et d'une idéalisation qui vise à inserer les données observées dans une cosmologie." Toëlle, "Vents, pluies liquides," 193.

89. "Et iste ardens uentus interdum in otioso et nulla solicitudine occupato homine surgit et in pectus eius flat et hominem illum aliquantum letum facit et ita de pectore eius in cerebrum ascendit et illud totum ac uenas eius ardente calore implet, atque tunc etiam pulmonem et cor tangit et sic ad genitalia loca uadit, in uiro in lumbos in muliere in umbilicum." CC, II, 180:9–13.

90. "Et cum sic homo aut de bonis aut de malis, que sibi placent, letatur, tunc predictus uentus interdum ex medulla exiens femur illius primum tangit et ita splen occupat atque uenas eiusdem splenis implet et se ad cor extendit et iecur replet ac ita hominem ridere facit. . . ." CC, II, 189:8–12.

91. This is not to suggest that Hildegard had any knowledge of *De ventis*, which was a rhetorical exericise, not a medical text, written in the third century B.C. and translated into Latin only in the fifteenth century. *De ventis* does, however, express ideas that were incorporated into medicine very early on, and did survive, sotto voce, in such techniques as *ventosa*. For an edition, see Jacques Jouanna, ed., *Hippocrate, Des vents, De l'art* (Paris: Les Belles Lettres, 1988). For a review of the textual tradition, see Jouanna, "Notice," in *Hippocrate, Des vents*, 11–98.

92. This is Jouanna's translation; "Le souffle (*pneuma*) à l'extérieur des corps s'appelle air (*aer*) et à l'interieur des corps, vent (*phusa*)"; see Jouanna, *Des vents*, 12, from Part I, ch. 3–5.

93. "Puis il montre que l'air est cause pour les êtres vivants de la vie (c. 4) et des maladies (c. 5)." Jouanna, *Des vents*, 12. "Conclusion (c. 15): l'air est la cause principale des maladies; tout le reste est secondaire." Jouanna, *Des vents*, 13. This is Jouanna's summary of *De ventis*.

94. See the series of passages explaining the different kinds of diseases as a result of problems with air and wind in Jouanna, *Des vents*, 115–124.

95. My summary of Bar-Sela, *Asaf on Anatomy and Physiology*, 379–380, an English translation, not the Hebrew original. Asaph's winds are related to the five internal winds of Ayurvedic medicine—*apana vayu*, located in the rectum and urethra; *samana vaya* in the stomach, responsible for the activity of digestion; *udana vayu*, in the throat, responsible for speech and singing; and, finally, *beana vayu*, in fluids, especially in the movement of the blood that is the pulse. See Thomas Alexander Wise, *Commentary on the Hindu System of Medicine* (Amsterdam: Oriental Press, 1901 [1860]), 43–44.

96. *Ventosa* is known in English as "cupping" and is surprisingly cross-cultural; an identical technique can be found in Chinese, Asian, and Ayurvedic cultures. It has changed little over the centuries. Thus the description in Dechambre, published in 1886 does not differ from that of Celsus, Albucasis, Hildegard, or, for that matter, from what a traditional Chinese practitioner will perform today in Chinatown. Dechambre describes: "Ventosa,

de ventus, vent. Nom qui a été donné à un petit appareil destine à faire momentanement le vide sur une surface plus ou moins circonscrite du corps, de manière à y provoquer un appel fluxionnaire dans un but thérapeutique. . . . On applique le plus hermetiquement possible, par une assez forte pression, le rebord mousse de l'orifice de la *ventouse* sur la peau, de manière à intercepter toute pénétration d'air . . . pas géneralement plus de quelques minutes." See A. L. Dechambre and L. Lereboullet, *Dictionnaire Encyclopédique des Sciences Médicales* (Paris: G. Masson, 1886), 570–571.

97. Cupping is still used in folk medicine as well as in traditional Chinese medicine, where it continues to be understood as a technique for withdrawing bad wind. See Maoshing Ni, *The Yellow Emperor's Classic of Medicine* (Boston: Shambala, 1995), 117.

98. "Et qui in oculis seu in auribus seu in toto capite dolet, cornu uel *uentosam* in confinio colli et dorsi ponat. Qui uero in pectore dolet, cornu in scapulis figat, aut si in lateribus dolet, id in utroque brachio et, ubi manus finitur, ponat, uel si in cruribus dolet, illud ad ylia figat, aut si in yliis torquetur, id inter nates et poblitem ponat, id est in coxas. In locum autem in quem *uentosa* uel cornu ponitur, non plus quam ter uel quater una eadem hora, qua sanguis ibi extrahitur, ponatur." CC, II, 166:11–17.

99. "Homo, qui in cruribus et in pedibus suis podagram sentit et inde dolet, cum idem dolor recens est, ille plurima cornua uel uentosas cruribus suis circumponat, scilicet a talo incipiens absque incisione cutis, ut humores sibi attrahant, et deinde a loco illo auferat et ea superius ponat quatinus iterum ibi subteriores humores sibi attrahant. Et sic faciat cutem non inscidendo nec rumpendo, usque dum ad nates suos perueniat; postquam autem hoc modo ad nates peruenerit, ligamen in superiori parte genu sibi circumponat, ne humores, quos cornibus illuc contraxit rursus descendant, et mox in confinio dorsi et posteriorum cum cornibus aut uentosis scarificando sanginem et malos humores emittat. Sic autem faciat, et dolor podagre illius cessabit." CC, III, 226:13–23.

100. For a description, see A. Corn. Celsi, "De sanguinis detractione per cucurbitas," in *De medicina* (Studiis Societatis Bipontinae, 1786), 90–91. "Cucurbitularum vero duo genera sunt: aeneum et corneum. . . . Cornea per se corpori imponitur; deinde, ubi ea parte qua exiguum foramen est, ore *spiritus* adductus est, superque cera cavum id clausum est, aeque inhaerescit" (90). I translate *spiritus* as air; see OLD, s.v. "spiritus."

101. "Ubi inhaesit, si concisa ante scalpello cutis est, sanguinem extrahit; si integra est, spiritum." Celsus, *De medicina*, 90.

102. "Corrupta materia, sive spiritu male habente. . . ." Celsus, *De medicina*, 91.

103. Both Indian and Chinese medicine accept and use the idea that an excess of wind, or wind trapped in the wrong place, can cause disease, especially rheumatic disease. On wind in Ayurvedic medicine, see Daniel C. Tabor, "Ripe

and Unripe: Concepts of Health and Sickness in Ayurvedic Medicine," *Social Science and Medicine* 15B (1981): 446–447; and Francis Zimmermann, *Le Discours des Remèdes au Pays des Épices: Ênquete sur la Médecine Hindoue* (Paris: Payot, 1989). Such wind diseases are usually localized and treated with topological practices like bloodletting and cupping. See Kenneth Zysk, *Asceticism and Healing in Ancient India: Medicine in the Buddhist Monastery* (New York: Oxford Press, 1991), 110. A similar concept exists in Chinese medicine; see Kuriyama, "Imagination of Winds," 26.

104. "The way to apply the cupping-vessels: the vessel is first put in position empty; then you suck moderately; do not long hold the cupping-vessel in place but apply it quickly and remove it quickly so that the humours gather evenly at the place. . . ." Spink, *Albucasis on Surgery and Instruments*, 662.

105. "Utatur eger calidis et humidis et ventositatem generantibus, sicut sunt ciceres, caro recens, fabe. . . . Bartholomew, *Practica*, 401.

106. "Satiriasis est immoderata virge erectio . . . hujus autem passionis causa est sola ventositas, que nervos inflat et virgam erigit." Bartholomew, *Practica*, 401.

107. "Suffocatio matricis est compressio spiritualium ex vitio matricis stomace mediante, unde inspirandi difficultatem patiuntur . . . et precipue in viduis que viris non utuntur. In matrice enim sperma retinetur et retenta diu corrumpitur, et venenosam naturam recipit, unde humores matricis in *ventositatem* resolvit . . ." Bartholomew, *Practica*, 404.

108. Maurmann writes that wind was "the dominant and primary constituent of the microcosmic man" for Hildegard, and so he has the name, "homo ventosus" (73). The quote from the *LDO* is: "For man with his arms and legs rules and sustains himself and is *ventosus* in the same way that the four principal winds with their collaterals hold up the sky. . . ." ("Nam et homo cum brachiis et cruribus suis totum se regit et sustentat, et ventosus est, quemadmodum quatuor principales venti cum collateralibus suis omne firmamentum in illud positi tenent. . . ." *LDO*, PL197: col. 842A).

109. "Le mauvais air est des plus redoutables, car il ne cesse d'aller et venir par les mutiples pertuis du corps." Marie-Christine Pouchelle, *Corps et Chirurgie à L'apogée du Moyen Age: Savoir et Imaginaire du Corps chez Henri de Mondeville, Chirurgien de Philippe Le Bel* (Paris: Flammarion, 1983), 256.

110. "Nam firmamentum est uelud capud hominis, sol luna, et stellae ut oculi, *aer ut auditus*, uenti uelud odoratus. . . ." CC, I, 33:1–2.

111. Remarkably, in the so-called Berlin fragment—perhaps a part of *Causes and Cures*—wind acts like an intermediary and modifies the qualities of the body. "Das Berliner fragment geht diesen Zusammenhangen in breiter Form nach; für die Heilkunde ergibt sich daraus, dass die Windkräfte als Vermittler der Labilität des Säftenhaushaltes fungieren." Schipperges, *Heilkunde*, 72.

112. As summarized and translated by Dronke, *LDO*, xliv.

113. "Deinde etiam vidi quia per diversam qualitatem ventorum et aeris, cum sibi invicem concurrunt, humores qui sunt in homine commoti et inmutati qualitatem illorum suscipiunt," in *LDO*, Part I, Vision III, Third Part, 114–115:23–25, as translated by Dronke.

114. This is Dronke's summary of the passage; see *LDO*, xlvii. The full passage will be examined in more detail in the Conclusion.

NOTES TO CHAPTER FOUR

1. "Quattuor sunt venti, quattuor anguli caeli, quattuor tempora, . . . quattuor humores in humana corpora." M. Wlaschky, "'Sapientia artis medicinae.' Ein frühmittelalterliches Kompendium der Medizin," *Kyklos* 1, no. 1 (1928), 104.

2. "But birds and animals and beasts, which are used by people, take their life according to the disposition of God from the air, and live on top of the earth. Worms and reptiles take their life from the juice of earth, and so live in the earth. And fish take their life from the watery air of flowing waters, whence they live in the water. . . ." ("Volucres autem et animalia et bestie, que in usus hominum uenire possunt et sunt, uitam de aere secundum dispositionem dei sumunt, unde super terram uersantur. Vermes uero et reptilia uitam sumunt de succo terrae et ideo in terra et subtus terram libenter sunt. Pisces autem uitam accipiunt de aquoso aere fluminum, unde etiam in aquis sunt. . . .") CC, II, 62:1–6.

3. "Ex uno autem humore aut ex duobus aut ex tribus homo constare non potest, sed ex quatuor, ut inuicem temperentur, sicut mundus ex quatuor elementis constat, que sibi inuicem concordant." CC, II, 91:5–7. Also, "Thus man lives because of the four humors, just as the world is made of the four elements." ("Sic homo, ut predictum est, ex quatuor humoribus uiget, ut etiam mundus ex quatuor elementis constat.") CC, II, 67:10–11.

4. "Deus ita creauit hominem, quod omnia animalia ad seruitutem eius subiecta sunt; sed cum homo preceptum dei transgressus est, mutatus est etiam tam corpore quam mente. Nam puritas sanguinis eius in alium modum uersa est, ita quod pro puritate spumam seminis eicit. Si enim homo in paradyso mansisset, in inmutabili et perfecto statu perstitisset. Sed hec omnia post transgressionem in alium et amarum modum uersa sunt." CC, II, 59:1–8.

5. "Tunc etiam flecmata et humores in eo crescunt, secundum quod naturam seminis eius fuit, quoniam secundum quod triticum aut siligo aut ordeum seminatur, secundum hoc etiam et grana naturaliter proferunt." CC, II, 101:18–20. (For now I leave *flecmata* untranslated to make clear that, at least here, Hildegard differentiates them from humors, but places them on the same level.)

6. This is my summary of Hildegard's ten-page discussion of the four different types of men, identified in the rubrics (but not in the text) as the choleric, sanguine, phlegmatic, and melancholic (CC, II, 106–114), and the different types of women (CC, II, 126–128). Dronke discusses Hildegard's material on women in "Hildegard of Bingen," 181–183; Cadden looks at the humoral types of both men and women in *Meanings of Sex Difference in the Middle Ages: Medicine, Science and Culture* (Cambridge: Cambridge University Press, 1993), 185–186 and 205.

7. "Sed iecur est in homine quasi uasculum, in quod cor et pulmo atque stomachus succos suos effundunt, quos illud iterum in omnia menbra refundit uelut uas aliquod iuxta fontem positum, quod aquam de fonte acceptam iterum in alia loca diffundit. Sed cum iecur perforatum est et marcidum, ut predictum est, bonos succos a corde et a pulmone atque a stomacho recipere non potest, et ita succi et humores illi ad cor et ad pulmonem et ad stomachum reuertentes uelut quandam inundationem faciunt." CC, II, 137:17–24.

8. "Sepe contingit, quod tempestates et procellae malorum humorum in aliquod membrum hominis cadunt et ibi insania sua meatum sanguinis uenarum eius claudunt, ita quod ille ibi in uenis inundare non potest, unde et uenae illae arescunt, quia meatu sanguinis carent. Et sic homo ille claudicare incipit." CC, II, 201:16–20.

9. "Et sepe contingit hominibus, qui diuersis cibis utuntur, quod inde facile infirmantur. . . . Mali humores superhabundanter in eis inundant et crescunt, ita quod contineri non possunt, quin hac et illac in hominibus illis inordinate fluant et sic tandem ad inferiora descendunt et in cruribus ac in pedibus eorum insanire incipiunt. . . . Et sic homo ille in cruribus et in pedibus suis podagram sentit et dolet. . . ." CC, II 141:9–18.

10. As becomes clear in the final section of this chapter, there was a near-identity between the juice (*sucus* or *succus*) and the *humor* of plants, both of which signified sap. Here Hildegard seems to use the two words almost interchangeably in the body as well. For a later elaboration of this concept, see Piero Camporesi, *The Juice of Life: The Symbolic and Magic Significance of Blood*, trans. Robert R. Bauer (New York: Continuum, 1995).

11. "Cum homo comedit et bibit, uitalis tractus rationalitatis, qui in homine est, gustum et subtiliorem succum et odorem eorum sursum ad cerebrum eius ducit et uenulas eius replendo calefacit. Sed cetera eorundem ciborum et potuum, que in stomachum perueniunt, cor et iecur et pulmo calefaciunt, et de eodem gustu et subtiliori succo et odore in uenulas suas trahunt, ita quod inde replentur et fomentantur et nutriuntur, uelut si aridum et siccum intestinum in aquam ponitur, quod inde mollificatur et intumescit et repletur. Ita cum homo manducat et bibit, de succo ciborum et potuum uene eius replentur et fomentantur, ita quod succus ille in uenis sanguinem et tabem fomentat et de succo uenarum sanguis, qui in carne est, rubedinem sibi attrahit." CC, II, 153:31–154:7.

12. "Homo post cibum statim non dormiat, antequam gustus et succus ac odor eorum ad loca sua perueniant. Sed post cibum per aliquam breuem moram a dormitione se contineat, ne, si repente post cibum dormierit, eadem dormitio gustum et succum ac odorem ciborum in iniusta et contraria loca ducat et ne uelut puluerem hac et illac in uenis dispergat." CC, II, 154:28–32.

13. "Humores . . . de eodem corpore per sudorem expellit." CC, II, 203:8–9.

14. "Et ita de humoribus illis, qui in homine sunt, uelut amarus fumus per tristitiam egrediens circa cor spargitur. . . ." CC, II, 187:2–3.

15. "Nam cum in hiis hominibus diuersi humores flecmatum excitantur, ita quod eadem flecmata de immoderato cibo et potu ac de inepta letitia, tristitia et ira ac de immoderata libidine in eisdem hominibus concutiuntur. . . ." CC, I, 66:1–4.

16. "Nam calor agrimonie et calor galange et uirtus storacis et calor polipodii ac calor celidonie maioris frigidos humores, de quibus flecma in homine nascitur, superant, et frigiditas fenigreci ac frigiditas storkesnabeles frigus eorundem humorum dissipant." CC, IV, 234:10–14.

17. "Si autem urina hec pura et lucida est secundum aquam, morietur, quoniam sanguis eiusdem hominis in frigus uersus est, et ideo ceteri humores simul confluunt, uelut lac coagulatur, quia calorem et sanguinem non habent; unde etiam eadem urina pura et lucida est, quoniam cum humoribus mixta non est, quia illi officia sua non exercent." CC, V, 277:11–15.

18. " . . . quoniam calor et splendor solis humores cerebri in inustum modum forte educerent." CC, IV, 235:3–4.

19. "Et sic nociui humores, qui in homine sunt, commouentur et febres accrescunt. . . ." CC, II, 202:28–29.

20. "Nam qui molles carnes habent, de nimiis potationibus mali humores, qui in eis sunt, in aliquod membrum eorum repente cadunt et illud dissipant. . . ." CC, II, 201:9–10.

21. Numerous authors have discussed Hildegard's humors, including Liebeschütz, *Das allegorische Weltbild*, 97–106; Maurmann, *Himmelsrichtungen*, 57; Cannon, "The Medicine of Hildegard of Bingen," 76–115; Jacquart, "Hildegarde et la physiologie," 125–131; Berger, *Hildegard's Medicine*, 35–37; Bäumer, *Wisse die Wege*, 320–321; and Sabina Flanagan, "Hildegard and the Humors: Medieval Theories of Illness and Personality," in *Madness, Melancholy, and the Limits of the Self*, eds. Andrew Weiner and Leonard Kaplan (Madison, Wisconsin: 1996), 14–23.

22. *Spuma* usually means foam, froth or scum; see Lewis and Short, s.v. "spuma."

23. *Tepidum* is an idiosyncratic term Hildegard uses probably for the quality usually called warm or hot, although its translation has been problematic. For instance, Flanagan translates it as "cool" and Cannon as "tepid" (and equivalent to warm), so making Hildegard's humors conform to the Galenic

qualities. However, in addition to "tepid," *tepidum* has the meaning of "the warmth natural to a living body;" see OLD, s.v. "tepidum." So, perhaps it signified the "vital heat" (*calor naturalis*)—that heat that distinguishes the live from the dead body. The *calor naturalis* will be discussed in Chapter Five.

24. " . . . uelud cum aliqua olla, ad magnum ignem posita repente ebullit et spumam eicit." CC, I, 22:19–20.

25. "Et cum homines isti sic in libidine ardent, ita quod sanguis eorum frequenter inordinate concutitur, ita quod nec rectus sanguis est nec recta aqua nec recta spuma, tunc in parvum [pravum in Kaiser] liuorem et in tabem uertitur. . . ." CC, II, 200:28–70.

26. "Et hii de tribus flecmatibus, scilicet de sicco et humido atque tepido flecmate uelut aquosam spumam sibi attrahunt, que de eisdem flecmatibus oritur et que uelut periculosas sagittas in uenas et medullas ac in carnem eorum emittit, ut feruens aqua que feruentem spumam eicit." CC, II, 65:26–30.

27. "Masculus autem a quinto decimo anno etatis sue gustum delectationis in se habet, et tunc etiam spumam humani seminis de uanis cogitationibus facile in eo exsudat. . . ." CC, II, 178:12–14. Also, "Et hec omnia de nimietate superfluitatis quasi tempestatem faciunt et de sanguine uenenosam spumam, quod semen est. . . ." CC, II, 95:6–8.

28. "Puella a duodecimo aetatis sue anno gustum delectationis in se sentit, et tunc etiam de lasciuis cogitationibus spuma delectationis de illa facile exsudat, quamuis eadem delectatio nondum matura ad semen sit." CC, II, 179:1–3.

29. "Et ignis medulle hominis incendit delectationem, que gustum peccati habet, et tunc delectatio cum gustu feruorem in sanguine uelut tempestatem facit, ita quod sanguis spumam facit et spumam illam similem lacte ad cauernas genitalium in dulcedine ducit. . . ." CC, II, 177:4–8.

30. "Sed alii sunt alterius generis eiusdem morbi, ita quod inconstantes et leues in moribus suis sunt et inpatientes, quorum anima, dum in ipsis supra modum hiis moribus fatigatur, se multotiens subtrahit et succumbit, et ita corpus uelut uiribus anime subtractis in terram cadit ac sic quasi mortuum iacet, usque dum iterum anima uires suas recipit. Isti autem blandam faciem et lenes gestus habent, et dum a morbo isto deiecti in terram cadunt, interdum aliquam uocem, sed tamen lugubrem et naturalem, emittunt et multam spumam de ore eiciunt. . . ." CC, II, 196:1–8.

31. "There are four humors, two good ones called *flecmata* and the two following, *liuores*." ("Nam cum quatuor sunt humores, duo precellentes flecmata dicuntur, et duo subsequentes dicuntur liuores.") CC, II, 83:14–15. Even this apparently clear definition has left much room for opposing interpretations. For example, Flanagan and Cannon have argued that Hildegard meant that *flecmata* and *livores* were defined by their *amount* relative to the other humors—the two in greatest quantity in any particular situation

would be called *flecmata*, and the lesser two would be called *livores* (Cannon, "The Medicine of Hildegard of Bingen," 88 and Flanagan, "Hildegard and the Humors," 16–17). In contrast, Berger interprets this passage to mean that there were two *intrinsically* good humors called the *flecmata* (the dry and the wet), and two *bad* ones called *livores* (*Hildegard's Medicine*, 134).

32. "Unusquisque autem humor, qui supereminens est, proximum se subsequentem quarta parte et dimidia parte tertie partis superat; et ille, qui subiectus est, duas partes ac reliquam partem tertiae partis temperat, ne modum suum transeat. Qui enim primus humor est, hoc modo secundum superat, qui duo flecmata dicuntur; et secundus tertium, ac tertius quartum, qui duo, uidelicet tertius et quartus, liuores uocantur. Et superiores in habundantia sua minores transeunt, et minores habundantiam illorum sua uacuitate temperant. Et cum homo sic est, in tranquillitate est. Cum autem quilibet humor modum suum excedit, homo ille in periculo est." CC, II, 84:15–24. Cannon constructs a polynomial equation for Hildegard's mathematics and a pie graph for these weird mathematics; see Cannon, "The Medicine of Hildegard of Bingen," 84–86; the pie graph is on 92.

33. Hildegard's humoral ratios are "a garbled version of the cosmic proportions first set out in *Timaeus* 35b and repeatedly invoked by medical writers. . . ." (Berger, *Hildegard's Medicine*, 135).

34. The actual system of Galen as presented in *De complexionibus* was more complicated, since, for each person, there was a proper balance of qualities, not a single absolute one for the human race. See McVaugh, "The Development of Medieval Pharmaceutical Theory," 10.

35. These consequences are alluded to but not fully worked out in Pozzi Escot, "Hildegard's Christianity: An Assimilation of Pagan and Ancient Classic Tradition," in *Wisdom Which Encircles Circles, Papers on Hildegard of Bingen*, ed. Audrey Ekdahl Davidson (Kalamazoo: Medieval International Publishing, 1996), 53–61. Escot noted that "for Hildegard, especially, all must be balance, all must be ordered, and all must reflect a mathematical mean" (56).

36. Hildegard's prescribed ratios can be interpreted in a number of ways. Here I have mainly used the first line: that the each humor exceeds the next by 1/4 and 1/6, leading to a ratio of 9:6:4:1.

37. "Whenever *siccum* is greater than *spuma*, and *spuma* than *humidum* and *tepidum*, the person will be sometimes irritable and sometimes happy, robust, and he can live a long time." ("In quo autem siccum transcendit spumam et spuma humidum et tepidum hic in stultitia interdum est iracundus ac interdum in stultitia letus. Et debilis non est et aliquantum robustus et diu uiuere potest . . . .") CC, II, 87:5–7.

38. "But if *spuma* exceeds *siccum* and *siccum*, *humidum* and *tepidum*, then the person will be gentle, tender, and will not live long." ("Sed si spuma excedit siccum et siccum humidum et tepidum, tunc homo ille bonos mores habet et beniuolus est ac tener corpore, sed diu non uiuit.") CC, II, 87:16–18.

39. Schipperges relates this idiosyncratic emphasis on humoral hierarchy to Hildegard's parallel preoccupation with hierarchy in politics (*Heilkunde*, 122). Glaze comments that "Hildegard, ever hierarchical, divides the humors, like the elements, into dominant and subordinate pairs; of the elements the 'higher' are celestial and immaterial, while the lower are terrestrial and material. Similarly, the humors within each person are divided into two classes, the dominant *flegmata*, and the subordinate, occasionally obstreperous, *livores*" (Glaze, "Medical Writer," 135).

40. For a list of all the different humoral permutations and their corresponding disease states, see Cannon, "The Medicine of Hildegard of Bingen," 95–105.

41. "Pestis, que dicitur *yctericia*, de superfluitate *fellis* nascitur. . . ." CC, II, 194:21. Here I assume that the yellow skin of jaundice, which comes, Hildegard tells us, from *fel*, meant that *fel* itself was yellow.

42. "The bitterness of *fel*" ("amaritudine *fellis*"). CC, II, 186:9–10.

43. "*Fel* increases with sweet foods, just like vinegar from a good wine." ("Et quemadmodum acetum forte et acer de bono uino fit, sic etiam fel de bonis et suauibus escis crescit.") CC, II, 186:13–14.

44. ". . . quia de infirmis humoribus et de febribus ac de magna et frequenti ira fel effluit." CC, II, 194:21–22.

45. ". . . the blackness of melancholia" (". . . nigredine melancolie"). CC, II, 186:10.

46. "Quod si etiam ipsa egestio nigra et arida est, signum mortis est, quia melancolia eandem digestionem in nigredinem et in ariditatem euertit. . . ." CC, V, 280:24–25.

47. "Melancolia autem de bonis et suauibus cibis decrescit; de malis uero et de amaris et de inmundis et de male paratis cibis . . . infirmitatum crescit." CC, II, 186:15–17.

48. "Vnde in hiis crescit flecma, quod nec humidum nec spissum est, sed tepidum, et quod est ut liuor, qui tenax est, et qui se ut gummi in longum protrahit. . . ." CC, II, 66:14–16.

49. "Si uero, cum in aestate ualde calidus interius est, ualde frigidos cibos comedit, *flecma* in se parat." CC, II, 157:32–158:1.

50. "If blood and *flecma* dry out. . . ." ("Cum enim sanguis et flecma in homine exaruerint. . . .") CC, II, 196:14–15.

51. "Sed flecma aliquando minoratur, aliquando augetur, et facile curari et depelli potest." CC, II, 133:21–22.

52. "Nam uinum est sanguis terre et . . . calorem suum ex uesica ad medullam ducit." CC, II, 181:17–19.

53. "Sanguis aridus est et non fluit absque humiditate aquae . . ." CC, II, 181:25–26.

54. "*Tabes*, that is, the water of blood . . ." ("Tabem scilicet aquam de sanguine. . . .") CC, II, 187:4.

55. "*Tabes* is separated from the blood and from the humors . . ." ("Et tabo . . . a sanguine et humoribus separatur. . . .") CC, V, 274:11–15.

56. This was also noted by Jacquart, who confirms that the humoral physiology assumed by Hildegard conformed to the traditions of "Arabic Galenism"; see the whole of Jacquart, "Hildegarde et la physiologie."

57. "Sed humores, qui a corde, a iecore, a pulmone, a stomacho et a ceteris interioribus procedunt, cum aliquando in iniustam diuersitatem et superfluitatem uertuntur, tunc etiam interdum tenaces et lubrici et tepidi fiunt. Et si tunc in homine remanent, eum in infirmitatem ducunt." CC, II, 194:5–9.

58. "Qui enim aut nimis pingues aut nimis macri sunt, malis humoribus sepe habundant, quoniam rectam institutionem et temperantiam in se non habent. Vnde aliquando de corde, de iecore, de pulmone, de stomacho et de uisceribus eorum mali humores exsurgunt, qui ad melancoliam tendunt. . . ." CC, II, 197:1–5.

59. "When someone eats raw food not tempered by spices, then its bad humors rise to the spleen and cause pain . . ." ("Cum homo aliquando crudum cibum comedit, mali humores eorumdem cibum, quia per nullum condimentum temperati sunt, ad splen interdum ascendunt et illud dolere faciunt. . . .") CC, II, 220:1–3.

60. "When there are too many bad humors in the spleen or other organs, they bring problems to the heart . . ." ("Si mali humores in uisceribus et in splene hominis superhabundauerint et cordi multas passiones per melancoliam intulerint. . . .") CC, III, 216:16–17.

61. "And thus from those humors in the body, something like a bitter smoke from sadness is scattered about the heart." ("Et ita de humoribus illis, qui in homine sunt, uelut amarus fumus per tristitiam egrediens circa cor spargitur . . .") CC, II, 187:2–3.

62. "And their anger stirs up their blood like a storm, and from this storm a certain smoke and humor rises to the brain" ("Et ira, que in ipsis est, totum sanguinem eorum in sanguineam et in magna inundationem saepe mouet, et sic de inundatione ista quasi quidam fumus et humor cerebrum eorum tangit.") CC, II, 195:18–20.

63. "When certain bad and fetid humors are stirred up, they send a kind of noxious smoke to the brain . . . and the brain then sends a kind of smoke to the lung." ("Cum autem quidam mali et fetidi humores in homine suscitantur, quemdam noxium fumum ad cerebrum emittunt. . . . Vnde cerebrum conmotum eundem fumum per quasdam uenas ad pulmonem ducit.") CC, II, 135:23–26.

64. "Agrimony, hyssop, and aserum make her sweat and menstrual blood oppresses her like an enemy because it takes out of women many diverse humors." ("Agrimonia enim et ysopus ac aserum eam exsudare faciunt, menstrualis autem sanguis eam opprimit et concutit, uelut inimicus

inimicum suum, quia ista de diuersis humoribus mulierum egreditur.")
CC, IV, 260:22–24.

65. "And if a person who is ill drinks water frequently then he will become
well, because the water will draw off the smoke, fetor, and rottenness of the
bad humors." ("Et homo, qui infirmus est, si aquas illas frequenter biberit,
sanitatem recipit, quia fumum, fetorem et putredinem malorum humorum
uelut bonum unguentum ab eo auferunt.") CC, I, 48:7–9.

66. "Homines autem, qui corpore sani sunt, ita quod mobiles humores, scilicet
qui hac et illac moueri possunt, in se non habent. . . ." CC, II, 202:6–8.

67. That is, *mali*. ("Et cum etiam mali humores superhabundant in homine. . . ."
CC, II, 191:25–26.)

68. That is, *pravi*. ("Cum frigiditate bibinelle temperatus prauos humores con-
mouet in homine. . . ." CC, III, 222:8–9.) (Moulinier has *parvos*, small.)

69. That is, *nocivi*. ("Et sic nociui humores qui in homine sunt, conmouentur et febres accrescunt. . . ." CC, II, 202:28–29); also, they may be *noxii*
("quia omnia noxios humores in eo augerent et accenderent. . . ." CC, IV,
250:23–251:1).

70. "Qualiter etiam pericula inquietudinum mali humores in hominibus faciunt, cum eos in egritudinem deiciuunt." CC, I, 26:5–6.

71. "Sicut quoque inundatio pluuiarum et tempestatum interdum exsurgit et
interdum silet, et sicut mustum nunc per feruorem ebullit nunc cadit, sic etiam
mali humores aliquando in homine eriguntur, aliquando tabescunt. . . ." CC,
II, 115:15–18.

72. "Dolor enim renum et lumborum multotiens ab iniustis humoribus exsurgit." CC, III, 223: 21–22. ("Injustus" means wrong, unjust, or even excessive, according to Lewis and Short, s.v. "injustus.")

73. "Sepe contingit, quod tempestates et procelle malorum humorum in aliquod membrum hominis cadunt et ibi insania sua meatum sanguinis uenarum eius claudunt, ita quia ille ibi in uenis inundare non potest, unde et
uenae ille arescunt, quod meatu sanguinis carent. Et sic homo ille claudicare
incipit." CC, II, 201:16–19.

74. "If anyone suffers from a fistula arising from bad and excessive humors, let
him purge himself, so that the excessive humors may diminish." ("Si quis
fistulam de malis et superfluis humoribus ortam in corpore suo alicubi patitur, purgatorias potiones sepe accipiat, quatinus superhabundantes humores
in se minuat.") CC, III, 226:26–28.

75. "If someone suffers from a pustule, let him protect himself from heat, wet,
cold, and the wind, and avoid hot, roasted, and fatty food, wine, and raw oil
and fruits, because all of these things increase the harmful humors." ("Sed
interim dum homo tumorem pustule huius patitur, ab igne et a frigore et
a uento et ab humido aere se muniat atque ab omni calido et asso et grosso
cibo et a uino se abstineat et cruda olera et cruda poma deuitet, quia omnia
noxios humores in eo augerent et accenderent.") CC, IV, 250:20–251:1.

And "Bad humors are evacuated when pustules drain." ("Et per hoc ulcus suavius mollificatur et suauius rumpitur et humores extrahuntur et suauius sanatur.") CC, III, 227:3–5.

76. "If the bad humors erupt in a rash, let him await its maturity." ("Si uero mali humores in scabie per totum corpus hominis erumperint, tunc iterum ad maturitatem et ad emanationem eorum aliquam diu exspectet.") CC, II, 194:16–18.

77. "If meanwhile from bad humors the testicles swell and cause pain, let the patient take fennel." ("Si interdum a malis humoribus in uirilibus uiri inflatio pessimi tumoris insurgit, que illum ibi dolere facit, ille feniculum accipiat.") CC, III, 224:15–17.

78. "Also bad humors in certain people make a kind of smoke which rises to the brain and makes them forgetful and dull." ("Mali quoque humores in quibusdam hominibus interdum fumum faciunt, qui usque ad cerebrum eorum ascendit et illud ita inficit, quod ex hoc stulti et obliuiosi ac sensu uacui fiunt.") CC, II, 184:21–23. Also, "Foul breath and bad humors can get to the brain and tire patients, often making them forgetful." ("Et fetidus halitus eorum atque mali humores, qui in ipsis sunt, usque ad cerebrum eorum pertingunt et illud in infirmitate fatigant, quod multotiens uelut in obliuione sui ducuntur.") CC, III, 218:12–14.

79. "For if the face is swollen, this is due to the different weak humors." ("Nam quod caro faciei hominis huius inflata est, hoc de diuersis infirmis humoribus est.") CC, V, 271:1–2.

80. "And thus the harmful humors that are in the body are excited and cause fevers, because the *anima* restrains their vital movements." ("Et sic nociui humores, qui in homine sunt, conmouentur et febres accrescunt, quia anima uitales motus suos contraxit.") CC, II, 202:28–29. Also, "When the bad fevers are in the body and when the bad humors in it begin to overflow. . . . ("Cum autem male febres in homine sunt, et cum mali humores in eo inundare incipiunt. . . .") CC, II, 184:16–17.

81. "Et cum hec alterato calore uini temperantur et ita sumuntur, bonos humores per inmoderatum risum istum destructos restituunt." CC, IV, 245:23–25.

82. "And these people have a delightful humor in them, which is not depressed by sadness or bitterness, and which the bitterness of melancholia flees and avoids." ("Sed et delectabilem humorem in se habent, qui nec tristitia nec acerbitate oppressus est et quem acerbitas melancolie fugit et deuitat.") CC, II, 108:22–24.

83. "Quidam uero homines, qui temperata corpora habent, ita quod in carne nec nimis pingues nec nimis macri sunt, temperatos etiam humores sepius in se habent. . . ." CC, II, 196:28–30.

84. "Worms do not arise from the usual humors nor from those that are sharp and acid." ("Sed uermes in hominibus a consuetudinalibus humoribus

non nascuntur nec ab humoribus illis, qui acres sunt ut acetum.") CC, II, 198:1–3.

85. "For the heat of agrimony, galangal, storax, polypoidy, and chelidony will suppress the cold humors from which *flecma* comes." ("Nam calor agrimonie et calor galange et uirtus storacis et calor polipodii ac calor celidonie maioris frigidos humores, de quibus flecma in homine nascitur, superant.") CC, IV, 234:10–12.

86. "For the cold of cinnamon, *bibinella*, and egg whites, along with the heat of pepper and flour . . . will balance the too hot or too cold humors . . . that cause nausea." ("Nam cum frigiditas cimini et frigiditas bibinelle et frigiditas uitelli oui cum calore piperis et cum calore farine . . . iniuste calidos et injuste frigidos humores, qui nausiam homini inferunt, conpescunt.") CC, IV, 246:15–24.

87. "Si autem in cibo suo supra modum bibit, malam inundationem procellarum in humoribus corporis sui facit, ita quod recti humores in eo disperguntur." CC, II, 154:18–20.

88. "Si homo amaritudine fellis et nigredine melancolie careret, semper sanus esset." CC, II, 186:9–10.

89. "Et si interim infrigidate fuerint, eas iterum in predicta aqua calefactas in eisdem locis sibi circumponat et hoc faciat, quamdiu in balneo illo sedeat, ut per humores herbarum istarum cutis et caro illius exterius et matrix interius mollificetur et ut uene eius, que clause sunt, aperiantur." CC, IV, 231:9–13.

90. "Lapides autem ignem et diuersos humores in se habent." CC, V, 283:6–7.

91. "Et homo, sive senex, sive juvenis sit, si caliginem in oculis quocunque modo patitur, accipiat in verno tempore folia arboris illius, antequam fructus ipsius anni proferat, cum in prima eruptione verni temporis sunt, quia tunc suavia et salubria sunt, velut juvenes puellae, antequam prolem gignant, et eadem folia tundat, et succum eorum exprimat. . . . Sed cum in verno tempore in affaldra primae eruptionis, id est ceppini procedunt, ramusculum unum absque ferro abscisionis abrumpe, et corrigiam cervi in pisam eruptione arboris et rami hac et illac trahe ut de succo ejus madida fiat, et cum ibi nichil plus maditatis senseris, tunc cultello ipsam eruptionem minimis ictibus incide, id est hacke, ut plus humoris profluat. . . ." *Physica*, cols.1215A–1217B.

92. "Quod si aliquando humor in ulceribus exsurgescit rubam comedat et ulcu compescitur." *Physica*, col.1164C.

93. " . . . tunc terra ista contra has infirmitates amplius non valet, quia humor ejusdem terrae et succus ipsius arboris jam in fructus illos ascendit. . . ." *Physica*, col.1217C.

94. " . . . spirituu sancto stillavit faciens maximum et plurimum fructum sicut purus humor super gramen descendens illud ad multum germen fecundat." *Scivias*, PL197, col.445C.

95. "Et terra habet sudorem atque humorem et succum. . . . Ita humor ejus utiles herbas, quae comedi possunt et quae ad alios usus hominis valent." *Physica*, col.1126B.

96. "Si quis diversos cibos immoderate et indiscrete sumpserit et sic de diversis humoribus ciborum illorum iecur eius leditur et indurescit, hic minner hufladecha accipiat. . . ." CC, III, 219:3–5.

97. "Cum autem interdum homo nouum et incognitum cibum comedit et nouum et incognitum uinum aut alium no uum potum bibit, tunc de nouis humoribus illis alii humores in eo mouentur et liquefacti per purgationem de naribus effluunt, sicut etiam nouum uinum in uas fusum sordes et feces se purgando eicit." CC, II, 174:31–175:3.

98. "Homo post cibum statim non dormiat, antequam gustus et succus ac odor eorum ad loca sua perueniant." CC, II, 154:28–29.

99. "A diuersis quoque humoribus, tam bonis quam malis, caro et uene hominis ingrossescunt, uelut a fermento farina eleuatur et intumescit." CC, II, 194: 4–5.

100. Certainly, standing blood does settle into three or four clearly distinct layers, and the bodily fluids and excretions can be made to fit into a four (or three or five) color classification. Actually this derivation was proposed at least as far back as Nemesius, who wrote that "phlebotomy demonstrates that all four humors reside in the blood, where sometimes a watery phlegm is in abundance, sometimes black choler or red choler." ("Inveniuntur autem multotiens quattuor humores in sanguine, ut est videre in phlebotomiis, aliquando quidem aquoso phlegmate abundante in ipso, quandoque autem cholera nigra vel rubea.") Alfanus of Salerno, *Nemesii episcopi Premnon Physicon*, ed. Carolus Burkhard (Leipzig: Teubner, 1917), 60:2–5.

101. Even its earliest developer, Polybius, noted this, according to Maria Michela Sassi, *The Science of Man in Ancient Greece*, trans. Paul Tucker (Chicago and London: University of Chicago Press, 2001), 153. Yet Jacques takes issue with this idea; black bile, he argues, does not correspond to a physical fluid, despite attempts by Galen to identify it with the sediment of blood, or Sigerist to identify it with melena (219). It was added merely to satisfy the requirements of the tetradic theory. See Jean-Marie Jacques, "La bile noire dans l'antiquité grecque: médecine et littérature," *Revue des Etudes Anciennes* 100, no. 1–2 (1998): 217–234.

102. Yet the Pneumatics also accepted the concept of the four humors whose balance was to be maintained through a strict regime; see Wellmann, *Die pneumatische Schule*, 146–148.

103. See Smith, "The Modern Hippocratic Traditions," in *The Hippocratic Tradition*; and also Owsei Temkin, *Galenism: The Rise and Decline of a Medical Philosophy* (Ithaca, N.Y. and London: Cornell University Press, 1973). Temkin pointed out that Galen wrote about 12,000 pages and was often self-contradictory (6). Moreover, "the doctrine of the four humors was not Galenic, it was

Hippocratic. . . . But the emphasis on these four humors as the Hippocratic humors, the linking of these with the Aristotelian qualities, and with the tissues of the body, was largely Galenic (103)." Thus Galen wrote that the four bodily humors were best understood to work not in themselves but primarily as carriers of the four qualities (Smith, 87). For an extended treatment of the theory of qualities and the workings of medicines, see McVaugh, "The Development of Medieval Pharmaceutical Theory," 13–136.

104. Maria Sassi even labelled her section on humoral theory, "Schematic and More Schematic." She described the process as "a kind of gradual and unbroken transition to an increasingly schematic version of the original schemas" (*The Science of Man*, 160). See also the summary by Ayoub, "Old English Waeta," 332–333. But perhaps it is our own understanding of the theory that is schematic. As we shall see, particular texts present particular versions of the theory, which are often, in their details, contradictory; it is precisely the variations within the schema that proves that it was not, mainly, schematic.

105. This is in Polybius' *Peri phusis anthropown* (c. 400 B.C.), according to Klibansky, *Saturn and Melancholy*, 8.

106. According to Schöner, *Viererschema*. He also noted this variability of a one- or two-element doctrine; a two- or three-quality doctrine; and a four-quality and four-humoral doctrine (57), as noted in Ayoub, "Old English Waeta," n. 5.

107. See the whole argument of Chapter Three in Smith, *Hippocratic Tradition*.

108. I am thinking here of such texts as the *Conservatio flebotomiae*, which, although it has nothing *per se* on the humors, does relate health *via* the humors to the times of year; see Karl Sudhoff, "Lateinische Texte über den Rhythmus der Säftebewegung im Gleichklang mit den Jahreszeiten, samt Nachtgleichen- und Sonnwends-Diätetik," in *Archiv für Geschichte der Medizin* 11 (1919), 206–211.

109. For a review, see Henry Sigerist, "The Latin Medical Literature of the Early Middle Ages," *Journal of the History of Medicine and Allied Sciences* 13 (1958): 127–145.

110. For texts, see Isidore of Seville, *Isidori Hispalensis episcopi Etymologiarum sive originum libri XX*, ed. W. M. Lindsay, 2 vols. (Oxford: Clarendon Press, 1911); William of Conches, *De philosophia mundi*, in PL172, cols. 39–102; and Charles Burnett, ed., *Pseudo-Bede: De Mundi Celestis Terrestrisque Constitutione. A Treatise on the Universe and its Soul* (London: Warburg Institute, 1985). Although found only in late-twelfth-century manuscripts, *Pseudo-Bede* was written in the eleventh century and so reflects early medieval views.

111. Glaze surveys the field in her dissertation; see "Perforated Wall," 10–104. Sudhoff's discovery of the list of medical manuscripts in Bishop Bruno's library, c. 1160, is very useful in that it provides some idea of the *maximum* amount of literature that could have been known to Hildegard by 1160, but not, probably, the usual texts for a medical practitioner. For the list, see Chapter Two.

112. In Kibre there are fifty-five Hippocratic texts in early medieval manuscripts. Of the fifty-five, however, most are only found in one or two manuscripts— probably not available to our student infirmarian. And even these texts were mainly fragments. For instance, the passage from the *De natura hominum* (*DNH*) is only a short passage and in a single ninth-century manuscript (Kibre, *Hippocrates Latinus*, 193).

113. According to McVaugh, it was not until the late eleventh century that the "teachings of antiquity began to pass into Latin medicine," and these first "translations" of Galen were "paraphrases," not Galen's work itself (Michael R. McVaugh, "The 'Humidum Radicale' in Thirteenth-Century Medicine," *Traditio* 30 [1974], 259 and 262). Strohmaier goes so far as to say that Galen "left virtually no trace in Latin literature." Gotthard Strohmaier, "Reception and Tradition: Medicine in the Byzantine and Arab World," in *Western Medical Thought from Antiquity to the Middle Ages*, ed. Mirko D. Grmek (Cambridge, Mass.: Harvard University Press, 1998), 141.

114. The list is from Vivian Nutton, "Medicine in Late Antiquity and the Early Middle Ages," in *The Western Medical Tradition: 800 B.C.–1800 A.D.*, ed. Lawrence I. Conrad, et al. (Cambridge: Cambridge University Press, 1995), 71–88. For the classic article on the texts that made up the twelfth-century medical student's "must-have" list, see Paul Oskar Kristeller, "Bartholomaeus, Musandinus, and Maurus of Salerno and Other Early Commentators of the Articella, with a Tentative List of Texts and Manuscripts," *Italia Medioevale e Humanistica* 19 (1976): 57–87.

115. There seem to be two separate camps, those who consider her knowledge representative of monastic medicine, and those who, like Dronke and Moulinier, find traces of the "new medicine." For one side of the argument, see Singer, "Scientific Views and Visions," 228, and Moulinier; for the other, see Heinrich Schipperges, "Einflüsse Arabischen Medizin auf die Microkosmosliteratur des 12. Jahrhundert," *Miscellanea Mediaevalia* 1 (1962): 139–42.

116. It is said to be a synopsis of the "Galenic" system of medicine, c. 1150, and was commonly found in Northern Europe, according to Temkin, *Galenism*, 107. For an edition, see Johannitius, "Introductio ad artem parvam Galieni," in Diego Gracia Guillén and José Luis Vidal, "La Isagoge de Joannitius," *Asclepio* 26–27 (1964–65): 267–382. I have chosen to look at the *Isagoge* because it was popular, early, and provides an "average" look at the new material; it is also readily available today in print. What follows is my translation and abridgment of the Latin version.

117. This is according to Temkin, *Galenism*, 81; and Baader, "Early Medieval Latin Adaptations of Byzantine Medicine in Western Europe," 259.

118. Although "it is clear to us today that virtually the whole of medieval medical theory derived, directly or indirectly, from the teachings of Galen . . .

this was not at first (if ever) clear to the Middle Ages. The first important works of medical theory that the West encountered may not actually have ventured far from Galenic principles, but they did not bear his name. . . ." (McVaugh, "Humidum Radicale," 259).

119. I have been unable to locate this text, which seems to have been printed only once, in 1528. The first book in Kühn's edition called *Ad Glauconem* is long and detailed but does not discuss the humors, and so does not seem to be the same text mentioned by these scholars; see C. G. Kühn, ed., *Klaudiou Galenou Apanta. Claudii Galeni Opera omnia*, vol. 11 (Hildesheim: Olms, 1964–1965 (1821–1833), 1–14.

120. For an edition, see Anonymous, "Ad Maecenatem," in *Marcelli de medicamentis*, ed. George Helmreich (Leipzig: Teubner, 1889), 9–13.

121. Anonymous, "Epistula ad Antiochum," in *Marcelli de medicamentis liber*, ed. Maximillianus Niedermann, Corpus Medicorum Latinorum (Leipzig, Berlin: Teubner, 1916), 10–13; the text exists in two ninth-century manuscripts (Thorndike, *A Catalogue of Incipits*, col. 480). Although purportedly written by Hippocrates to the king of Antioch, this was not a text by Hippocrates that reached the Middle Ages; rather "Hippocrates" was used as a medical authority for otherwise anonymous texts. Kibre provides several pre-twelfth-century manuscripts, which she refers to as slightly different versions (*Hippocrates Latinus*, 145–148). In fact she includes under this *De quattuor humoribus* and the "Sapientia" (157).

122. Vindician is found in more than twelve manuscripts, only one of which is from the twelfth century. For a printed edition, see Vindician, "Epitome altera. Epistula ad Pentadium," in *Theodori Prisciani Euporiston libri III*, ed. Valentin Rose (Leipzig: Teubner, 1894), 485–492. As an example of the textual problems alluded to above, it was first printed as a work of Galen.

123. Also known as the *Epistula ypogratis* (according to Thorndike, *A Catalogue of Incipits*, col.118), this was an early (sixth century, according to Wlaschky, "Sapientia," 113) popular text. It is found in more than a dozen pre-twelfth-century manuscripts (Kibre, *Hippocrates Latinus*, 151) and printed in Marcellus, *Corpus Medicorum Latinorum* V (1968), 26–32.

124. The letter by Diokles survived only in a fourteenth-century manuscript and was in Greek (Wlaschky, "Sapientia," 110) and so is not included in our canon. *De ciborum*, the classic cooking text, might be considered a quasi-medical text since it relies on the humors and qualities for its explanations of the medicinal uses of foods, but it does not provide any *explicit* passage on humoral theory. *Ad Glauconem*, in its "modern" edition in Kühn, is very long (146 pages), but during the early Middle Ages only the first chapter of the first Book (Baader, "Early Medieval Latin Adaptations," 283) was available. As printed, the first chapter of the first Book has nothing on the

humors, which are only much later mentioned, in Book 1, Chapter 4, as the temperaments called hot and cold, and wet and dry (79).

125. The *Liber Aurelius* is probably an early abridgment of a Methodist text and may provide a Methodist take on the humors. For an edition and commentary, see Charles Daremberg, "'Aurelius de Acutis Passionibus,' texte publié pour la première fois d'après un manuscrit de la bibliothèque de Bourgogne à Bruxelles, corrigé et accompagné de notes critiques," *Janus* 2 (1847): 468–499, 690–731. The actual text takes up only seven pages. It must have been fairly popular, since it is found in nine early (9th–11th c.) manuscripts. It, too, was printed as a work of Galen (Thorndike, *A Catalogue of Incipits*, col. 998). Daremberg thought that it was either an abridgment (468) or a plagiarism (469) of an earlier Methodist text, *De morbis acutis* of Caelius. In the text the humors are not the only causes of disease; Aurelius also diagnoses such problems as constriction [*astrictum*] and laxation [*solutum*] (481), as would be expected from a Methodist author.

126. "Quatuor humores sunt in humano corpore, id est sanguis, phlegma, cholera rubea, cholera nigra. . . . Sanguis in arteriis habitat et in venis, phlegma in cerebro, fel rufum in jecore, nigrum vero in splene. Sanguinis alia pars in corde, alia est in jecore; cholera rubes . . . in dextero latere . . . ; ubi splen jacet, cholera nigra dominatur; phlegmatis alia pars in vesica, alia pars in pectore dominatur. Sanguis colorem habet rubicundum, phlegma candidum, fel rufum roseum, fel nigrum et spissum. . . . Igitur sanguis natura amarus est, phlegma salsum et dulce, fel rufum acerbum, fel nigrum forte et acre. . . ." Anonymous, *De quattuor humoribus*, in *Collectio Salernitana*, vol. 2, ed. Salvatore da Renzi, (Naples: Dalla typografia del filiatre-sebezio, 1852–1859), 411–412.

127. This is a tantalizing allusion to a possible source for Hildegard's use of *viriditas* for a liquid in the body. Another allusion may be that "every food having been ground first in the stomach, goes out of the stomach and is sent throughout the body; so the liver sends the *greenest* and most whole into the body." ("Omnis autem materia ciborum detunsa per gurgulionem excipitur primo ab stomacho, que tenuissima itincra per totum corpus dispensatur. Ergo *viridissima* et integerrima in sanguinem dimittit iecor.") Vindician, "Epitome Altera," 475:2–5.

128. "Virtus sane ipsorum umorum talis est, sanguis est fervens umidus et dulcis, cholera xanthe id est rubea amara viridis ignea et sicca, melena cholera id est nigra acida frigida et sicca, flegma frigidum salsum et umidum." Vindician, "Epistula ad Pentadium," 486:14–487:2.

129. "Pulsus autem suos habent hi quattuor umores. . . ." Vindician, "Epistula ad Pentadium," 489:6.

130. "Haec omnia crescunt suis temporibus . . ." Vindician, "Epistula ad Pentadium," 487:3.

131. Asaph even added a piece that Hildegard only suggests: the tastes and flavors
      of foods, which come from the land, *also* carry the qualities of hot and cold,
      wet and dry into the body just like the winds do, but in a different way. "In
      particular, everything which is sweet is warm and wet; everything which is bit-
      ter is warm and dry; everything which is sour is cold and wet; and everything
      which is sharp is cold and dry." "Ausser diesen sphärischen Einflüssen hängt
      die Elementen-Harmonie des Körpers von der Naturkraft der Nahrungsmittel
      ab. Diese sind zumeist schon an ihrem Geschmack zu erkennen: alles Süsse ist
      warm und feucht, alles Bittere ist warm und trocken, alles Sauere ist kalt und
      feucht und alles Scharfe ist kalt und trocken." Venantianer, *Asaf Judaeus*, 78.

132. "Haec omnia habent respirationes suas per singulas partes corporis, sanguis
      per nares, cholera rubea per aures, cholera nigra per oculos, flegma per os."
      Vindician, "Epistula ad Pentadium," 487:17–19.

133. "Per omnia enim viscera venae currunt quae sanguinem regunt." Anon.,
      "Ad Maecenatem," 9:24–25.

134. "Sanguine vero in hepate praedicto modo generato, ad omnia transit membra
      per venas, quorum calore digestus, in eorum similitudinem transit." William
      of Conches, *De philosophia mundi*, PL172, Book 4, Chapter 19, col. 93.

135. "Corpus igitur homini ex quattuor umoribus constat. Namque habet in
      se sanguinem choleram rubeam choleram nigram et flegma. Qui quattuor
      umores habitant vel dominantur in suis locis. Sanguis dominatur in dextro
      latere, in epate quod iecur vocamus, sed et cholera rubea ibidem domina-
      tur. In sinistro vero latere, id est in splene, cholera nigra dominatur. Flegma
      autem in capite et alia pars eius in vesica. Sanguinis tamen pars dominatur
      in corde." Vindician, "Epistula ad Pentadium," 486:5–13.

136. "Quorum calore digestus, in eorum similitudinem transit." William of
      Conches, *De philosophia mundi*, col. 93; see full quote above.

137. "Superfluitas vero per sudores exit. Alia vero pars ad hepar revertitur, ibi
      decocta descendens, exit cum urina, semenque vocatur." William of
      Conches, *De philosophia mundi*, col. 93.

138. "Hi humores sunt commixti, quibus, si mensura non excesserint, perpetua
      sanitas est hominibus. Si autem imminuantur, aut suprabundant, aut spis-
      sentur, aut detenuentur, aut evadant a natura, aut acerbiores fuerint, aut
      sedes suas relinquent, aut incognita occupaverint, varie emergunt valetudi-
      nes in hominibus." Anon., *De quattuor humoribus*, 411.

139. "Homini sani urina mane alba erit, ante prandium rufa, pransi rursus can-
      dida, item ante cenam russea." Anon., "Ad Maecenatem," 10:21–23.

140. "Oportet vere hos humores alios nutrire, alios extenuare, alios compensare,
      alios temperare." Anon., *De quattuor humoribus*, 411.

141. For a discussion of the differences between Soranus and the Hippocrat-
      ics as evidenced by his *Gynecology* (which did survive), see Green, "The
      Transmission of Ancient Theories of Female Physiology and Disease
      Through the Early Middle Ages," especially 23–36. Green argues that "like

the Methodists in general, Soranus completely rejected the theory of four humors." If this is the case, then the *Liber Aurelius* cannot be a Methodist text. In particular, it is not a translation or even paraphrase (as argued by Daremberg) of Soranus, since Aurelius accepted the four-humor theory as bodily fluids. But Mudry argues that Soranus was Aurelius' ultimate, if not proximate source; see Philip Mudry, "Caelius Aurelianus ou l'anti-Romain: un aspect particulier du traité des maladies aigues et des maladies chroniques," in *Maladie et maladies dans les textes latins antiques et médiévaux*, ed. Carl Deroux (Bruxelles: Latomus, 1998), 313–329.

142. "Quae sunt valitudines acutae, quae oxea Graeci dicunt, fiunt vel nascuntur ex sanguine vel ex fele rubeo, nam aut cito transeunt aut celerius occidunt; veteres vero causae quae cronia graece dicuntur, fiunt vel nascuntur ex flegmate et ex nigro felle." Caelius, *Liber Aurelius*, 479. There is no new edition of this text, according to Gerhard Bendz, ed., *Caelii Aureliani Celerum passionum libri III. Tardarum passionum libri V*, vol. 6, Corpus Medicorum Latinorum (Berlin: Akademie Verlag, 1990–1993).

143. "Ex sanguine autem et felle acutae passiones nascuntur, quas Graeci oxea vocant. Ex phlegmate vero et melancholia veteres causae procedunt, quas Graeci chronia dicunt." Isidore, "De quatuor humoribus corporis," in *Etymologiarum*, PL82, col. 185 B.

144. "Omnibus hominibus generantur aegritudines ex quatuor humoribus unde et homo factus est. . . . Omnes humores, si amplius extra cursum naturalem plus creverunt, aegritudines faciunt aut ex sponte digeruntur. . . ." Caelius, *Liber Aurelius*, 479.

145. "Cum enim a capite morbus oritur, solet capitis dolor conparere; tunc et supercilia gravantur, tempora saliunt, aures sonant, oculi lacrimant, nares repletae odorem non sentiunt. Cum ex his ergo aliquid accidit, caput purgari oportet hac ratione. Hysopi aut cunilae bubulae fasciculum cum aqua deferve facies, inde aquam ore continebis, sed caput calide tractabis, ut fluat pituita." Anon., "Ad Maecenatem," 10:29–11:2.

146. According to the OLD, *pituita* had two meanings. In the body it was diseased material, phlegm or rheum, but it was also used for plants, to mean the viscous, gummy moisture that exudes from trees; that is, sap. See OLD, s.v. "pituita." There is the same overlap between humor and sap, as we shall see shortly.

147. "Cum autem a thorace morbus nascitur, incipit caput sudare linguaque fit crassior aut os amarum aut tusillae dolent. . . . Plus autem prodest, si ieiunus bilem eieceris; eam enim dicimus matrem morborum." Anon., "Ad Maecenatem," 11:7–16.

148. "Phlebothomia quoque initium est sanitatis . . . cum cautela fleri debet. . . ." Anon., *De quattuor humoribus*, 412.

149. "Chirugia est ferramentorum incisio . . . cucurbita, a suspiriio *ventosa* vocatur. . . . Omne quod intra cutem vel alitus aestuat, sive humore, sive sanguinem." Isidore, *Etymologiarium*, PL82, col. 195A.

150. "Nam si sit causa sanguinis, qui est sanguis est dulcis, humidus, et calidus: occurrendum est sic ut adhibeatur e contrario quod sit frigidum, amarum, siccum." Vindician, "Epistula ad Pentadium," 491:1–3.

151. "Omne itaque corpus hominum, pecudum alitumque, ex quattuor generibus constat, sed praecipue hominum: calido, frigido, sicco et umido." Anon., "Ad Maecenatem," 9:18–20.

152. "Frigido enim continentur viscera, unde spiramus: calore continetur anima, qua vivimus, id est qua vitam sentimus. Sicca sunt ossa, quae vires faciunt ad sustinendum laborem; umidus est sanguis quo alitur vita." Anon., "Ad Maecenatem," 9:20–24.

153. "Haec omnia crescunt suis temporibus. Sanguis crescit verno tempore, ab VIII. id. febr. usque in VIII. id. mai et sunt dies XCI. Cholera rubea aestate, ab VIII. id. mai usque in VIII. id. aug. et sunt dies XC. Cholera nigra autumno, ab VIII id aug. usque in VIII. id novemb. et sunt dies XCII. Flegma vero hieme ab VIII. id. novemb. usque in VIII id. febr. et sunt dies XCII." Vindician, "Epistula ad Pentadium," 487:3–9.

154. "Hyems enim, utpote longius sole remoto, frigida est et humida. Ver, illo super terras redeunte, humidum et calidum. Aestas, illo superfervente, calida et sicca. Autumnus, illo ad inferiora decedente, siccus et frigidus." The Venerable Bede, *De temporum ratione*, ed. J. A. Giles, *The Miscellaneous Works of Venerable Bede* (London: Whitaker and Co., 1843), 217–220.

155. "Terra namque sicca et frigida; aqua frigida et humida; aer humidus et calidus. Ignis est calidus et siccus: ideoque haec autumno, illa hyemi: iste veri, ille comparatur aestati." Bede, *De temporum ratione*, 218.

156. "Sanguis siquidem qui vere crescit, humidus et calidus. Cholera rubea, quae aestate, calida, et sicca. Cholera nigra, quae autumno, sicca et frigida. Phlegmata, quae hyeme, frigida sunt, et humida". Bede, *De temporum ratione*, 218.

157. "Humor: the fluid occurring in plants, sap." OLD, s.v. "humor."

158. "*Chymos:* The juice of plants, or animal juices (humours); juice in a wider sense. Also flavours." Henry George Liddell and Robert Scott, *A Greek-English Lexicon*, 1 ed. (Oxford: Clarendon Press, 1966 [1879], s.v. "chymos."

159. According to the OLD, *reuma* can mean the sea, a liquid, or a flowing; and according to Lewis and Short, s.v. "reuma," it meant a flowing. Here it seems to be something like phlegm, or, perhaps, sap; in this context it is hard to know whether reuma refers to bodies or plants.

160. "Mense autem Martio a XV kal. exurgent omnes humores et morbositates et reuma et radices herbarum. Vexantur corpora humana, in tumore erunt. Per venas morbositates detrahendum erit per flebotomum; per catarticum corpora purgantur. Dominantur humores usque in mense Iulio. Post III idus Iulii suspende, medice, flebotomum et catarticum, quia caniculares dies ingrediuntur, et cinocaumata sunt dies LXXIII. Post III idus septembres exurgent colera rubea et flegma nimia, et caliditas corporis accedit.

Autumnus autem erit dies LXIIII. Suspendite a Venere, quia vitiata. Ante III kal. decembres exurget colera nigra, salsa et humecta et utere Venere. Hieme saepius distillatio uvulae, flegma et reuma nimia amplius exurgent; usque in kal. Martii reumatidius erit." Anon., "Sapientia," 105:4–14.

161. "Ab hieme igitur salubrius fiet exordium. Ex die VIII kal. Ianuarias corporibus humor adcrescit usque veris aequinoctium, quod incipit VIII kal. Apriles. Hoc ergo tempore utendum calidissimis cibis et optimis, vino aliquatenus indulgendum, usus etiam venerius minime respuendus. Ex aequinoctio incipit fleuma crescere, id est concretus humor et frigidus, usque in ortum pleiadum, qui incipit pridie idus Maias. Ergo tum adhiberi convenit bene odoratos, sed et acres cibos. Exhinc sanguis ardentior febribus augmenta suppeditat nec minus humor obibitur usque ad aestivi temporis commutationem, quae ab VIII kal. Iulias inchoatur, a quo die venere abstinendum ieiunandumque. Ab octavo enim kal. Iulias aestiva progressio incrementa fellis exsuscitat usque in autumni aequinoctium, quod est VI kal. Octobris. Erit igitur utendum acutis et sapidis et bene olentibus cibis, parcius ieiunandum et venerio usu ab VIII kal. Iulias certe per dies XII penitus abstinendum. Post autumni aequinoctium, id est post diem VI kal. Octobris, acres biles vires et augmentum sortiuntur in occasum pleiadum, qui est IIII idus Novembris, umores etiam graves serpunt. Erit igitur utendum acidis et acribus cibis, venere quoque abstinendum, parcius etiam ieiunandum. Ab hoc rursus tempore id est a IIII idus Novembris, sanguis deficit; congruit igitur levioribus cibis uti, vino indulgere nec usu venerio abstinere in VIII kal. Ianuarias, ex quo incipit tempus hibernum." Anon., "Epistula ad Antiochum," 13:6–27.

162. "Similiter arboribus et herbis sucus quidam a terra radicibus additur qui, per omnes partes diffuses et naturali quodam calore induratus, transit in robur." Burnett, *Pseudo-Bede*, 18:4. It is important to note here that Burnett translates *robur* as "woody substance," its most literal meaning, although, clearly, most plants do not make woody substances. I have used here its later and more conceptual meaning of "strength, vigor, power;" see OLD, s.v. "robur." Similarly he translates *sucus* (also *succus*), as "moisture," while I have used its earlier and more common, literal meaning as "juice, sap"; see OLD, s.v. "succus."

163. "Jecur, nomen habet eo quod ignis ibi habeat sedem, qui in cerebro subvolat. Inde ad oculos caeterosque sensus et membra diffunditur, et calore suo ad se succum ex cibo tractum vertit in sanguinem. . . ." Isidore, *Etymologarium*, PL 82, col. 412C. "Semen est quod jactum sumitur aut a terra, aut ab utero ad gigendum, vel fructus vel fetus. Est enim liquor ex cibi et corporis decoctione factus, ac diffusus per venas atque medullas, qui inde desudatus in modum sentinae concrescit in renibus ejectusque per coitum, et in utero mulieris susceptus, calore quodammodo viscerum et menstruali sanguinis irrigatione formatur in corpus." Isidore, *Etymologarium*, PL82, col. 414B.

164. "Modi ciborum sunt duo. Est enim cibus qui bonum gignit humorem, est malus qui malum gignit humorem. Bonum autem gignit qui bonum

sanguinem creat et aequalem in commixtione et opere, ut est panis mundus.
. . . Malus autem contraria operatur. . . . Est et olerum genus quod malum
generat humorem scilicet choleram rubeam, ut est nasturtium, sinapis et
allium. Sunt et quae generant melancholiam, sicut lens. . . . Et sunt quae generant phlegma, ut est caro porcellorum. . . ." Johannitius, *Isagoge*, 333.

165. "Miu medicinarum ciborum et potuum complexiones possunt sciri per
      sapores. Quia omnis res aut ex calida ut frigida et humida, aut sicca et
      frigida, aut humida et frigida, frigida et sicca. Quarum qualitates per
      saporem perpendi possunt. Quae sapores sunt viii. Dulcis, unctuosus, ace-
      tosus, salsus, amarus, acutus, stipticus, ponticus. Sunt quaedam horum
      quae sensui passiones faciunt cum delectatione, et quaedam cum noco-
      mento. Quae delectatem facem est dulci et unctuosus. Cetera nocumenta
      prantunt. Nam dulcis cum temperat sicut naterci. Unctuoissi est calidus et
      humidus, in secondo gradus. Amarus calidus et siccus in iii gradu., stipti-
      cus frigus et siccus in vvvi gr. Ponticus similiter. Sed tamen quaedam qua
      nichil sensuis faciunt quae sunt insipidus quae friguida et humida sunt
      in primo gradu." My transcription from Zurich's Zentral bibliothek MS
      C128, fol. 98v.

166. Bar-Sela, *Asaph on Anatomy and Physiology*, 361. The difficulties of under-
      standing these correlations in Asaph are confounded not only by the lack
      of a complete translation of the text, but by the apparent contradictions
      within the text itself. Thus on p. 366 of Bar-Sela, sweet is linked to cold and
      wet while ten pages later, it is linked to the qualities of hot and wet. I have
      resolved this by using the most consistent connections.

167. "Et ideo sapor est, qui certissimum dat experimentum virtutis plantarum."
      See E. Meyer, and C. Jessen, eds., "Tract. II," in *Albertus Magnus, De Vegeta-*
      bilibus, Libri VII (Berlin, 1867), chapter III, section 69.

168. The distinction between a plant's qualitative effect and its specific effect was
      usual in medieval pharmacy. "Every compound medicine must have at least
      one medicinally active ingredient (more than one if the compound is meant
      to cure a multiple illness), its root [*radix*, Bacon]; it is this that is to be
      opposed to the peccant humor or quality (35). . . . Tastes derive directly
      from a medicine's complexion: medicines which taste sharp, bitter, salty, or
      sweet, are hot, while those which taste pontic, styptic or sour must be cold"
      (McVaugh, "The Development of Medieval Pharmaceutical Theory," 120).

169. "Potestas herbarum, id est vis et possibilitas. Nam in herbarum cura vis ipsa
      dynamis dicitur. Unde et Dynamidia nuncupatur, ubi eorum medicinae
      scribuntur." Isidore, *Etymologarium*, PL82, col. 194C.

NOTES TO CHAPTER FIVE

1. "Herbarum namque atque arborum uita uiriditas uocatur." Gregorius Mag-
    nus, "Moralia in Iob," *Cl* 1708, SL 143, lib. : 6, par. : 16, linea : 17. And:

"Vivunt dico, non per animam sed per viriditatem." Gregorius Magnus, "xl homiliarum in evangelia libri duo," *Cl* 1711 lib. 2, hom. 29, cap. 2, linea 7.

2. I have mainly left *viriditas* untranslated since any translation—greenness, greening, greening power, the greening power of Nature, or the Web of Greening Life Energy—implies a political, cultural, or even religious stance. Also, I will not continue to treat it as a foreign concept by italicizing it.

3. Not necessarily what we think of as cancer, however. According to Lewis and Short, *cancer* was more like a chronic, non-healing wound: a "crawling, eating, suppurating ulcer, a malignant tumor, a cancer," s.v. "cancer."

4. "Accipe uiolas et succum earum exprime ac eum per pannum cola et ad ter-tiam partem succi huius oleum olyue pensa et deinde ad quantitatem succi uiole hyrcinum sepum pondera et hec simul in olla noua fac feruere, et sic unguentum fit. Tunc locum corporis, ubi cancer aut alii uermes hominem comedunt, circumunge et etiam desuper unge, et morientur, cum de eo gus-tauerint." CC, IV, 251:11–16.

5. "Succum aristologie longe ad pondus duorum nummorum." CC, III, 221:14–15.

6. That is, "que ouum recte moderationis includere potest." CC, II, 159:14.

7. The following tables are not meant to be definitive, since identifying medi-cinal plants whether by their Latin or German names is notoriously unreli-able. I have assumed here that the Latin and German names do represent what are called today by the same names. My decision as to whether a rec-ommended herb was grown in the garden, found in the wild, or bought at market is based on a number of facts: the kinds of herbs known to be planted in monastic infirmary gardens; the kind of plants found wild in the Rhineland today; and the kind of herbs and spices that were foreign to the Rhineland and must, therefore, have been bought at market. On infirmary plants, see John H. Harvey, "Westminster Abbey: The Infirmarer's Garden," *Garden History* 20, no. 2 (1992): 97–115; Paul Meyvaert, "The Medieval Monastic Garden," in *Medieval Gardens*, ed. E. B. MacDougall (Washing-ton, D.C.: Dumbarton Oaks, 1986), 23–53, as well as Horn, *Plan of St. Gall*; and Francis J. Green, "The Archaeology and Documentary Evidence for Plants from the Medieval Period in England," in *Plants and Ancient Man: Studies in Paleoethnobotany. Proceedings of the Sixth Symposium of the International Work Group for Paleoethnobotany*, eds. W. Van Zeist and W. A. Casparie (Rotterdam: A. A. Balkema, 1984), 99–114. For the usual herbs on apothecary shelves, see M. Jenkins, "Medicine and Spices with Special Reference to Medieval Monastic Accounts," *Garden History* 4 (1976): 47–49. Of course, identification is limited by the difficulty of knowing whether what a plant is called today corresponds to its twelfth-century equivalent, as Müller noted in her "Verfasserfrage." Given these caveats, the tables do provide some idea of what Hildegard expected to be on hand. I have left untranslated plant names for which I could find no translation.

8. "Nam diuerse et nobiles herbe et pulueres atque condimenta nobilium herbarum inordinate comesta sanis hominibus non proderunt sed potius lesionem eis inferunt. . . ." CC, III, 228:12–14. What Hildegard meant by noble plants is not exactly clear. She hints that they are those which have a good smell. ("Et mox modicum calidi vini bibat, et alias nobiles herbas, quae bonum odorem habent." *Physica*, col. 1154B).

9. "Qui nausiam patitur, ciminum accipiat et ad eius tertiam partem piper atque bibinelle uelut quartam partem cimini et hec puluerizet et puram farinam simile accipiat et puluerem istum farine huic inmittat. Et sic cum uitello oui ac cum modica aqua tortellos aut in calida fornace aut sub calidis cineribus faciat et tortellos istos comedat, sed et predictum puluerem super pannum positum manducet. Nam cum frigiditas cimini et frigiditas bibinelle et frigiditas uitelli oui cum calore piperis et cum calore farine simile temperantur et cum suauitate aque componuntur ac cum suaui calore fornacis, ut predictum est, coquuntur, iniuste calidos iniuste frigidos humores, qui nausiam homini inferunt conpescunt." CC, IV, 246:10–19.

10. "Cum homo sanus in corpore est, si oculos puros et perlucidos habet, cuiuscumque coloris sint, signum uite tenet; ita, si oculi eius hoc modo perlucidi sunt, quemadmodum candida nubes est, per quam interdum uelut weithden nubes apparet, et iste uiuet nec cito morietur. Nam uisus anime in oculis hominis huius potens est, cum oculi ipsius puri et perlucidi sunt, quoniam anima in corpore eius potenter sedet, quatinus plura opera in ipso operetur. Oculi enim hominis fenestre anime sunt. Qui autem turbidos oculos habet, cum sanus est, ita quod oculi eius non sunt perlucidi, cuiuscumque coloris sint, signum mortis tenet; item si oculi eius hoc modo turbidi sunt, uelut aliqua nubes est, que superius ita spissa est, quod sub ea quasi weithden nubes considerari non potest, et iste cito infirmabitur et mors subsequetur. In uisu enim oculorum hominis istius anima potens non est, quia pauca opera ibi operabitur et quasi obnubilata sedet, uelut homo, qui in opinione et in dubio est, quando sedem suam deserat quando domum suam egrediatur." CC, V, 269:2–17.

11. "Homo etiam, cum sanus est, cuius color in genis sub cute rubeus seu moderate rubeus est, ita quod color iste sub cute cernitur, ut in pomo, quod perlucidum et purum est: signum uite habet, cum rubeus color in genis eius sub cute uidetur, ut in candida nube euenit, per quam interdum weiten nubes apparet. Et iste uiuet nec cito morietur. Nam rubeus color, qui sub cute in genis hominis apparet, ut predictum est, igneum spiramen uite, scilicet anime, est, quod anima ignis est. Et ideo in genis ostenditur, quoniam anima in corpore illo secura sedet et cito exitura non est. Cum autem homo sanus est, si tunc rubeus seu moderate rubeus color in genis eius super cutem iacet, ita quod nulla cutis sub eadem rubedine in genis considerari potest: signum mortis habet, cum rubedo in genis eius tam fortiter pressa super cutem iacet, quod sub ea cutis uideri non ualet ut in rubeo

pomo, sub cuius rubedine nulla cutis uidetur, sed tantum superius quedam rubedo; et iste cito infirmabitur, et mors subsequetur. Rubeus enim color iste in genis super cutem iacens igneum spiramen uite anime est, quoniam anima in homine illo uim suam extra corpus suum demonstrat et se in corpore debilem et incertam ostendit, uelut homo, qui interdum ad ianuam domus sue procedit, cum de ea egressurus est." CC, V, 269:18–270:11.

12. "Quicquid homo bibet, siue uinum, siue cereuisiam, siue medonem, siue aquam, quelibet urina qualitatem siue sanitatis seu infirmitatis ostendit. . . ." CC, V, 273:16–18.

13. "Et quia idem humores in homine illo feruent, digestio eius recta et naturalis esse non potest, unde et succus eiusdem urine aliquantum conmiscetur, et ideo est spissa. Sed idem homo ob hanc infirmitatem diu languet, sed non moritur, quia humores in eo ab inuicem se non separant. Et cum eadem urina in urinale nouiter excepta est et rubea est, si tunc deinde aliquantum in palorem mutatur, et si etiam in ea uaria liniamenta ut uenule apparent, scilicet rubee et aquose ac turbide: signum mortis est, et morietur." CC, V, 277:26–278:5.

14. "Qui griseos oculos habet, interdum leuis, interdum preceps aut ualde lasciuus aut tardus aut in moribus suis incompositus est, sed tamen omnia que facit, probiter perficit. Qui igneos oculos habet, qui similes sunt nigre nubi iuxta sole site, iste prudens et acutus ingenio ac iracundus est." CC, V, 283:16–284:2.

15. In the manuscript the lunary is certainly a separate sixth Book because it begins, as do the other Books, with a large majuscule, although Kaiser printed it as part of Book Five. Schipperges maintained that it was spurious (*Heilkunde*, 41–42). For a review of lunaries, see Laurel Means, *Medieval Lunar Astrology: A Collection of Representative Middle English Texts* (Lewiston: E. Mellen, 1993).

16. For instance, a twelfth-century nun wrote the *Codex Guta-Sintram*, which is said to contain medical passages; see Gérard Cames, "Un joyau de l'enluminure Alsacienne: le Codex Guta-Sintram (1154) à Strasbourg," *Bulletin de la Societé nationale des antiquaires de France* (1978–1979): 255–261.

17. The frequency of hits was determined by using a scanned-in copy of *Causes and Cures*, the CLCT4 *Cetedoc* editions, and the *Patrologia Latina* versions of *Scivias, Epistolarium, Physica*, the *Liber vitae meritorum*, and the *Liber divinorum operum*. A rough idea of the frequency per page of text can be had by calculating the frequency of viriditas per page of text, although pages are not exactly equal since in some instances I had to use the CCCM version of a text, (where footnotes are at the bottom of the page); and in most others, the *Patrologia*, except for *Causes and Cures*. Viriditas occurred most frequently in her last work, the *LDO* (87); in the 1163 *LVM*, it occurred 44 times; in *Scivias*, 60, and in the medical *Physica*, 58 times, but only 39 times in *Causes and Cures*. There is no word for viriditas in the *Lingua ignota*.

18. That this is not simply a question of subject is proven by the fact that *Physica's* Book One, whose subject is the same as Books Three and Four of *Causes and Cures*, namely, the medicinal use of herbs, has the second *highest* frequency of hits, just before the *Liber divinorum operum*.

19. "Si sanguis et aqua in oculis hominis aut pre senectute aut pre aliqua infirmitate supra modum attenuantur uadat ad uiride gramen et illud tam diu inspiciat, dum oculi eius uelut lacrimando madefiant, quia uiriditas graminis illius hoc, quod in oculis eius turbidum est, aufert et eos claros et lucidos facit." CC, III, 211:10–14.

20. "Nam calor ficus et frigiditas erlen nature huius sunt, quod humiditatem ad se trahunt. Rore autem perfundentur, ut suauitate eius uiriditas eorum temperetur, et etiam calore solis, qui succum eorum ad lenitatem trahit, ne oculos ledat." CC, III, 214:17–20.

21. "In hyeme autem hec puluerizet et puluerem istum item cum escis suis manducet, quia uiriditatem earundem herbarum tunc habere non potest." CC, IV, 239:22–24.

22. "Et hee herbule uirides esse debent, quoniam uirtus earum in uiriditate precipue uiget. . . ." CC, IV, 235:13–14.

23. Müller also suggested that viriditas was a concept for explaining the stronger effect of green over dried plants. That is, by viriditas Hildegard meant that when the herbalist gathered herbs, their (actual) greenness needed to be taken into account. Without viriditas, a plant was powerless ("Stellung," 31). Moulinier argued the contrary—that Hildegard preferred green plants (and green food, e.g., green crab's liver) because they possessed a *metaphorical* viriditas (*Manuscrit Perdu*, 267).

24. A number of articles are exclusively devoted to Hildegard's viriditas; see Gabrielle Lautenschlager, "'Viriditas': Ein Begriff und seine Bedeutung," in *Hildegard von Bingen: Prophetin durch die Zeiten. Zum 900. Geburtstag*, ed. Edeltraud Forster (Freiburg: Herder, 1997), 224–237; Miriam Schmitt, "Hildegard of Bingen: Viriditas, Web of Greening Life Energy I," *American Benedictine Review* 50, no. 3 (1999): 253–276 and Miriam Schmitt, "Hildegard of Bingen: Viriditas, Web of Greening Life Energy II," *American Benedictine Review* 50, no. 4 (1999): 353–380. As for the Internet, there are more than 9000 references to viriditas, and counting.

25. On Hildegard's use of color in general, see Christel Meier, "Die Bedeutung der Farben im Werk Hildegards von Bingen," *Frühmittelalterliche Studien Jahrbuch des Instituts für Frühmittelalterforschung der Universität Münster* 6 (1972): 245–355; and Roland Maisonneuve, "Le symbolisme sacré des couleurs chez deux mystiques médiévales: Hildegarde de Bingen, Julienne de Norwich," *Sénéfiance* 24 (1988): 253–272.

26. "Hier ist 'viriditas' das Lebensprinzip einer höchsten Potenz. . . ." Schipperges, *Heilkunde*, 307.

27. "Unter dem Bild des Grüns gibt Hildegard eine umfassende Vorstellung der lebensfrischen Naturkraft, wie sie uns als das vegetative und animalische Lebensprinzip vor Augen tritt. Es ist die Farbe alles Keimens und Spriessens, alles Bluhens und Wachsens. . . . Das Grün ubertragt jene Energie, die allem sichtbaren Leben seine Straffheit und Farbigkeit sein Frische und Fulle zufuhrt." Schipperges, *Heilkunde*, 304.

28. For instance, Dronke found that "greenness, *viriditas*, is perhaps the most frequent and most central image in Hildegard's writings" ("Tradition," 83). For Feldman, "the central concept of health, for Hildegard, is viriditas." ("Dieser Zentralbegriff ihrer Heilkunde, die viriditas, die grüne Lebensfrische, spricht noch nach achthundert Jahren unmittelbar, ohne grosse Erklärungen") (*Hildegard von Bingen*, 126). For Müller, "Eine zentrale Bedeutung hat die 'viriditas'" ("Stellung," 31). For Pernoud, viriditas is one of Hildegard's favorite notions, designating internal energy ("l'une des notions favorites d'Hildegarde . . . designant cette énergie interne. . . ") (*Hildegarde de Bingen*, 106). For Berger, it is "a key notion throughout Hildegard's works (*Hildegard's Medicine*, 128); for Flanagan, "one of the first principles in her scientific works" (*Hildegard of Bingen*, 114); for Schmitt, "a vision of the cosmos which she consistently characterizes by one word, viriditas (greenness)" ("Viriditas," 283). For Baird, it "enters into the very fabric of the universe in Hildegard's cosmic scheme of things" ("Letters," 7). For Behling, "one of the most common images is viriditas—greening power—the living principle of nature, as in the springing of plants from the earth, the organized development of humans, as well as in the soul's flowering" (*Die Pflanzenwelt der Mittelalterliche Kathedralen*, 40).

29. "Eine von Hildegard postulierte Natur- und Lebenskraft" (Müller, "Stellung," 31).

30. "Hildegards theologische Anthropologie entspricht hingegen dem biblischen Befund. . . . Das unsichtbare Prinzip der Seele wird in diesem Zusammenhang als 'grünende Lebenskraft' beschrieben (230). . . ist die Natur für Hildegard eine Lehrmeisterin. Die Impuls hierzu gibt jedoch keine weltimmanente Kraft, sondern die Gottheit selbst" (Lautenschlager, "Viriditas," 234). Schmitt asserts that "since neither Scripture, the Rule nor other writers are the direct source for the multivalent meanings attached to this key concept by the poet-mystic, perhaps viriditas is Hildegard's singular idea to illuminate the dynamic interconnectedness and intercommunion of all levels of created life with each other and with God as Source" (257). At the literal level viriditas designated nothing less than the "vigor, life-energy and fertility of the cosmos" (Schmitt, "Viriditas I," 259).

31. Flanagan explored a similar subject, the attraction Hildegard's medicine holds for New Age adherents, in "Zwischen New Age und wissenschaftlicher Forschung: die Rezeption Hildegards von Bingen in der englischsprachigen Welt," in *Prophetin durch die Zeiten: zum 900. Geburtstag,*

ed. Edeltraud Forster (Freiburg: Herder, 1997), 476–484. The meanings culled by this school from Hildegard's material recall the "vegetation school" of Gawain studies. Indeed, there are authors who propose that Hildegard's viriditas originated in pre-Christian nature religion; see Robert Olson, "The Green Man in Hildegard of Bingen," *Studia Mystica* 15, no. 4 (1992): 3–18. That there are political stakes in seeing Hildegard's viriditas as a pagan remnant, as God in Nature, or as a physical substance, can be approached by means of the analogous historiography of the Greenness of Sir Gawain's Green Knight. But that is another story.

32.   On vitalism, holism, and science, see Ann Harrington, *Re-enchanted Science: Holism in German Culture from Wilhelm to Hitler* (Princeton: Princeton University Press, 1996); Catherine Velay Vallentin, "Le folklore et l'histoire: Le congrès internationale de folklore de 1939," *Annales*, Mars-Avril (1999): 481–506; Joachim Wolschke-Bulmahn and Gert Goenig, "From Open-mindedness to Nationalism: Garden Design and Ideology in Germany during the Early 20th Century," in *People-Plant Relationships: Setting Research Priorities*, eds. Joel Flagler and Raymond P. Poincelot (New York: Haworth Press, 1994), 133–150; and Heinrich Marzell, "Der Zauber der Heilkräuter in der Antike und Neuzeit," *Sudhoffs Archiv* 29, no. 2 (1936): 3–26. According to Harrington, the vitalists were responding to the bold nineteenth-century proclamation by Virchow that there was "no spiritual rector, no life-spirit, water-spirit or fire-spirit . . . everywhere there is mechanistic process only" (*Reenchanted Science*, 7). With its political agenda, vitalism was brought into the twentieth century by the German *Holismus* movement, which provided the philosophic underpinnings for both socialism and fascism. Thus *Holismus* supported "organic" socialistic communities like the kibbutz, as well as fascism and the concept of the Jew as prototypical rootless city dweller, cold scientist, and materialist banker (*Reenchanted Science*, xx). Ironically, Jewish kibbutzniks used *Holismus* in support of a Jewish return to an agricultural life in Palestine, while Nazi philosophers used it in support of a "natural" racism that was "rooted in the soil." There was therefore, a vitalism of the Left and a vitalism of the Right, just as there was a folklorism of the Left and of the Right. Both parties contrasted the vitalist principles of Holism with the values engendered by science, the city, and the machine. Both might have agreed with the idea that "a detachment of man from nature . . . leads to his annihilation . . . This striving for connectedness with all of life, indeed with Nature in general into which we are born—that, so far as I can see, is the deepest purpose and true essence of National Socialist thinking" of Ernst Lehman (*Reenchanted Science*, 172). Nazi *Holismus* even affected accounts of medieval horticulture. Hans Hasler, an important historian of medieval German horticulture, was a Nazi. Plants were categorized into native German and southern foreign species and proper German gardens had no place for foreigners; see

Hermann Fischer, *Mittelalterliche Pflanzenkunde* (München: Münchner Drucke, 1929). Our ideas of medieval German horticulture, not to speak of viriditas, are tinged by this vitalist romanticism.

33. "Quoque claritate simili modo, ut supra dictum est, quasi per speculum aerem uidebam, puritatem super puritatem limpidissimarum aquarum habentem, et splendorem super splendorem solis emittentem de se; qui et flatum habebat, qui omnem uiriditatem herbarum et florum paradisi ac terre continebat." *Liber vite meritorum*, pars : 6, cap. : 30 (visio), linea : 585 XXX. In prefata.

34. "Tu namque qui in uiriditate beati uiri per instructionem nominis sui positus es." *Epistolarium* (*Epistula Hildegardis*), epist.: 27R, linea: 36.

35. "Tu ergo qui in uice Christi es, uirgam correptionis illi per admonitionem prouide, quia dies eius sine uiriditate securitatis spei aridi sunt." *Epistolarium* (*Epistula Hildegardis*), epist.: 67, linea : 8.

36. "O ipsi insipientes et infructuosi exsistunt, quia sunt inutiles absque exaratione legis Dei et sine uiriditate uerbi ipsius." *Sciuias*, CCCM 43, pars: 2, uisio : 5, linea: 1350.

37. "In omni uiriditate beatitudinis . . ." *Sciuias*, CCCM 43, pars : 2, uisio : 6, linea : 475.

38. "Viriditate uiventis spirationis . . ." *Sciuias*, CCCM 43, pars : 2, uisio : 6, linea : 518.

39. "Non sic Unigenitus meus natus est, sed in uiriditate integritatis exiuit." *Sciuias*, CCCM 43, pars : 2, uisio : 6, linea : 1028.

40. "In plena uiriditate ueritatis." *Sciuias*, CCCM 43A, pars : 3, uisio : 6, linea : 835.

41. "Et sic de suspiriis et lacrimis istis uiriditas penitentiae in eodem homine exurgit." *Liber diuinorum operum*, pars : 1, visio : 4, cap. : 32 (commentarii), linea : 29.

42. "Similiter per exempla iustorum conpunctio in fideli homine aliquando exoritur, que+ illi uiriditatem bonorum operum et ariditatem malorum infert." *Liber diuinorum operum*, pars : 1, visio : 4, cap. : 39 (commentarii), linea : 52.

43. "Homo a Spiritu Sancto interius doctus, tunc de corde suo uiriditatem abstinentie. . . ." *Liber diuinorum operum* pars : 2, visio : 1, cap. : 34 (commentarii), linea : 49.

44. In fact, Hildegard almost always uses viriditas as a metaphor, not as a substance, in her theological writing. In *Scivias* she writes only of the viriditas of the virgin, of the holy spirit, and of God. Of the many passages on viriditas in the *Liber vitae meritorum*, only a few times does it signify a substance. In her letters, viriditas is never used as a substance, and in the *Liber divinorum operum* it appears as a substance only once.

45. Of the thirty-nine passages in *Causes and Cures* where viriditas occurs, it is used as a metaphor—the viriditas of knowledge—only once. "For whenever

a person wishes to learn something, the Holy Spirit sprinkles the greenness of knowledge with dew, so that he may learn and remember what he wants." ("Nam cum homo quodlibet opus uel quamlibet artem per optionem et per desiderium scire uult, spiritus sanctus rore suo uiriditatem scientie illius perfundit, unde et discit et capit, quod discere uult.") CC, II, 103:4–7.

46. "Radices herbarum in hyeme uiriditatem in se habent, quam in estate in flores emittunt." CC, II, 120:15–16.

47. "Swertula calida et sicca est et omnis vis ejus in radice est, et viriditas ejus in folia ascendit." *Physica*, col. 1177.

48. "Per radicem quoque arboris, que+ uiriditatem in se continet, flores et poma enutriuntur." *Liber diuinorum operum*, pars : 2, visio : 1, cap. : 43 (commentarii), linea : 65.

49. "Sed et cum sol calorem suum ad se trahit, aque maximum frigus habent et spumam suam, scilicet niuem, emittunt, que terram operit et uiriditatem terre inpinguat. . . ." CC, II, 54:1–3.

50. "Et sol totum firmamentum confirmat atque splendorem suum per totam terram spargit, unde ipsa uiriditatem et flores profert." CC, I, 27:22–24

51. "Et ita cum eo incedunt usque ad signum libre, ubi uiriditas et ariditas quasi in libra sunt, ita quod uiriditas abscedit et ariditas accedit." CC, I, 36:13–14.

52. "Cum uero sol ad terram declinatur, frigus terre de aqua ei occurrit et omnia uiridia arida facit . . . et hyemps est." CC, I, 27:25–27.

53. " . . . et sic in calore uiriditatem, in frigiditate ariditatem ostendit." CC, I, 54:22–23.

54. "Aqua . . . et lignis succum dat, pomiferis gustum, herbis uiriditatem . . ." CC, I, 44:16–18.

55. "Herbe etiam de fluente humore illius uiriditatem in se habent . . ." CC, I, 45:8–9. "Air has four powers: putting forth dew, exciting every viriditas, moving the wind by which flowers are brought out, and spreading heat, which ripens all. "Sed aer quatuor uires habet, uidelicet rorem emittendo, omnem uiriditatem excitando, flatum mouendo, cum quo flores educit, calorem dilatando, cumquo omnia maturescere facit. . . ." CC, I, 43:25–44:1.

56. "Stutgras contraria homini ad comedendum, ut unkrut, quoniam viriditas eorum mala est." *Physica*, col. 1184.

57. "De bertram. Et aliquantum sicci et eadem temperies pura est, et bonam viriditatem tenet." *Physica*, col. 1138C.

58. "Berwurtz calida est, et siccam viriditatem in se habet." *Physica*, col. 1184.

59. "Ringula frigida et humida est, et fortem viriditatem in se habet, et contra venenum valet." *Physica*, col. 1170.

60. "De asaro. Calidum existit et siccum, ac quasdam vires pigmentorum habet, quia viriditas ejus suavis et utilis est." *Physica*, col. 1206D.

61. "Apio. Quocunque autem modo comedatur, vagam mentem homini inducit quia viriditas ejus eum interdum laedit." *Physica*, col. 1159. Or, "whoever

wants to eat *ruba* needs to remove its skin, because it is sticky and its viriditas is harmful." ("De ruba. Et qui crudum comedere vult, exteriorem corticem totam auferat, quod spissa sit, quia viriditas illius hominem laedit.") *Physica*, col. 1164.

62. "De dille. Et hae herbae virides esse debent, quoniam virtus earum in viriditate praecipue viget. Quod si in hyeme est, herbas istas . . . haec pulverizet, et pulverem item cum escis suis manducet, quia viriditatem earundem herbarum tunc habere non potest." *Physica*, col. 1138.

63. "Et sol in igne radiorum suorum calorem super aromata et super flores illos misit, atque ros et pluuia illis humorem uiriditatis dedit." *Epistularium (Epistulae Hildegardis)*," CCCM 91, ep. : 85 R/B, linea : 3.

64. "Lac: Ipsa tunc humores et succos ac viriditatem in se habet." *Physica*, col. 1198.

65. "Gichtbaum viriditas ac succus ejus per se non valet." *Physica*, col. 1245.

66. "Et terra in qua idem uir a genibus suis ad suras suas erat, humorem et uiriditatem ac germen in se habuit." *Liber vite meritorum*, pars : 4 (visio), linea : 5.

67. "Crasso quam frigidus est, et etiam humidus existit, et plus de viriditate terrae quam de sole crescit." *Physica*, col. 1160D.

68. "Ruta de forti et de plena, id est queckin, viriditate terrae magis quam de calore crescit; et calorem temperatum." *Physica*, col. 1155B. Also mustard: "Sinape . . . vento illo crescit qui poma educit, et quia etiam de viriditate terrae crescit, et inde aliquantum succi habet." *Physica*, col. 1166C.

69. "In mane diei . . . herbae viriditatem tam fortiter sugunt, ut agnus qui lac sugit. . . ." *Physica*, col. 1249.

70. For a discussion of Hildegard's possible botanical sources, see Laurence Moulinier, "Deux jalons de la construction d'un savoir botanique en Allemagne aux xiie.-xiiie. siècles: Hildegarde de Bingen et Albert le Grand," in *Le Monde Végétal (XIIe. au XIIIe. siècles). Savoirs et Usages Sociaux*, eds. Allen J. Grieco, O. Redon, and L. Tongiorgi (Saint-Denis: Presses Universitaires de Vincennes, 1993), 89–105; and Laurence Moulinier, "Abbesse et Agronome: Hildegarde et le savoir botanique de son temps," in *Hildegard of Bingen: The Context of Her Thought and Art*, eds. Charles Burnett and Peter Dronke (London: The Warburg Institute, 1998), 135–56.

71. This is based on the medical texts available in the mid-twelfth-century Rhineland as laid out in Chapter Four. These include: Anonymous, "Epistula ad Antiochum"; Anonymous, "De quattuor humoribus"; Anonymous, "Ad Maecenatem,"; the Elder Pliny, *Plinii Secondii quae fertur una cum Gargilii Martialis medicina;* M. Wlaschky, "Sapientia artis medicinae"; Bartholomeus, *Practica;* Anthimus, *De observatio[ne] ciborum: Text, Commentary, and Glossary, with a Study of the Latinity*, ed. Shirley Howard Weber (Leiden: E. J. Brill, 1924). (N.B.: Since none of these texts is yet available as

a database, the methodology of this section differed from that in the rest of this chapter since it entailed reading the entire text.)

72. For a summary and timeline of the relevant treatises on pharmacy, see Rudolf Schmitz, *Geschichte der Pharmazie* (Eschbon: Govi, 1998). I have looked at the most popular herbals that were relatively available when Hildegard was writing her natural scientific texts, c. 1150; namely, the set of texts edited in Ernst Howald and Henry Sigerist, eds., *Antonii Musae de herba vettonica, liber Pseudo-Apulei herbarius, Anonymi de taxone, liber Sexti Placiti, liber medicinae ex animalibus=Corpus Medicorum Latinorum*, vol. 4 (Leipzig-Berlin: Teubner, 1927). The typical text simply provides a listing of plants (as well as stones in the lapidaries and animal substances in the bestiaries), along with their descriptions, synonyms, and medicinal properties.

73. Despite its promising title, the "De Herbis of Henry" is only a poem to plants; see Bernd Ruppel, "Ein verschollenes Gedicht des 12. Jahrhunderts: Heinrichs von Huntington 'De Herbis,'" *Frühmittelalterliche Studien* 31 (1997): 197–213.

74. For a review, see René Martin, *Recherches sur les Agronoms Latins et leurs Conceptions Economiques et Sociales* (Paris: Belles Lettres, 1971).

75. Horticultural material from Hildegard's era is scanty. I looked at the Carolingian *Cartulare*, the blueprint for the monastic gardens at St. Gall, and late material on gardens and gardening. For a literature review, see Geneviève Sodigné-Costes, "Simples et jardins, une histoire," in *Vergers et jardins dans l'univers médiéval*, Sénéfiance (Aix-en-Provence: C.U.E.R.M.A., 1990), 329–342. See also John H. Harvey, *Medieval Gardens* (Oregon: Timber Press, 1981); Meyvaert, "Monastic Infirmary Gardens" and Harvey, "Westminster Abbey." There are no references to viriditas in any of the essays in Allen J. Grieco, ed., *Le monde végétal (XIIe.-XVIIe. siècles)* (Saint-Denis: Presses Universitaires de Vincennes, 1993).

76. For instance, Strabo simply describes the medicinal effects of plants. Walter of Henley and Godfrey do convey practical material on gardening but do not discuss plant physiology. Of course, it may have been that plant physiology, and, therefore, Hildegard's understanding of viriditas, was part of oral culture. Braekman pointed out that "the treatise of Bollard is no doubt an example illustrative of a much vaster body of practical know-how, the majority of which was passed from generation to generation by word of mouth" ("Bollard's Middle English book of planting and grafting and its background," 26). Whether viriditas was part of oral culture is impossible to know. Certainly most gardeners will have some idea of the function of sap, however.

77. This is based on a computer search of Jerome's *Vulgate*. There are, of course, numerous comparisons in the Bible between grass and the short, fleeting life of human beings, but only this one reference to viriditas. Although Schmitt referred to seventy references to viriditas in the Old Testament and five in the New Testament, she references the *Douay* and *New American*

Bible concordances ("Viriditas I," 257) not the Vulgate, and they do not appear in a search of the *Vulgate* database. Moreover, there is only one reference to viriditas in Fischer's *Concordantiae;* see Bonifatius Fischer, *Novae concordantiae bibliorum sacrorum iuxta vulgatam versionem critice editam,* vol. 5 (Stuttgart: Frommann-Holzboog, 1977), col. 5625.

78. This is my awkward but literal translation. The King James version is more elegant: "The children of the ungodly shall not bring forth many branches, but are as unclean roots upon a hard rock. The weed growing upon every water and bank of a river shall be pulled up before all grass" (Ecclus. 40:15–16). The *Vulgate* has: "Nepotes impiorum non multiplicabunt ramos, et radices immundae super cacumen petrae sonant. *Viriditas* super omnem aquam et ad oram fluminis ante omnem faenum evelletur." Ecclus. 40:15–16.

79. "Non inmerito auctoris sui participans nomen collega est cum uiriditate gemmarum. . . ." Cassiodorus, "Uariarum libri duodecim," *Cl.* 0896, lib. : 2, epist. : 39, linea : 48.

80. "Smaragdi nitens viriditas. . . ." Isidorus Hispalensis, "De Lapidibus et Metallis," *Etymologarium,* PL82, col. 577A.

81. "Beryllius in India gignitur, gentis suae lingua nomen habens, viriditate similis smaragdo, sed cum pallore." Isidorus Hispalensis, *Etymologiarum siue Originum,* libri XX *Cl.* 1186, lib. : 16, cap. : 7, par. : 5.

82. "Species autem terrae est germinatio et viriditas agri." Ambrosius Mediolanensis, "Exameron," *Cl.* 0123, dies : 3, cap. : 6, par. : 25, pag. : 76, linea : 3.

83. Also "viriditas ubertasque silvarum," in Lactantius, PL6, Liber Septimus: "De Vita Beata," Capt. III, col. 745C.

84. "Vidisti amoenitatem arborum herbarum viriditatem" in Ambrosius Mediolanensis, "Explanatio psalmorum xii," *Cl.* 0140, psalmus 1, cap. 24, pars 4, pag. 18, linea 20.

85. "Quod nominatur in hac terra herba quaedam 'semper uiuit,' ibi uerum locum habet: ibi semper uiuit, uita sola est ibi, corruptio nulla, indigentia nulla est, uiriditas aeterna obtinetur, ariditas non timetur." Augustinus Hipponensis, "Sermones," *Cl.* 0284, sermo.: 25A, ed. : SL 41, linea : 73. Also,"herbidum dicimus locum, in quo herbarum viriditas nunquam cessat, herbosum autem, qui facile herbam generat, et ad tempus arescit." Isidorus Hispalensis, "De differentiis uerborum," *Cl.* 1187, par. : 287, col. : 39, linea: 37.

86. Thus, "in viriditate auri, ut quidam legunt. . . ." in Origenes sec. transl. Rufini, "Commentarium in Canticum canticorum," *Cl.* 0198 2 (A), lib. : 4, pag. : 234, linea : 4; and "utrum propterea ibi est auri uiriditas id est sapientia et caritas . . ." in Augustinus Hipponensis, "Enarrationes in Psalmos," *Cl.* 0283, SL 39, psalmus : 67, par. : 18, linea : 15. Du Cange was aware of this double meaning of viriditas even in his early dictionary, defining it as "rutilans," that is, red and shiny. Charles Du Cange, Du Fresne Sieur, *Glossarium mediae et infimae latinatis* (Niort: L.

Favre, 1886), s.v. "viridis." Green and gold come together in the alchemical "greened bronze" of the twelfth-century alchemist, Senior: "You have asked about viriditas, assuming that bronze is sick because of its viriditas, but, in fact, the only thing in the bronze which *is* perfect is its viriditas." Quoted by Carl G. Jung, *Psychology and Alchemy* (Princeton: Princeton University Press, 1968), 159.

87. "Ortus est uiror calami, id est uiriditas praedicationis, et uiror iunci, id est uiriditas rectae fidei." Heiricus Autissiodorensis, "Homiliae per circulum anni, pars hiemalis, hom.: 65, linea: 276 (in *Cetedoc*). Another example: "Quia plebem eius inutilem fidelis emergens populus gentium et pertinax in uiriditate fidei et usque ad consummationem temporum manens in fide patriarcharum protrusit." Paschasius Radbertus, "Expositio in Matheo. Libri xii," *CCCM*, 56B lib. : 9, linea : 3340.

88. "Sed quia viriditatem intimi amoris perdidit, iam stipula sicca est." Gregorius Magnus, *Moralia in Job. Cl.* 1708, SL143, lib. 8, par, 42, linea 93.

89. "In hortis sancta ecclesia in hortis unaquaeque anima habitat quae iam uiriditate spei et bonorum operum est repleta." Beda Uenerabilis, In *Cantica canticorum*, libri vi. *Cl.* 1353, lib. : 6, linea : 599.

90. "Planta in me radices verarum virtutum, et germina sanctae meditationis cum viriditate boni operis fac excrescere et sursum pullulare. . . ." Thomas à Kempis, *De elevatione mentis*, vol. : 2, pag. : 415, linea : 21.

91. "Ergo paradisus est plurima ligna habens, sed ligna fructifera, ligna plena suci atque uirtutis, de quibus dictum est: exultabunt omnia ligna siluarum ligna semper florentia uiriditate meritorum." Ambrosius Mediolanensis, "De paradiso." *Cl.* 0124, cap. : 1, par. : 3, pag. : 266, linea : 19.3.

92. "Et viriditatem vitae spiritalis accipiunt." Lathcen, *Ecloga de Moralibus Job quas Gregorius fecit, Cl.* 716, lib. 12, line 52.

93. "Ibi et lapis prasinus, qui uiridis est, et significat uiriditatem uitae quam debemus habere, ut non simus marcidi nec mortui, sed semper in uiriditate sanctae conuersationis." Aelredus Rieuallensis, *Sermones i-xlvi*, sermo.: 39, linea: 125.

94. "Neque enim hoc otiosum aut ab re fuit quod hoc uestis sacerdotalis uirore beati uiri discessio praeoccupari meruit qui aeternae uiriditatis fructum suis in actibus non dehabuit." Onulfus, *Uita Popponis Stabulensis*, pag. : 301, linea : 29.

95. For instance, the "viriditate dilectionis" of Gregorius Magnus, "Regula pastoralis." *Cl.* 1712, pars 3: cap 23, linea 22) or of youth ("viriditatem pueritiae") in Ambrosius Mediolanensis, "Exameron," *Cl.* 0123, dies: 3 cap. 17, par: 71, pag. 108, linea: 18.

96. "Si florentem vineam per inaequalitatem aeris immoderatum frigus attigerit protinus ab omni humore viriditatis arefacit." Guillelmus de S. Theoderico, *Excerpta ex lib. Gregorii papae super Cantica*, col.: 469, linea: 46.

97. *Laetari* can mean to be fertilized, according to Lewis and Short, s.v. "laeto."

98. "Arborem, quam florere uides, quam summa conspicis uiriditate laetari, subterraneo suco fecunditatis animatur, reddens in superficiem quod continet in radice." Cassiodorus, "Variarum libri duodecim," *Cl.* 0896, lib. : 9, epist. : 2, linea : 12.

99. "Liquet enim quia frumentum cum seminatur in terra, in prima quidem facie deficit, sed unde putrescit prius in pulvere inde post modum viriditatem recipit in renovatione." Heiricus Autissiodorensis, *Homilae per circulum anni*, Pars Aestiva, Hom.: 49, linea 76.

100. "In tanta namque unius grani paruitate et paene nulla sui dissimilitudine, ubi latet ligni duritia, et ligno tenerior uel durior medulla, asperitas corticis, uiriditas radicis, sapor fructuum, suauitas odorum, colorum diuersitas, mollities foliorum?" Gregorius Magnus, "Moralia in Iob," *Cl.* 1708, SL 143, lib. : 6, par. : 15, linea : 53.

101. "Et cum ad aedificium arbusta succidimus ut prius viriditatis humor exsiccari debeat." Gregorius Magnus, "Registrum epistularum," *Cl.* 1714. SL 140. lib. 5., epist. 58, linea 63.

102. "Hiems est; intus est uiriditas in radice." Augustinus Hipponensis, "In Iohannis epistulam ad Parthos tractatus," *Cl.* 0279, tract. : 9, col. PL : 2050, linea : 42.

103. "Haec est uiriditas in aestate arborum per hyemem quasi arescentium, sed in occulto uirentium." Augustinus Hipponensis, "Sermones," *Cl.* 0284, sermo.: 25A, ed. : SL 41, linea : 67

104. "Consideremus nunc ubi in illo paruo grano seminis latet fortitudo ligni, asperitas corticis, saporis odoris que magnitudo, ubertas fructuum, uiriditas foliorum." Gregorius Magnus, "xl homiliarum in euangelia libri duo," *Cl.* 1711, lib.: 2, hom. : 26, cap. : 12, linea : 28.

105. "The palm tree bears late, staying in its viriditas for a long time." ("Palma enim tarde proficit, sed diu in viriditate subsistit.") Gregorius Magnus, "Moralia in Job," *Cl.* 1708, SL 143A, lib. 19, par. 27, linea 22. By implication, its viriditas becomes fruit.

106. "Sed ille in exiguo grano mirabilior praestantior que vis est, qua valuit adiacens humor conmixtus terrae tamquam materies verti in ligni illius qualitatem, in ramorum diffusionem in foliorum viriditatem ac figuram, in fructuum forma." Augustinus Hipponensis, "De Genesi ad litteram libri duodecim," *Cl.* 0266, lib. 5: par. 23, pag. 167, linea 22.

107. "Infructuosae quidem arbores salicum, sed tamen tantae viriditatis sunt." Gregorius Magnus, "Moralia in Job," *Cl.* 1708, SL 143B, lib. 33, par: 5, lines 1.

108. "Quasi faenum germinat viriditatem in specie, non in fructu soliditatem." Ambrosius Mediolanensis, "Exameron," *Cl.* 0123, dies 3, cap. 7, par. 29, pag. 78, linea 17.

109. "Palmes, qui de vite procedit, succum non habet in semetipso, sed succum viriditatis vitis transfundit in palmites, ut de succo illius palmites viridescant

et effloreant." Godefridus (siue Irimbertus?) Admontensis, *Homiliae festi-uales hom.*: 41, col. : 835, linea : 53.

110. There is not a lot of material on medieval botany, but see Karen Reeds, *Botany in Medieval and Renaissance Universities* (New York: Garland, 1991); Bernard Ribémont and Geneviève Sodigné-Costes, "Botanique médiévale: tradition, observation, imaginaire. L'exemple de l'encyclopédisme," in *Le Moyen Age et la Science: Approche de quelques disciplines et personalités scientifiques médiévales*, ed. Bernard Ribémont (Paris: Klinksieck, 1991), 153–169; Mauro Ambrosoli, *The Wild and the Sown: Botany and Agriculture in Western Europe, 1350–1850*, trans. Mary McCann Salvatorelli (Cambridge: Cambridge University Press, 1997); Agnes Arber, *Herbals, Their Origin and Evolution: A Chapter in the History of Botany (1470–1670)* (Cambridge: Cambridge University Press, 1912); Agnes Arber, "From Medieval Herbalism to the Birth of Modern Botany," in *Science, Medicine, and History : Essays on the Evolution of Scientific Thought and Medical Practice Written in Honour of Charles Singer*, ed. Edgar Ashworth Underwood (London, New York: Oxford University Press, 1953), 317–336; Charles Joret, "Les incantations botaniques des manuscrits f. 277 de la bibliothèque de l'école de médecine de Montpellier et f. 19 de la bibliothèque académique de Breslau," *Romania* 17 (1888): 337–354; A. G. Morton, *History of Botanical Science* (London: Academic Press, 1981); Laurence Moulinier, "La botanique d'Hildegarde de Bingen," *Médiévales* 16–17 (1989): 113–129; John Riddle, "Medieval Medical Botany," *Journal of the History of Biology* 14, no. 1 (1981): 43–81; and Katherine E. Stannard and Richard Kay, eds., *Pristina Medicamenta: Ancient and Medieval Medical Botany* (Aldershot: Variorum, 1999).

111. Little is known about *De plantis'* Latin translator and commentator, Alfred Sareshal (or Shareshill), but see Roger French, "The Use of Alfred of Shareshill's Commentary on the 'De Plantis' in University Teachings in the Thirteenth Century," *Viator* 28 (1997): 223–252; H. J. Drossaart Lulofs and E. L. J. Poortman, eds., *Nicolaus Damascenus De Plantis: Five Translations* (Amsterdam: North Holland Publishing Company, 1989); and R. James Long, "Alfred of Sareshel's Commentary on the Pseudo-Aristotelian 'De Plantis': A Critical Edition," *Medieval Studies* 47 (1985): 125–167. Each author proposes a different hypothetical biography for Alfred. Long's and Lulofs' Alfred was Jewish and probably a doctor, with dates as early as 1160–1200, while French's was from the West Midlands, and lived somewhat later. The text was popular (preserved in more than 170 manuscripts) and, until supplanted by the derivative text of Albertus Magnus, was the sole source of written botanical information in medieval Europe.

112. Because it mentions ancient Greek authors, it is thought to be an originally Greek treatise, written between the fourth century B.C. and the first century A.D. It was then translated into Syriac in the ninth century, and

       into Arabic somewhat later. "The prevailing opinion is that *De plantis* is a compilation of extracts from Aristotle and Theophrastus" (Lulofs, *De plantis*, v). However, the earliest surviving manuscripts are in Latin; the Arabic and Hebrew manuscripts come from the fourteenth century, and the existing Greek text, which provides the basis for the *De plantis* printed in the Loeb edition of Aristotle, is, in fact, a fifteenth-century Greek rendering of Alfred of Shareshill's Latin text! (Lulofs, 563).

113.   I use the Latin edition of Lulofs, although I have also examined Long's version. The text is lengthy, and I have summarized, translated, excerpted, and rearranged the material. Unless otherwise noted, the citations refer to the page and/or passage number in Lulofs.

114.   "Quia generari, nutriri, augeri, iuventute virescere senioque dissolvi conspexerunt." Lulofs, 517.

115.   "Calor et humor naturalis." Lulofs, 525. "Quando consumabuntur, infirmabitur et veterascet et corrumpetur et arefiet." Lulofs, 525.

116.   "Et [quaedam] partes arborum simplices ut humor inventus in ea et nodi et venae; et quaedam partes sunt compositae ex his ut rami, virgae et similia." Lulofs, 525, section 61.

117.   "Sicut pili hominis et ungues." Lulofs, 570, section 70.

118.   "Et sicut est in ovo vis generandi pullum et materia cibi eius usque ad horum sui complementi et sui exitus ab eo . . . ita et planta." Lulofs, 523, section 46.

119.   "Tota siquidem planta quattuor indiget . . . semine terminato, loco convenienti, aqua moderata, aere consimili." Lulofs, 551, section 99. As we saw in Chapter Three, "air" meant climate.

120.   "Et ipsa indiget temporibus anni et sole et temperantia et vere plus omni re." Lulofs, 523, section 43.

121.   "Plantarum quaedam est domestica, quaedam hortensis, quaedam silvestris." Lulofs, 531, section 93.

122.   "Et quaedam plantarum nascuntur in locis siccis et quaedam maribus et quaedam in fluminibus . . . , in locis siccis . . . , in montibus . . . , in planitie . . . , in locis aridissimis . . . , in locis altis . . . , etc." Lulofs, 531, sections 96 and 97.

123.   This is a summary of Lulofs, *De plantis*, 528, section 78.

124.   "Quia humor qui est in magnis arboribus in quibusdam est ut lac, ut in ficubus, et in quibusdam similis est pici, ut humor qui est in vite, et quidam est origanalis, ut qui est in origano, et in planta quae dicitur opigaidum." Lulofs, 528, section 75.

125.   "Estque principium cibi plantarum a terra et principium generationis fructuum a sole, quamvis Anaxagora dixit quod earum frigus est ab aere. Et ideo dixit Lecineon quod terra est mater plantarum et sol pater." Lulofs, 523, section 44.

126.   "Et sunt qui putant plantam completam et integram esse propter duas vires quas habet, et propter cibum qui adaptatus est ad cibandum illam,

et longitudinem suae existentiae et temporis sui. Et quando fronduerit et fructifecaverit durabit vita eius et vertetur ad illam iuventus eius non fiet in ea aliquid superfluum." Lulofs, 524, section 50.

127. "Locus vero et labor huic rei conferunt et maxime tempus anni in quo plantatur." Lulofs, 536, section 126.

128. "Et planta terrae affixa non separatur ab ea. Quidam quoque loci meliores sunt quibusdam." Lulofs, 532, section 99.

129. "Planta autem quae provenit in montibus altis, si fuerit species, erit perniciosior et aptior medicinae." Lulofs, section 200. "Loca vero a sole remota non erunt multarum plantarum . . . nec habebit planta vires folia et fructus producendi." Lulofs, section 201.

130. This is Long's interpretation of "vita occulta" (131).

131. "Nec sensum habent nec desiderium: desiderium enim non est nisi ex sensu, et nostrae voluntatis finis ad sensum converititur." Lulofs, section 11.

132. "Non habet autem planta motum ex se: terrae enim affix est." Lulofs, section 22.

133. "Et non habet planta spiritum . . . et omne animal habet animam, sed planta est res imperfecta." Lulofs, 521, section 33.

134. "Et nodi et venae et caro totius plantae ex quattuor sunt elementis." Lulofs, 529, section 86.

135. "Et cortices et lignum et medulla arboris nascuntur ab humore." Lulofs, 529, section 84.

136. "Et radix plantae mediatrix est inter plantam et cibum. Et ideo vocant eam Graeci radicem et causam vitae plantarum, quia ipsa causam vitae plantis adducit." Lulofs, 529, section 81.

137. "Plantae vero secundo modo inest motus, [et est attracti] et est vis terrae quae attrahit humorem; eritque in attractione motus venitque ad locum." Lulofs, 540, section 142. And "humorem ad extremitates plantae trahit dispergiturque materia per omnes partes eius." Lulofs, 540, section 147.

138. "Est autem prima digestio desub planta et secunda quae est in medulla quae exit a terra quae est in media planta." Lulofs, 555, section 222.

139. Here I use the edition of French for my translation. "Hec est ratio viriditatis extrinsecus in arboribus sicut et in omnibus terre nascentibus quia calor aeris plurimum habundans extrinsecus nutrimentum arboris scilicet humorem vel terreum vaporem ut plurimum trahit ad exteriora in quo humore cum sit multum aquositatis et debilis agens caliditas est color viridis et quod materia sit plurimum terea, patet quia desiccata humiditate remanet cortex nigra. In foliis autem quibusdam maior habundat aquositas cum aliquantula terreitate. Unde et in eis non est adeo intensa viriditas quod patet cum arescunt fiunt citrina. Interius autem melior celebratur digestio propter temperatam humiditatem non superhabundantem. Et ideo arbor alba est interius, vel secundum naturam nutrimenti." "The Use of Alfred of Shareshill's Commentary," 243. Lulofs gives: "Viriditas vero plantae debet esse res communissima

in arboribus. Videmus quod communius est albedo et viriditas exterius. Et hoc est quia materia utuntur propinquiori. Opportet ergo ut sit viriditas in omnibus arboribus, quia materiae attrahuntur et rarificant lignum arboris, fluitque calor per parvam digestionem remanetque ibi humor. Apparetque exterius: erit ergo viriditas. Et hoc est in foliis nisi quia maior inest digestio." Lulofs, 554, section 215. Long paraphrases the text as: "Greenness must be the most common characteristic of plant life . . . [because] in every life form that draws its nutrient from the earth there is present in dominant measure the *earthy humor*. Now the *color* of this humor, since it has an admixture of *moisture* and *heat* (though the latter only minimally) is *green*. That the earthy humor predominates, moreover, is evident because the desiccated cortex turns black. In some leaves, however, there is a greater percentage of moisture and a correspondingly smaller percentage of the earthy element; this means that their green will be less intense. Indeed we note that when such leaves dry out they turn yellow" (Long, "Alfred of Sareshel's Commentary on the Pseudo-Aristotelian 'De Plantis': A Critical Edition," 135–136). [Emphasis added.]

140. "Ergo in processu anni retinebit illa aqua colorem propter frigiditatem acris. Et quia accessit calor ad frigus, impulit calor humiditatem exterius cum eo quod tinxit colore caloris, et apparet ideo color in apparenti arboris. Consequenter vertuntur frigus et siccitas ad actum, et humor retinet calorem, et ideo apparet alius color." Lulofs, 559, section 245.

141. "Eritque sapor fructus acidus in suctione, et quando plus digestum fuerit, dissolvetur paulatim aceositas, donec consumatur apparebitque dulcedo. Erit ergo fructus dulcis, foliaque suae et extremitates acida. Cumque perfecta fuerit maturitas, erit amarus, et hoc est propter superfluum calorem cum pauco humore." Lulofs, 559, section 248.

142. "According to the agronomists, the least digested sap produced leaves, a food that is often referred to as fit for animals or for the poorest urban and rural population. The better digested sap produced flowers. . . . Best of all, the most perfectly digested sap produced fruit." Allen J. Grieco, "The Social Politics of pre-Linnaean Botanical Classification," *I Tatti Studies: Essays in the Renaissance* 4 (1991), 149.

143. "Succorum vero qui in fructibus sunt quidam sunt potabiles, ut succus uvarum, malorum [malorum granatorum], mororum et myrti; quidam vero unctuosi, ut succus olivae, nucis pinei; quidam dulces mellares, ut dactyli et ficus; quidam calidi et acuti, ut origani et sinapis et quidam amari, ut absinthium et centaurea." Lulofs, 533, section 105.

144. According to Pastoureau, Bartholomew's passage sums up twelfth-century ideas about green. There is still no complete new edition of the *De proprietatibus rerum*, although several chapters have been recently edited, and there are modern editions of Trevisa's English translation and of the medieval French. See Heinz Meyer, *Die Enzyklopädie des Bartholomaeus*

Anglicus: Untersuchungen zur Überlieferungs- und Rezeptionsgeschichte von 'de proprietatibus rerum' (Munich: Fink, 2000); and M. C. Seymour and Colleagues, *Bartholomaeus Anglicus and His Encyclopedia* (Aldershot: Variorum, 1992).

145. "Color viridis generatur actione calidi in materia mediocri, vergente tamen ad dominium humidi, ut patet in foliis, fructibus, et herbis, et ideo generabitur color multum habens de nigredine, non tamen perringens totaliter ad nigredinem. Ex admixtione enim remissi albi ut glauci, et intensi nigri in superficie humidi, viriditas generatur, quando calor agens in materia non potest adurere humidum, nec ad plenum docoquere, ut totaliter in nigrum convertatur. Unde viriditas in herbis et in fructibus signum est crudi humoris et indigesti, ut dicit Aris. Et hoc patet, quia color viridis in plantis et in fructibus mutatur in glaucum in autumno, quia in foliis et in herbis est multa materia humida et grossa, quae per actionem calidi paulatim consumitur, nec omnino destituitur a colore in materia elevante, licet frigus dominetur. Et ideo quaedam arbores virescunt in vere et in aestate, in hyeme vero vel in autumno pallescunt, quia adveniente caliditate vernali, provocatur humor ad exteriora, qui tactus calore fit viridis, sed advenient frigiditate repercutitur humor, et abundat siccitas, et fit color glaucus, et dicit Comment super Arist de plant infime." Bartholomaeus Anglicus, *De proprietatibus rerum*, 1154–1155.

146. "Est itaque color viridis medius inter rubeum et nigrum generatus et hoc patet per transitum cholerae rubeae in melancholiam innaturalem. Cholera enim cum sit rubea, transit in melancholiam, quae est nigra, mediante cholera innaturali, scil. aeruginosa et praesina, quae viridis invenitur. Et ideo color viridis maxime delectat visum, propter concursum partium ignearum et terrestrium. Nam luciditas ignes, quae in viridi est temperata, delectat visum. Obscuritas etiam terrea sive nigredo, cum non sit in extremo, mediocritet spiritum visibilem aggregat et confortat. Et ideo nullus color est ita delectabilis visui sicut viror, ut patet in smaragdo, qui oculos sculpentium gemmas et metalla maxime reparat et confortat." Bartholomaeus, *De proprietatibus rerum*, 1155.

147. "Sunt autem viridia folia planta et gramina et caetera terrae nascentia ex dominio partium terrestrium, in quibus radicatur, tamquam in materia, et ex virtute ignea tanquam ex causa effectiva, quae dissoluit terrea et subtiliat et rarefacit, et sic attrahendo earum fumos ad extrema tingit herbae superficiem tali colore non nigro nec rubeo sed virore. Nam nigredo temperat rubedinis disgregativam claritatem et claritas incoporata nigredivis, ipsam reducit ad mediocritatem. Et dominio ergo partium terrestrium et ignearum viridis color generatur. Et quamvis folia fructus et gramina virescant, flores tamen virides raro vel nunquam inveniuntur, quod accidit propter subtilitatem materiae florum, in qua si praedominantes partes fuerint aquae e acetae, erit color albus. Si vero aquae cum igneis prevalu-

erint, erit color glaucus vel pallidus aut citrinus. Si autem praeabundeverint igneae partes aereis erit color roseus et si fuerint prevaleres partes aquae cum terrestribus erit color blavius vel violaceus. Si autem aeque prevaluerint partes igneae cum terrestribus posse quidem fieri color viridis vel niger sed talem commixtionem subtilitas materia floris vel rarefacto non admittit. Et ideo flores non virescunt generalitur nec nigrescunt." Bartholomaeus Anglicus, *De proprietatibus rerum*, 1155–1156.

148. "Et ergo color viridis inter rubedinem medius et nigredinem, delectans visum et ad sui aspectum oculorum attractivus, aciei visus confortativus et reparativus. Unde cervi et animalia agrestia alia loca virentia diligunt et frequentant, non tantum propter pastum, verum etiam propter visum. Et ideo venatores viridibus vestimentis semper se induunt, quia propter aspectum viroris quem bestia naturalitur diligunt, minus venatorurm insidia expavescunt ut dicit galen." Bartholomaeus Anglicus, *De proprietatibus rerum*, 1156.

149. This bit is not in the Latin edition of 1601, but is in the French manuscript published by Salvat. ". . . Et pour ce il a moult de noir en la couleur verte, et est la verdure des fruis et des herbes signe de humeur crue et mal digeree, sicomme dit Avicenne. Et ce appert car tant comme l'umeur des fruis est plus digeree et plus meure, tant plus appetice la verdure et vient autre couleur sicomme blanc ou rouge ou noir ou jaune." See Michel Salvat, "Le traité des couleurs de Barthelemi l'Anglais (XIIIe. s.)," *Sénéfiance* 24 (1988), 379.

150. On colors in the Middle Ages, see Michel Pastoureau, "Le temps mis en couleurs: Des couleurs liturgiques, aux modes vestimentaires XIIe.-XIIIe. s.," *Bibliothèque de l'Ecole des Chartes* 157 (1999): 111–135; Michel Pastoureau, "Voir les couleurs au XIIIe. siècle," *Micrologus* 6 (1998): 147–165; Michel Pastoureau, *Couleurs, Images, Symboles: Etudes d'histoire et de l'anthropologie* (Paris: Le Léopard d'Or, 1989); Michel Pastoureau, *Figures et Couleurs: Etudes sur la symbolique et la sensibilité médiévale* (Paris: Le Léopard d'Or, 1986). See also Peter Dronke, "Tradition and Innovation in Medieval Western Colour-Imagery," *Eranos Jahrbuch* 41, no. 4 (1972): 51–107; and Ulrich Ernst, "Farbe und Schrift im Mittelalter unter Berücksichtigung antiker Grundlagen und neuzeitlicher Rezeptionsformen," in *Testo e immagine nell'alto Medioevo* (Spoleto: Centro italiano di studi sull'alto Medioevo, 1994), 343–415. See also Roland Maisonneuve, "Le symbolisme sacré"; and Sylvie Fayet, "Le regard scientifique sur les couleurs à travers quelques encyclopédistes Latins du xiie. s." *Bibliothèque de l'École des Chartes* 150 (1992): 51–70. Most research, however, focuses on late literary sources like the romances, theological tracts, sumptuary laws and heraldry and may not reflect popular attitudes.

151. Although perhaps the character of green also came in part from how it was made, in practice, by painters and dyers. For instance, according to Theophilus, the painter could make green either from saps and/or metallic

copper; and there were four types: sap (*succus*) green (from buckthorn, iris, leek, and elderberry), Spanish green (from copper and vinegar), salt green (from copper, vinegar, and salt), and emerald green; see Presbyter Theophilus, *Theophilus presbyter schedula diversarum artium*, ed. Albert Ilg (Osnabruck: O. Zeller, 1970[1874]); and Theophilus Presbyter, *On Divers Arts: The Treatise of Theophilus*, trans. John G. Smith Hawthorne and Cyril Stanley (Chicago: Chicago University Press, 1963). Notably, when a modern stained-glass designer followed Theophilus' instructions, he found that "iron produces green with an *equilibrium* of oxidation and reduction," suggesting that the painter's green was, in fact, created only when two opposing forces were in balance. See Donald Royce-Roll, "The Colors of Romanesque Stained Glass," *Journal of Glass Studies* 33 (1994), 75, emphasis added.

152. According to Hugh of St. Victor, green was the most beautiful of colors, because it meant the new life in seeds. "Above all beauty—green. How it takes the spirit as we look upon the new life of seeds as they come forth." ("Postremo super omne pulchrum viride. Quomodo animos intuentium rapit, quando vere novo, nova quadam vita germina prodeunt et erecta sursum in spiculis suis quasi deorsum moret calcata ad imaginum futurae et resurrectionis in lucem pariter erumpunt.") Hugh of St. Victor, *Disdascalion*, PL176, cols. 820D–821B.

153. Green was in the middle of the *heraldic* color scheme (white/yellow/red/green/blue/black), although it could sometimes be found near black or even off by itself in other formulations. Likewise it was a tempering, mediating color for liturgical use, between the extremes of white, black and red. According to Pope Innocent III, "viridis color medius est inter albedinem et nigritiam et ruborem." Pastoureau, "Le temps mis en couleurs," 117.

154. Indeed, the devil wore green and green was the color of Islam and of jealousy; D. J. Robertson, "Why the Devil Wears Green," *MLN* 69 (1954): 470–472.

155. "Riuulus autem menstrui temporis in muliere est genitiua uiriditas et floriditas eius, que in prole frondet, quia ut arbor uiriditate sua floret et frondet et fructus profert, sic femina de uiriditate riuulorum menstrui sanguinis flores et frondes in fructu uentris sui educit." CC, II, 145:6–9. (In modern French menstruation is still a woman's "flowers" [*les fleurs*].)

156. "Sed ut arbor, que uiriditate caret, infructiferum lignum dicitur, ita et femina, que uiriditatem floriditatis sue in forti etate non habet infertilis nuncupatur." CC, II, 145:9–11.

157. "Et idem Adam de uiriditate terre uirilis et de elementis fortissimus erat. . . ." CC, II, 77:5.

158. "Sed si quis masculus hiis duabus uiribus forte naturaliter per defectum aut per abscisionem caret, uirilem uiriditatem non habet. . . ." CC, II, 140:9–11.

159. Schipperges also noted that viriditas could be within the body ("Ach des Menschen Fleisch und Blut sind grün, weil sie die Offenheit und Bereitshaft des Leibes darstellen. . .") *Heilkunde*, 303.

160. "Et uiriditas anime spumam et humiditatem ad caput, scilicet in cerebrum misit. . . ." CC, II, 72:1–2.

161. "Aqua enim in homine est, cum sanguis in eo non deest. Et hec humiditatem in homine facit, ita quod etiam uiriditas in eo uiget et quod coagulatio ossium in eo perdurat." CC, II, 74:6–8.

162. "Homo enim de frigiditate et caliditate fecundus est . . . quia calor uiriditas eius ac frigiditas ariditas eius est; ac per hec omnia germinat." CC, II, 75:10–12.

163. "Et quia hunc defectum in corporibus habent, ideo etiam tardi in ingenio sunt, et uene temporum eorum non sunt plene in uiriditate, sed ad similitudinem calamorum et quarundam herbarum fragiles uenas habent. . . ." CC, II, 113:13–16.

164. "Sed postquam alicui homini crines in caluitio ceciderint, amplius nullo medicamine restaurari poterunt, quoniam humectatio et uiriditas, quam in cute capitis, id est in hirneschedele suo, prius habebat, iam exaruit, nec amodo ulla uiriditas ibi exsurgere poterit, unde etiam capilli amod ibi non renascentur." CC, II, 129:11–15.

165. "Et quia etiam aliquantum uiriles sunt propter uiriditatem, quam in se habent, aliquantum lanuginis circa mentum interdum emittunt." CC, II, 127:11–13.

166. "Sic etiam forte et pretiosum uinum uim uesice hominis torret, ita quod ipsa rectam uiriditatem medulle eius afferre non poterit." CC, II, 181:14–16.

167. "Et etiam cerebrum hominis tangit: quia in viribus suis non solum terrena sed etiam caelestia sapit, cum Deum sapienter cognoscit: ac se per omnia membra hominis transfundit, quoniam viriditatem medullarum ac venarum et omnium membrorum toti corpori tribut, velut arbor ex sua radice sucum et viriditatem omnibus ramis dat." *Scivias*, CCCM 43, pars: 1, visio 4, linea 565.

168. "Et simul conmixta in locum, ubi talpa fodit, ponenda sunt, quia eadem terra sanior quam alia est, quatinus ibi succum et uiriditatem suam cum succo et uiriditate terre accipiant, quia etiam prius succo terre perfusa sunt. . . ." CC, IV, 253:19–20.

169. "Viriditatem et fructus quibus homines nutriri deberent." *Liber vite meritorum*, pars 3, cap. 26, linea 510.

170. "Id est in uiriditate calorem ei infundens, quoniam terra est carnalis materia hominis, nutriens eum suco suo sicut mater lactat filios suo." *Scivias*, pars : 2, visio : 1, cap. : 7 (commentarii), linea : 213.

171. "Terra nutrit viriditatem; viriditas, fructum, fructus, animalia." *Scivias*, pars 2, visio 1 cap. 2, linea 128.

172. "Dynames herbarum quas ut crescentibus nemeris lunae observes, dum tollis et conponis, curato . . . cerebra etiam hominum augentur crescente luna." Anonymous, *Ad Maecenatem*, 13.

173. In *Cure que ex hominibus fiunt*, a *viridis* urine signified internal illness. "Urina viridis in visceribus morbum esse significat." (Roberto Simonini, "Herbolarium et materia medici: Codex MS n296 della Biblioteca Governmental di Lucca." *Atti e memorie della R. Acc. di sc. lett ed arti di Modena* 5, no. 1 (1936), 208.)

174. In Vindician, bile was called *viridis:* "Virtus sane ipsorum talis est . . . cholera xanthe id est rubea amara viridis ignea et sicca" ("The power of . . . yellow coler is red, bitter, *viridis*, fiery and dry.") "Ad Pentadium," 486.

175. Although found no earlier than texts from the thirteenth century, all three concepts were ancient ideas and may have been common knowledge. Certainly, it is simply a matter of experience that the live body differs from the corpse in very particular ways—it is hot, not cold; moving, not still; bleeds if injured, and repairs itself. These experiential aspects may have been an important source for the concepts of *calor innatus, humidum radicale*, and *vis medicatrix naturae.*

176. See Michael R. McVaugh, "The 'Humidum Radicale' in Thirteenth-Century Medicine," *Traditio* 30 (1974): 259–283. *Humidum* can mean either "moist" or "full of sap," according to the OLD, s.v. "humidus." Perhaps the *humidum radicale* should be translated as *rooty sap*, a direct equivalent to viriditas.

177. Since semen was considered to be the most refined end-product of the digestion of food, it may have provided a conceptual analogue for Hildegard's discovery of a bodily viriditas. Although this is pure speculation, such a leap does occur in Ayurvedic medicine (with *rasa*) and in Chinese medicine (with *ching*).

178. On this metaphor, see Peter H. Niebyl, "Old Age, Fever, and the Lamp Metaphor." *Journal of the History of Medicine* 26 (1972): 351–368.

179. See Michael Stolberg, "Die Lehre vom 'calor innatus' im lateinischen Canon medicinae des Avicenna," *Sudhoffs Archiv* 77, no. 1 (1993): 33–53; Richard J. Durling, "The Innate Heat in Galen," *Medizinhistorisches Journal* 23 (1988): 210–221; Friedrich Solmsen, "The Vital Heat, the Inborn Pneuma, and the Aether," *Journal of Hellenic Studies* 77 (1957): 119–123.

180. Stolberg states that it is equivalent to the healing power of Nature; see below.

181. Stolberg called it the "unrenewable, life-preserving faculties of the 'innate heat'"(53).

182. Again, the physician's role was to prescribe a regime that would both minimize the wasting of the *calor innatus* and replenish it. There are parallels to

Eastern doctrines, especially to tantric techniques for conserving and acquiring *chi*.

183. The *Isagoge* mentions still other powers and spirits in the body. There was the natural power (*virtus naturalis*), which allows the body to take in, retain, digest, transform, and expel the exterior world; the breathy power (*virtus spiritalis*), which moves the heart and arteries; and the animating power (*virtus animalis*), which causes the body to move and to sense (Johannitius, *Isagoge*, 319–323). Each of these powers expresses one of the aspects of the living, growing, breathing, moving body. There were also the three "spirits" of the three main organs: the *spiritus naturalis* of the liver, the *spiritus animalis* of the brain, and the *spiritus vitalis* of the heart (Johannitius, *Isagoge*, 335). Although these concepts do overlap with Hildegard's use of viriditas in some ways, they are not substances but powers. See also Hill, "The Grain and the Spirit in Medieval Anatomy," and James Bono, Jr., "Medical Spirits and the Medieval Language of Life," *Traditio* 40 (1984): 91–130.

184. See Max Neuburger, "A Historical Study of the Concept of Nature from a Medical Viewpoint," *Isis* 35 (1944): 16–28; Max Neuburger, *The Doctrine of the Healing Power of Nature throughout the Course of Time*, trans. Linn J. Boyd (New York: s.n., 1932); and John Harley Warner, "The Nature-Trusting Heresy: American Physicians and the Concept of the Healing Power of Nature in the 1850s and 1860s," *Perspectives in American History* 11 (1977): 291–324.

185. This "nature" is not Mother Nature nor external nature, but *physis*, the live body's *physiologic* ability to repair itself.

186. Neuburger claimed that later authors postulated a complete identity between the *calor innatus* and the *vis medicatrix naturae*, according to Stolberg in "Vital Heat," fn. 13.

187. For example, the vegetation analogy is notable in Galen's "to flesh a healing wound is possible for *Nature* but impossible for the physician, as it is also to ripen what is half-mature or unripe." Cited in Neuburger, *Doctrine of the Healing Power of Nature*, 19.

NOTES TO CHAPTER SIX

1. "Vidi, et ecce ventus orientalis ventusque australis cum collateralibus suis, per flatus fortitudinis sue firmamentum moventes, illud ab oriente usque ad occidentem super terram circumvolui faciebant; ibique ventus occidentalis necnon et ventus septentrionalis et collaterales ipsorum, illud suscipientes spiraminibusque suis impellentes, ab occidente usque ad orientem sub terra reiciebant." *LDO*, 114:1–8.

2. "Vidi quoque quod a die quo dies prolongari incipiunt prefatus australis ventus cum collateralibus suis idem firmamentum in australi plaga sursum

versus septentrionem usque in diem, quo ultra non prolongantur, quasi ful-
ciendo paulatim attollebat; et quod ab eodem die quo dies adbreviari incipiunt septentrionalis ventus cum collateralibus suis ipsum firmamentum,
claritatem solis abhorrens, a septentrione ad austrum repellendo paulatim
deprimebat, usque dum auster illud iterum a longitudine dierum erigere
incipiebat." *LDO*, 114:8–16.

3. "Sed et vidi quod in superiori igne circulus apparebat, qui totum firmamentum ab oriente versus occidentem circumcingebat, de quo ventus ab
occidente progrediens septem planetas ab occidente ad orientem contra circumvolutionem firmamenti ire compellebat; et iste sicut alii prefati venti,
in mundum flatus suos non emittebat, sed tantum cursum planetarum, ut
predictum est, temperabat." *LDO*, 114:17–22.

4. "Deinde etiam vidi, quia per diversam qualitatem ventorum et aeris, cum
sibi in invicem concurrunt, humores qui sunt in homini commoti et
inmutati qualitatem illorum suscipiunt. Unicuique enim superiorum elementorum aer qualitati illius conveniens, per quem illud, scilicet elemen-
tum, vi ventorum ad circumvolutionem inpellatur, inest; alioquin non
moveretur, et de quolibet istorum cum ministerio solis, lune et stellarum
aer qui mundum temperat exspiratur. . . . Et iterum vidi, cum quispiam
ventorum omnium predictarum qualitatum in qualibet plaga mundi aut
diverso cursu solis et lune aut iudicio Dei, ut predictum est, excitatur, et
ut illic aere conmoto sibique contemperato flatum suum emittat, quod
idem aer, per mundum spirans et ea que in mundo sunt temperando con-
servans, secundum eundem flatum hominem in humoribus suis aliquantum mutabilem reddit; quoniam cum ille, scilicet homo, cuius naturalis
qualitas eidem flatui convenit, aerem hunc sic inmutatum in se inducit
et emittit, eo quo anima illum suscipiens ad interiora corporis transmit-
tat, humores qui in ipso sunt etiam inmutantur eique aut infirmitatem
aut sanitatem, ut supra demonstratum est, multociens inducunt." *LDO*,
114:23–115:50.

5. For an excellent essay on the diagram, see Peter S. Baker, "Byrhtferth of
Ramsey, "De concordia," (www.engl.virginia.edu/OE/Editions/Decon.pdf).
On Byrhtferth, see Peter S. Baker and Michael Lapidge, eds. *Byrhtferth's Enchiri-
dion*, *EETS* (Oxford: Oxford University Press, 1995).

NOTES TO THE APPENDIX

1. Hagen describes it as a "s. XII membr. 4* f. 97. Scriptus est liber a 1143 vel
1144 in claustro Disibodenberg propre Kreuznacii teste Fiala Solodurensi.
1. f. 1A–f. 47B: Calendarium romanorum ecclesiasticum, mutilo ab initio.
Deest januarios et februarios. 2. f. 47B–f. 50A: Tabula calendaria Henrici
monachi. 3. f. 50B–f. 97A: Regula monachorum." See Hermannus Hagen,
Catalogus Codicum Bernensum (Bern: Haller, 1875), 276.

2. In the catalogue at Bern Burgerbibliothek the texts are identified as: the *Martyrologium Usuardi*, fols. lr–47v, a *Tabulum computi*, fols. 48r–50r, and the *Regula sancti Benedicti*, fols. 50r–97v. Schrader and Führkötter observed that two scribes worked on the codex; one wrote the Rule and the other the martyrology (*Echtheit des Schrifttums*, fn. 17, p. 77).

3. The numeration refers to Franciscus Buecheler and Alexander Riese, eds., *Anthologia Latina*, vol. 1 (Leipzig: B.G. Teubner, 1894–1926). What has happened is that the composer of Disibodenberg's poem has taken the text by Ausonius (AL 640) and used its lines as the first line of each stanza. The provenance of each of the two lines that follow, however, is not known.

4. For an edition, see Monachus Usuardus, *Usuardi martyrologium*, in PL123, col. 453 to end, and PL124, cols. 10–860.

5. According to Hennig, various lines from Ausonius' AL 640 are also part of the calendrical poem in Bede's *De temporum ratione*; see Bede, *De temporum ratione*, in *The Miscellaneous Works of Venerable Bede*, ed. J.A. Giles (London: Whitaker and Co., 1843), 217–220.

6. For information on the martyrologies, see Dom Jacques Dubois, *Les Martyrologes du Moyen Age Latin*, Typologie des sources du moyen age occidental, fas. 26 (Abbeville: F. Paillart, 1990).

7. Of the traditional martyrologies, only Hrbanus Maurus, once Archbishop of Mainz, gives information on Disibod: "Et in suburbanis moguntiacensis, ecclesia natale sancti disibothi confessioris. . . ." See Hrbanus Maurus, *Liber de computo*, PL110, col. 1167.

8. My translation; see Usuard, *Usuardi martyrologium*, PL124, cols. 55–56.

9. "Henric monachus istam collegit tabulam verbis indoctis ad modum consulens. oportet igitur quemlibet studiose istum retinere nunc cum litterarum docentem cunctos cristiani nominis quando usualiter debeant cureis carnibus septem hebdomadis jejeunis. populi episcopi abbates decim canonici diaconi subdiaconi cantores cuncti prepositi custodes quivis magistri indocti pueri moniales monachi presbiteis inclusi domine abbatisse prelati socii prounce idiothe . . . scolares iuventes adultes decrepiti . . . principes diversi reges palatines comites . . . agricoles bubulci recitati pariter venite. . . . Discite mirabilem. . . . Audite Henricem pauperem certe monachum." My transcription.

10. On Hildegard and the Rule, see Constable, "Hildegard's Explanation of the Rule of St. Benedict."

Bibliography

Acht, Peter, ed. *Mainzer Urkundenbuch*, Vol. 2, *Die Urkunden seit dem Tode Erzbischof Adalberts (1137) bis zum Tode Erzbischof Konrads (1200)*. Darmstadt: Selbstverlag der Hessischen Historischen Kommission, 1968–71.

Ackerknecht, Erwin. *A Short History of Medicine*. New York: Ronald Press, 1955.

Acklin-Zimmermann, Beatrice, ed. *Denkmodelle von Frauen im Mittelalter*. Freiburg: Universitätsverlag, 1994.

Ado, Archiepiscopus Viennensis in Gallia. *Vetus Romanum Martyrologium*. 143–180. *Patrologia Latina*. Vol. 123. Paris: Garnier, 1855.

Alfanus of Salerno. *Nemesii episcopi premnon physicon*. Edited by Carolus Burkhard, 59–72. Leipzig: Teubner, 1917.

Allers, Rudolf. "Microcosmus. From Anaximandros to Paracelsus." *Traditio* 2, no. 2 (1944): 319–408.

Alvarez-Millan, Cristina. "Practice versus Theory: Tenth-Century Case Histories from the Islamic Middle East." *Social History of Medicine* 13, no. 2 (2000): 293–306.

Ambrosoli, Mauro. "L'opus agriculturae di Palladio: volgarizzamenti e identificazione dell'ambiente naturale fra tre cinquecento." *Quaderni Storici* 18, no. 1 (1983): 227–254.

———. *The Wild and the Sown: Botany and Agriculture in Western Europe, 1350–1850*. Translated by Mary McCann Salvatorelli. Cambridge: Cambridge University Press, 1997.

Amundsen, Darrel W. "Medicine and Faith in Early Christianity." *Bulletin of the History of Medicine* 56 (1982): 326–350.

———. "Medieval Canon Law on Medical and Surgical Practice by the Clergy." *Bulletin of the History of Medicine* 52 (1978): 22–44.

Anders, Stefan. *Rheinland, Pfalz und Saar*. Frankfurt: Umschau Verlag, 1966.

Anderson, Frank J. *An Illustrated History of the Herbals*. New York: Columbia University Press, 1977.

Anderson, William. "The Green Man: Tracing a Vegetation Image in Medieval Art." *Parabola* 14, no. 3 (1989): 26–33.

Anderson, William, and Clive Hicks. *The Green Man: Archetype of Our Oneness with the Earth*. London: HarperCollins, 1990.

Anglicus, Bartholomaeus. *De genuinis rerum coelestium, terrestrium, et inferarum proprietatibus, libri XVIII. Opus incomparabile, theologis ureconsultis, medicis, omniumque disciplinarum et artium alumnis, utilissimum futurum. Cui accessit liber XIX de variarum rerum accidentibus*. . . . Frankfurt: Wolfgang Richter, Nachdruck Frankfurt a. M., 1964 [1601].

Anonymous. "Ad Maecenatem." In *Marcelli de medicamentis*. Edited by George Helmreich, 9–13. Leipzig: Teubner, 1889.

Anonymous. "Bodley Herbal and Bestiary (ca. 1100) with introduction by W. O. Hassal." Oxford: Oxford Microfilm Publications, 1978.

Anonymous. "De Quattuor Humoribus." In *Collectio Salernitana*. Edited by Salvatore da Renzi. Vol. 2, 411–412. Naples: Dalla typografia del filiatre-sebezio, 1852–1859.

Anonymous. "Epistula ad Antiochum." In *Marcelli de medicamentis liber*. Edited by Maximillianus Niedermann, 10–13. Leipzig, Berlin: Teubner, 1916.

Anthimus. *De Observatio[ne] Ciborum: Text, Commentary, and Glossary, with a Study of the Latinity*. Edited by Shirley Howard Weber. Leiden: E. J. Brill, 1924.

Arber, Agnes. "From Medieval Herbalism to the Birth of Modern Botany." In *Science, Medicine, and History: Essays on the Evolution of Scientific Thought and Medical Practice Written in Honour of Charles Singer*. Edited by Edgar Ashworth Underwood, 317–336. London, New York: Oxford University Press, 1953.

———. *Herbals, Their Origin and Evolution: A Chapter in the History of Botany (1470–1670)*. Cambridge: Cambridge University Press, 1912.

Aris, Marc-Aelko, Michael Embach, Werner Lauter, Irmgard Müller, Franz Staab, and Scholastica Steinle, eds. *Hildegard von Bingen. Internationale wissenschaftliche Bibliographie*. Mainz: Gesellschaft für mittelrheinishe Kirchengeschichte, 1998.

Armengaud, Auguste, and Claude Rivals. *Moulins à Vent et Meuniers des Pays d'Oc*. Portet-sur-Garonne: Loubatières, 1992.

Aristotle. *Meteorologica*. Translated by H. D. P. Lee. Cambridge, Massachusetts: Harvard University Press, 1952.

Atherton, Mark, ed. *Hildegard of Bingen: Selected Writings*. Suffolk: Penguin Books, 2001.

Augustodunensis, Honorius. *De Imagine Mundi*, in *Patrologia Latina*. Vol. 172, 119–186. Paris: Garnier, 1855.

Ayoub, Lois. "Old English Waeta and the Medical Theory of the Humours." *Journal of English and Germanic Philology* 3, no. July (1995): 332–346.

Baader, Gerhard. "Early Medieval Latin Adaptations of Byzantine Medicine in Western Europe." In *Dumbarton Oaks Symposium on Byzantine Medicine*. Edited by John Scarborough, 251–259. Washington, D.C.: Dumbarton Oaks Research Library and Collection, 1985.

———. "Mittelalterliche Medizin in bayerischen Klostern." *Sudhoffs Archiv* 57, no. 3 (1973): 275–296.

Baedeker, Karl. *The Rhine*. Leipzig: Baedeker, 1884.

Baehrens, Aemilius. *Miscellanea Critica*. Gröning: Woltersii, 1878.

———. *Poetae Latini Minores*. Vol. 1. Leipzig: Teubner, 1879.

Baker, Peter S. "Byrhtferth of Ramsey, De concordia mensium atque elementorum." www.engl.virginia.edu/OE/Editions/Decon.pdf.

Baker, Peter S., and Michael Lapidge, eds. *Byrhtferth's Enchiridion*. Oxford: Oxford University Press, 1995.

Ballestrasse, Flavio. *Medicina Monastica*. Vol. 93, *Scientia Veterum*. Pisa: Casa Editrice Giardini, 1966.

Bar-Sela, A., and Hebbel E. Hoff. "Asaf on Anatomy and Physiology." *Journal of the History of Medicine* 20 (1965): 358–389.

Barrière, Bernadette. "La place des monastères Cisterciens dans le paysage rural des XIIe.-XIIIe. s." In *Moines et Monastères dans les societés de rite grec et latin*. Edited by Jean-Loup Le Maître, Michel Dmitriev, and Pierre Gonneau, 191–210. Geneva: Droz, 1996.

Bartholomaeus. "Practica." In *Collectio Salernitana*. Edited by Salvatore da Renzi. Vol. 4, 321–406. Naples: Filiatre-Sebezio, 1855.

Basford, Kathleen. *The Green Man*. Ipswich: D. S. Brewer, 1978.

———. "A New View of 'Green Man' Sculptures." *Folklore* 102, no. 2 (1991): 237–9.

Bates, Don, ed. *Knowledge and the Scholarly Medical Traditions*. Cambridge: Cambridge University Press, 1995.

Bauer, Dieter R., and Klaus Herbers, eds. *Hagiographie im Kontext: Wirkungsweisen und Möglichkeiten historischer Auswertung*. Stuttgart: Franz Steiner, 2000.

Bäumer, Änne. *Wisse die Wege: Leben und Werk Hildegards von Bingen. Eine Monographie zu ihrem 900. Geburtstag*. Frankfurt am Main: Peter Lang, 1998.

Beaujouan, Guy. "Réflexions sur les rapports entre théorie et pratique au Moyen Age." In *The Cultural Context of Medieval Learning*. Edited by J. E. Murdoch and Edith Sylla, 437–484. Dordrecht/Boston: D. Reidel, 1975.

Beccaria, Augusto. *I codici di medicina del periodo presalernitano (secoli IX, X e XI)*. Vol. 53. Rome: Storia e Letteratura, 1956.

Beck, Bernard. "Jardin monastique, jardin mystique: ordonnance et signification des jardins monastiques médiévaux." *Revue d'histoire de la pharmacie* 48, no. 327 (2000): 377–394.

Becker, Gustavus. *Catalogi bibliothecarum antiqui*. Bonn, 1885.

Bede. *The Reckoning of Time: translated with introduction, notes and commentary by Faith Wallis*. Translated by Faith Wallis. Liverpool: Liverpool University Press, 1999.

Bede. *De temporum ratione*. Edited by J. A. Giles, *The Miscellaneous Works of Venerable Bede*. London: Whitaker and Co., 1843.

———. "De temporum ratione." In *Bedae Opera de temporibus*. Edited by Charles William Jones. Cambridge, Mass.: Medieval Academy of America, 1943.

———. *Didascalia Genuina. Patrologia Latina*. Vol. 90, 192–194 and 246–250. Paris: Garnier, 1854.

Behling, Lottlisa. *Die Pflanze in der mittelalterlicher Tafelmalerei.* Weimar: H. Bohlaus Nachfolger, 1957.

———. *Die Pflanzenwelt der mittelalterliche Kathedralen.* Köln: Bohlau, 1964.

Bell, David N. "The English Cistercians and the Practice of Medicine." *Citeaux: Commentarii cistercienses* 40 (1989): 139–174.

Bendz, Gerhard, ed. *Caelii Aureliani Celerum passionum libri III. Tardarum passionum libri V.* CML. Vol. 6. Berlin: Akademie Verlag, 1990–1993.

Benson, Larry D. "The Greenness of the Green Knight." In *Art and Tradition in Sir Gawain and the Green Knight,* edited by Larry D. Benson, 90–95. New Brunswick, New Jersey: Rutgers University Press, 1965.

Berger, Margret. *Hildegard of Bingen: On Natural Philosophy and Medicine.* Cambridge: D. S. Brewer, 1999.

Berman, Constance H. *Medieval Agriculture, the Southern French Countryside and the Early Cistercians. A Study of Forty-Three Monasteries.* Philadelphia: Transactions of the American Philosophical Society, 1986.

Berndt, Rainer, ed. *"Im Angesicht Gottes suche der Mensch sich selbst": Hildegard von Bingen (1098–1179).* Berlin: Akademie Verlag, 2001.

Bertelli, Sergio. *The King's Body: Sacred Rituals of Power in Medieval and Early Modern Europe.* Translated by R. Burr Litchfield. University Park, Pa.: Pennsylvania State University Press, 2001.

Besserman, Lawrence. "The Idea of the Green Knight." *ELH* 53, no. 2 (1986): 219–239.

Beuys, Barbara. *Den ich bin krank vor Liebe: Das Leben der Hildegard von Bingen.* Munich: Carl Hanser Verlag, 2001.

Beyer, Heinrich, Leopold Eltester, and Adam Goerz. *Urkundenbuch zur Geschichte der mittelrheinischen Territorien 2 Band.* Vol. 1. Koblenz, repr. Hildesheim: Scientia Verlag Aalen, repr. Georg Olms, 1860 (1974).

Biggs, Frederck M., Thomas D. Hill, Paul E Szarmach, and E. Gordon Whatley, eds. *Sources of Anglo-Saxon Literary Culture.* Electronic Edition: www.wmich.edu/medieval/saslc/volone/index.html.

Birch, Walter de Gray, ed. *Liber Vitae: Register and Martyrology of New Minster and Hyde Abbey.* London: Simpkin and Co., 1892.

Bird, Jessalynn. "Texts on Hospitals: Translation of Jacques de Vitry 'Historia Occidentalis 229' and Edition of Jacques de Vitry's 'Sermons to Hospitallers.'" In *Religion and Medicine in the Middle Ages.* Edited by Peter Biller and Joseph Ziegler, 109–134. Oxford: York Medieval Press, 2001.

Blaine, Bradford B. "Mills." In *Dictionary of the Middle Ages,* edited by Joseph Strayer, 394–395. New York: Scribners, 1982–1989.

Blumenfeld-Kosinski, Renate, and Timea Szell, eds. *Images of Sainthood in Medieval Europe.* Ithaca: Cornell University Press, 1991.

Blunt, Wilfrid, and Sandra Raphael. *The Illustrated Herbal.* New York: Metropolitan Museum of Art, 1979.

Bodarwé, Katrinette. "Pflege und Medizin in mittelalterlichen Frauenkonventen." *Medizinhistorisches Journal* 37, no. 2 (2002): 231–63.

Böhmer, J. Fr. *Regesten zur Geschichte der mainzer Erzbischof von Bonifatus bis Heinrich 2*. Innsbruck: Neudruck der Ausgabe, 1877.

Bois, Guy. *The Transformation of the Year One Thousand: The Village of Lournand from Antiquity to Feudalism*. Translated by Jean Birrell. Manchester: Manchester University Press, 1992.

Boitani, Piero, and Anna Torti, eds. *The Body and Soul in Medieval Literature*. Cambridge: D. S. Brewer, 1999.

Bono, James, J. "Medical Spirits and the Medieval Language of Life." *Traditio* 40 (1984): 91–130.

Borland, Jennifer. "Subverting Tradition: The Transformed Female in Hildegard of Bingen's 'Scivias.'" Paper presented at the Seeing Gender: Perspectives on Medieval Gender Conference, King's College, London 2002.

Borst, Arno. *The Ordering of Time: From the Ancient Computus to the Modern Computer*. Translated by Andrew Winnard. Chicago: University of Chicago Press, 1993.

Boudet, Jacques. *Chronologie Universelle*. Paris: Bondas, 1983.

Boulaine, Jean. *Histoire de l'agronomie en France*. London: Lavoisier, 1992.

Bouteiller, Marcelle. *Médecine Populaire d'Hier et d'Aujourd'hui*. Paris: Editions G. P. Maisonneuve et la Rose, 1987.

Bozóky, Edina. "Mythic Mediation in Healing Incantations." In *Health, Disease and Healing in Medieval Culture*, edited by Sheila Campbell, Bert Hall and David Klausner, 84–92. New York: St. Martin's Press, 1992.

Braekman, W. L. "Bollard's Middle English Book of Planting and Grafting and Its Background." *Studia Neophilologica* 57, no. 1 (1985): 19–39.

Brede, Maria Laetitia. "Die Klöster der heiligen Hildegard Rupertsberg und Eibingen." In *Hildegard von Bingen 1179–1979: Festschrift zum 800. Todestag der Heiligen*, edited by Anton Ph. Brück, 77–94. Mainz: Selbstverlag der Gesellschaft für mittelrheinische Kirchengeschichte, 1979.

Brown, Jane. *A Social History of Gardens and Gardening*. London: HarperCollins, 1999.

Brown, Michelle P. *A Guide to Western Historical Scripts*. Toronto: University of Toronto Press, 1990.

Brück, Anton Ph., ed. *Hildegard von Bingen 1179–1979: Festschrift zum 800. Todestag der Heiligen*. Mainz: Selbstverlag der Gesellschaft für mittelrheinishen Kirchengeschichte, 1979.

Bruder, Petrus. "Acta Inquisitionis de Virtutibus et Miraculis S. Hildegardis." In *Analecta Bollandiana*, edited by Carolus de Smedt, Gulielmus van Hoof and Josephus de Backer, Vol. 2. 116–129. Paris: Societé Génerale de Librairie Catholique, 1883.

Buecheler, Franciscus, and Alexander Riese, eds. *Anthologia Latina*. Vol. 1. Leipzig: B. G. Teubner, 1894–1926.

Buhler, Curt F. "Prayers and Charms in Certain Middle English Scrolls." *Speculum* 39 (1964): 270–8.

Bullough, Vern. L. "Training of the Nonuniversity-Educated Medical Practitioners in the Later Middle Ages." *Journal of the History of Medicine* 15 (1959): 446–58.

Burchkardt, Max, Pascal Lander, and Martin Steinmann, eds. *Katalog der datierten Handschriften in der Schweiz in lateinischer Schrift vom Anfang des Mittelalters bis 1550*. Dietikon-Zurich: Urs Graf, 1977–1991.

Burkholder, Kristen M. "'Attempree diete was al hir phisik': The Medieval Application of Medical Theory to Fasting." *Essays in Medieval Studies* 13 (1996): 15–29.

Burnett, Charles. "Hildegard of Bingen and the Science of the Stars." In *Hildegard of Bingen: The Context of Her Thought and Art*, edited by Charles Burnett and Peter Dronke, 111–120. London: The Warburg Institute, 1998.

———. "King Ptolemy and Alchandreus the Philosopher: The Earliest Texts on the Astrolabe and Arabic Astrology at Fleury, Micy and Chartres." *Annals of Science* 55 (1998): 329–368.

———, ed. *Pseudo-Bede: De Mundi Celestis Terrestrisque Constitutione. A Treatise on the Universe and Its Soul*. London: Warburg Institute, 1985.

Burnett, Charles, and Peter Dronke, eds. *Hildegard of Bingen: The Context of Her Thought and Art*. London: The Warburg Institute, 1998.

Burrow, John Anthony. *A Reading of Sir Gawain and the Green Knight*. New York: Barnes and Noble, 1966.

Buschinger, Danielle, and Andre Crepin, eds. *Les Quatre Elements dans la Culture Médiévale*. Goppingen: Kummerle, 1983.

Bynum, Caroline Walker. "Material Continuity, Personal Survival, and the Resurrection of the Body: A Scholastic Discussion in its Medieval and Modern Contexts." *History of Religions* 30 (1990): 51–85.

———. *Metamorphosis and Identity*. New York: Zone Books, 2001.

———. *The Resurrection of the Body in Western Christianity, 200–1336*. New York: Columbia University Press, 1995.

Bynum, W. F., and Roy Porter, eds. *Companion Encyclopedia of the History of Medicine*. 2 vols. London: Routledge, 1993.

Cadden, Joan. "It Takes All Kinds: Sexuality and Gender Differences in Hildegard of Bingen's 'Book of Compound Medicine.'" *Traditio: Studies in Ancient and Medieval History, Thought and Religion* 40 (1984): 149–174.

———. *Meanings of Sex Difference in the Middle Ages: Medicine, Science and Culture*. Cambridge: Cambridge University Press, 1993.

———. "Science and Rhetoric in the Middle Ages: The Natural Philosophy of William of Conches." *Journal of the History of Ideas* 56, no. 1 (1995): 1–24.

Cambell, Kate Hurd-Mead. *A History of Women in Medicine from Earliest Times to the Beginning of the Nineteenth Century*. Connecticut: Haddam Press, 1938.

Cameron, M. L. *Anglo-Saxon Medicine*. Vol. 7, *Cambridge Studies in Anglo-Saxon England*. Cambridge: Cambridge University Press, 1993.

Cames, Gerard. "Un joyau de l'enluminure Alsacienne: le Codex Guta-Sintram (1154) à Strasbourg." *Bulletin de la Societé nationale des antiquaires de France* (1978–1979): 255–261.

Campbell, Bruce M. "A New Perspective on Medieval and Early Modern Agriculture: Six Centuries of Early Norfolk Farming." *Past and Present* 141 (1993): 38–105.

Campbell, Bruce M., James A. Galloway, and Margaret Murphy. "Rural Land-Use in the Metropolitan Hinterland 1270–1339: The Evidence of Inquisitiones Post Mortem." *Agricultural History* 40 (1992): 1–22.

Camporesi, Piero. *The Juice of Life: The Symbolic and Magic Significance of Blood.* Translated by Robert R. Bauer. New York: Continuum, 1995.

————. "Plants as Symbols." In *The Anatomy of the Senses: Natural Symbols in Medieval and Early Modern Italy*, 26–36. Cambridge, U. K.: Polity Press, 1994.

Cannon, Sue Spencer. "The Medicine of Hildegard of Bingen. Her Twelfth-Century Theories and Their Twentieth-century Appeal as a Form of Alternative Medicine." Ph.D. diss., UCLA, 1993.

Cantimpré, Thomas de. *Liber de natura rerum von Thomas Cantimpratensis.* Berlin: W. De Gruyter, 1973.

Caraffa, Filippo, ed. *Bibliotheca Sanctorum.* Vol. 2. Rome: Società Graffia Roma, 1962.

Cavallo, Guglielmo. *Exultet: Rotoli liturgici del medioevo meridonale.* Rome: Istituo Poligrafico, 1994.

Caviness, Madeline H. "Artist: 'To See, Hear and Know All at Once.'" In *Voice of the Living Light*, edited by Barbara Newman, 110–124. Berkeley: University of California Press, 1998.

————. "Hildegard as Designer of the Illustrations to Her Works." In *Hildegard of Bingen: The Context of her Thought and Art*, edited by Charles Burnett and Peter Dronke, 29–62. London: Warburg Institute, 1998.

————. "Hildegard of Bingen: Some Recent Books—a Review Essay." *Speculum* 77, no. 1 (2002): 113–120.

————. *Stained Glass Windows.* Turnhout: Brepols, 1996.

Cazelles, Brigitte. "Introduction." In *Images of Sainthood in Medieval Europe*, edited by Renate Blumenfeld-Kosinski and Timea Szell, 1–17. Ithaca: Cornell University Press, 1991.

Celsi, A. Corn. *De Medicina:* Studiis Societatis Bipontinae, 1786.

Chamberlein, Marcia Kathleen. "Hildegard of Bingen's *Causes and Cures:* A Radical Feminist Response to the Doctor-Cook Binary." In *Hildegard of Bingen: A Book of Essays*, edited by Maud Burnett McInerney, 53–72. London: Garland Press, 1998.

Chambers, E. K. *The Medieval Stage.* Oxford: Oxford University Press, 1925 (1903).

Chapelot, Jean, and Robert Fossier. *The Village and House in the Middle Ages.* Translated by Henry Cleere. London: B. T. Batsford, 1985.

Chazin, Carol Anne. "The Planning of English Monastic Infirmary Halls in the Twelfth and Thirteenth Centuries." M. A., University of California, 1966.

Chevalier, Jean, and Alain Gheerbrant. *Dictionnaire des symboles*. Paris: Seghers, 1978.

Citrome, Jeremy J. "Bodies that Splatter: Surgery, Chivalry, and the Body in the *Practica* of John Arderne." *Exemplaria* 13, no. 1 (2001): 137–172.

Clerc, Daniel le. *Histoire de la médecine où l'on voit l'origine et le progrès de cet art de siècle en siècle*. La Hay: Isaac van der Kloot, 1729.

Collins, Minta. *Medieval Herbals: The Illustrative Tradition*. London: British Library, 2000.

Columella, Lucius Junius Moderatus. *On Agriculture, with a Recension of the Text and an English Translation*. Edited by Harrison Boyd Ash. 3 vols. *Loeb Classics*. Cambridge, Mass.: Harvard University Press, 1941–1955.

Conde, Linaje. "La enfermedad en la organizaciòn monastica Visigothica." *Asclepio* (1970): 202–213.

Conrad, Lawrence I., Michael Neve, Vivian Nutton, Roy Porter, and Andrew Wear, eds. *The Western Medical Tradition: 800 B.C.–1800 A.D.* Cambridge: Cambridge University Press, 1995.

Constable, Giles. "Hildegard's Explanation of the Rule of St. Benedict." In *Hildegard von Bingen in ihrem historischen Umfeld. Internationaler wissenschaftlicher Kongress zum 900 jährigen Jubiläum, 13–19 September 1998, Bingen am Rhein*, edited by Alfred Haverkamp and Alexander Reverchon, 163–187. Mainz: Philipp von Zabern, 2000.

Corvi, Antonio. "Il monaco 'pigmentarius.'" In *La farmacia monastica e conventuale*, edited by Antonio Corvi and Ernesto Riva, 36–39. Pisa: Pacini, 1996.

Corvi, Antonio, and Ernesto Riva. *La farmacia monastica e conventuale*. Pisa: Pacini, 1996.

Cotten, C. M. *Ethnobotany: Principles and Applications*. Chicago: Wiley, 1996.

Coulet, M. "Pour une histoire du jardin." *Le Moyen Age* 73 (1967): 239–70.

Craine, Renate. *Hildegard: Prophet of the Cosmic Christ*. New York: The Crossroad Publishing Company, 1997.

Crossgrove, William C. "Das landwirtschaftliche Handbuch von Petrus de Crescentiis in der deutschen Fassung des Bruder Franciscus." *Sudhoffs Archiv* 78, no. 1 (1994): 98–106.

———. "Medicine in the Twelve Books on Rural Practices of Petrus de Crescentiis." In *Manuscript Sources for Medieval Medicine*, edited by Margaret R. Schleissner, 81–103. New York: Garland Publishing Inc., 1995.

———. "The Vernacularization of Medieval Science, Technology and Medicine, Introduction." *Early Science and Medicine* 3, no. 2 (1998): 81–87.

Cylkowski, David G. "A Middle English Treatise on Horticulture: Godfridus super Palladium." In *Popular and Practical Science of Medieval England*, edited by Lister M. Matheson, 301–329. East Lansing: Colleagues Press, 1994.

D'Alverny, Marie-Thérèse. "Le cosmos symbolique du XIIe. siècle." *Archives d'histoire doctrinale et littéraire* 20 (1953): 31–81.

————. "Une Baguette Magique." In *Pensée Médiévale en Occident*, edited by Charles Burnett, 1–11. Brookfield, Vt.: Variorum, 1995.

Daaleman, Timothy. "The Medical World of Hildegard of Bingen." *American Benedictine Review* 44 (1993): 280–9.

Dales, Richard C., ed. *Marius: On the Elements*. Berkeley, Los Angeles, and London: University of California Press, 1976.

Daniell, Christopher. *Death and Burial in Medieval England: 1066–1550*. London, New York: Routledge, 1997.

Daremberg, Charles. "'Aurelius de Acutis Passionibus,' texte publié pour la première fois d'après un manuscrit de la bibliothèque de Bourgogne à Bruxelles, corrigé et accompagné de notes critiques." *Janus* 2 (1847): 468–499, 690–731.

————. *Histoire des sciences médicales comprenant l'anatomie, la physiologie, la médecine, la chirugie, et les doctrines de pathologie générale. Tome premier: Depuis les temps historiques jusqu'à Harvey*. Paris: Baitlere et Fils, 1870.

Davis, Scott. "The Cosmological Balance of the Emotional and Spiritual Worlds: Phenomenological Structuralism in Traditional Chinese Medical Thought." *Culture, Medicine and Psychiatry* 20 (1996): 83–123.

Davy, M. M. *Initiation à la Symbolique Romane*. Paris: Flammarion, 1977.

DeBlieu, Jean. *Wind: How the Flow of Air has Shaped Life, Myth and the Land*. Boston: Houghton Mifflin, 1998.

Dechambre, A. L., and L. Lereboullet. *Dictionnaire Encyclopédique des Sciences Médicales*. Paris: G. Masson, 1886.

Delatte, Armand. *Herbarius: Recherches sur le cérémonial usité chez les anciens pour la cuillette des simples et des plantes magiques*. 2 ed. Paris: Droz, 1938.

Delisle, Léopold. *Inventaire des manuscrits de la Bibliothèque Nationale*. Paris: Champion, 1884.

Delley, D. *Hildegarde de Bingen et les Plantes Medicinales*. Basle, 1988.

Derolez, Albert, ed. *Guiberti Gemblacensis epistolae quae in codice B. R. Brux. 5527–5534 inveniuntur*. Vol. 66 and 66A, *CCCM*. Turnhout: Brepols, 1988.

————. "The Manuscript Transmission of Hildegard of Bingen's Writings: The State of the Problem." In *Hildegard of Bingen: The Context of Her Thought and Art*, edited by Charles Burnett and Peter Dronke, 17–28. London: Warburg Institute, 1998.

Derolez, Albert, and Peter Dronke, eds. *Liber divinorum operum*. Vol. 92, *CCCM*. Turnhout: Brepols, 1996.

Detienne, Marcel. *The Garden of Adonis: Spices in Greek Mythology*. Translated by Lloyd, Janet. Sussex: Harvester Press, 1977.

Dickson, J. H., and R. R. Mill, eds. *Plants and People: Economic Botany in Northern Europe 800–1800*. Edinburgh: Edinburgh University Press, 1994.

Diepgen, Paul. "Die volkstümlichen und die wissenschaftlichen Grundlagen der Therapie in der Geschichte der Medizin." In *Medizin und Kultur*, edited by Paul Diepgen, 61–62. Stuttgart: F. Engke, 1938.

Diers, Michaela. *Hildegard von Bingen*. München: Deutscher Taschenbuch Verlag, 1998.

Dilg, Peter, ed. *Inter folia fructus: Gedenkschrift für Rudolf Schmitz (1918–1992)*. Frankfurt: Govi, 1995.

Donovan, Leslie A. *Women Saints' Lives in Old English Prose, Translated from Old English with an Introductory Note and Interpretative Essay*. Rochester, New York: D. S. Brewer, 2000.

Dronke, Peter. "The Allegorical World-Picture of Hildegard of Bingen: Revaluations and New Problems." In *Hildegard of Bingen: The Context of Her Thought and Art*, edited by Charles Burnett and Peter Dronke, 1–16. London: Warburg Institute, 1998.

———. "Bernard Silvestris: Natura and Personification." *Journal of the Warburg and Courtauld Institute* 43 (1980): 16–31.

———. "Hildegard of Bingen." In *Women Writers of the Middle Ages: A Critical Study of Texts from Perpetua (+203) to Marguerite Porete (+1310)*, 144–201, notes 306–15, edition of autobiographical setion, 231–241. Cambridge: Cambridge University Press, 1984.

———. "Hildegard's Inventions. Aspects of her Language and Imagery." In *Hildegard von Bingen in ihrem historischen Umfeld. Internationaler wissenschaftlicher Kongress zum 900 jährigen Jubiläum, 13–19 September 1998, Bingen am Rhein*, edited by Alfred Haverkamp and Alexander Reverchon, 299–320. Mainz: Philipp von Zabern, 2000.

———. "Liber Nemroth." In *Dante e le tradizioni latine medievali*, 179–187. Bologna: Il mulino, 1990.

———. *Poetic Individuality in the Middle Ages*. Oxford: Clarendon Press, 1970.

———. "Problemata Hildegardiana." *Mittellateinisches Jahrbuch* 16 (1981): 97–131.

———. "Tradition and Innovation in Medieval Western Colour-Imagery." *Eranos Jahrbuch* 41, no. 4 (1972): 51–107.

Drouin, Jean-Marc. *Reinventer la nature: l'écologie et son histoire*. Paris: Desclec de Brouwer, 1991.

Du Cange, Charles, Du Fresne Sieur. *Glossarium mediae et infimae latinatis*. Niort: L. Favre, 1886.

Dubois, Jacques. *Les Martyrologes du Moyen Age Latin, Typologie des sources du moyen age occidental, fas. 26*. Abbeville: F. Paillart, 1990.

Dubois, Jacques, and Geneviève Renaud, eds. *Le martyrologe d'Adon: Ses deux familles, ses trois recensions: texte et commentaire*. Paris: Editions du Centre national de la recherche scientifique, 1984.

Duby, Georges. *Rural Economy and Country Life in the Medieval West*. Translated by Cynthia Postan. Philadelphia: University of Pennsylvania Press,1998(1968).

Ducos, Joëlle, ed. *La météorologie en France au Moyen Age: XIIe.-XVe. siècles*. Paris: Honoré Champion, 1998.

Duff, J. Wight. *Minor Latin Poets*. Cambridge: Harvard University Press, 1934.

Dufour, Jean Yves. "Essai d'archéologie horticole en banlieue parisienne Saint-Denis et rueil-Malmaison (XIV-XIX s.)." *Histoire et Societés Rurales* 1, no. 7 (1997): 11–40.

Duft, Johannes. *Notker der Artz: Klostermedizin und Mönchsarzt im frühmittelalterlichen S. Gallen.* St. Gall: Verlag der Buchdruckerei Ostschweiz, 1970.

Dulieu, Louis. *La médecine à Montpellier. Le Moyen Age.* 3 vols. Vol. 1. No city: No publisher, 1976.

Durling, Richard J. "The Innate Heat in Galen." *Medizinhistorisches Journal* 23 (1988): 210–221.

Eastwood, Bruce S. *The Revival of Planetary Astronomy in Carolingian and Post-Carolingian Europe.* Ashgate: Variorum, 2002.

Echternach, Theodoric of. "Vita Sanctae Hildegardis=Leben der heiligen Hildegard von Bingen." In *Vita Sanctae Hildegardis=Leben der heiligen Hildegard von Bingen. Canonizatio Sanctae Hildegardis= Kanonisation der heiligen Hildegard,* edited by Monika Klaes, 79–235. Freiburg: Herder, 1998.

Eis, Gerhard. *Altdeutsche Zauberspruche.* Berlin: Walter de Gruyter, 1964.

Elliot, Dyan. *Fallen Bodies: Pollution, Sexuality, and Demonology in the Middle Ages.* Philadelphia: University of Pennsylvania Press, 1999.

Enders, Markus. "Das Naturverständnis Hildegards von Bingen." In *"Im Angesicht Gottes suche der Mensch sich selbst": Hildegard von Bingen (1098–1179),* edited by Rainer Berndt, 461–502. Berlin: Akademie Verlag, 2001.

Engbring, Gertrude. "Saint Hildegard, Twelfth Century Physician." *Bulletin of the History of Medicine* 8 (1940): 770–784.

Epler, D. C., Jr. "Bloodletting in Early Chinese Medicine and its Relation to the Origin of Acupuncture." *Bulletin of the History of Medicine* 54 (1980): 337–367.

Ernst, Ulrich. "Farbe und Schrift im Mittelalter unter Berücksichtigung antiker Grundlagen und neuzeitlicher Rezeptionsformen." In *Testo e immagine nell'alto Medioevo,* 343–415. Spoleto: Centro italiano di studi sull'alto Medioevo, 1994.

Escot, Pozzi. "Hildegard's Christianity: An Assimilation of Pagan and Ancient Classic Tradition." In *Wisdom which Encircles Circles, Papers on Hildegard of Bingen,* edited by Audrey Ekdahl Davidson, 53–61. Kalamazoo: Medieval International Publishing, 1996.

Etkin, Nina. "Consuming a Therapeutic Landscape: A Muticontextual Framework for Assessing the Health Significance of Human-Plant Interactions." In *People-Plant Relationships: Setting Research Priorities,* edited by Joel Flagler and Raymond P. Poincelot, 61–81. New York: Haworth Press, 1994.

————, ed. *Plants in Indigenous Medicine and Diet: Biobehavorial Approaches.* New York: Redigare, 1986.

Etkin, Nina, and T. Johns. "'Pharma foods' and 'Nutraceuticals': Paradigm shifts in Biotherapeutics." In *Plants for Food and Medicine: Proceedings of the Joint Conference of the Society for Economic Botany and the International Society for Ethnopharmacology,* edited by Nina Etkin and others, 3–16. London: Royal Botanical Gardens, 1998.

Faucher, D. "Les jardins familiaux et la technique agricole." *Annales* 14 (1979): 297–307.

Fayet, Sylvie. "Le regard scientifique sur les couleurs à travers quelques encyclo-pédistes latins du XIIe. s." *Bibliothèque de l'Ecole des Chartes* 150 (1992): 51–70.

Feher, Michel, Ramona Naddaff, and Nadia Tazi, eds. *Fragments for a History of the Human Body*. Vols. 1–3. New York: Zone, 1989.

Fehringer, Barbara. *Das Speyerer Kräuterbuch mit den Heilpflanzen Hildegards von Bingen. Eine Studie zur mittelhochdeutschen Physica-Rezeption mit kritischer Ausgabe des Textes*. Würzburg: Königshausen und Neumann, 1994.

Feldmann, Christian. *Hildegard von Bingen. Nonne und Genie*. Freiburg: Herder Spektrum, 1995.

Felten, Franz J. "'Novi esse volunt . . . deserentes bene contritam viam . . .' Hilde-gard von Bingen und Reformbewegungen im religiosen Leben ihrer Zeit." In *"Im Angesicht Gottes suche der Mensch sich selbst": Hildegard von Bingen (1098–1179)*, edited by Rainer Berndt, 27–86. Berlin: Akademie Verlag, 2001.

Ferrante, Joan. "Correspondent: 'Blessed Is the Speech.'" In *Voice of the Living Light: Hildegard of Bingen and Her World*, edited by Barbara Newman, 91–109. Berkeley: University of California Press, 1998.

———. "'Scribe quae vides et audis': Hildegard, Her Language and Her Secre-taries." In *The Tongue of the Fathers: Gender and Ideology in Twelfth-Century Latin*, edited by David Townsend and Andrew Taylor, 102–135: University of Pennsylvania Press, 1998.

Fery-Hue, Françoise. "Le Romarin et ses propriétés, un traité faussement attribué à Aldebrandin de Sienne." *Romania* 115, no. 1, 2 (1997): 138–192.

Fife, Austin. "The Concept of the Sacredness of Bees, Honey and Wax in Christian Popular Tradition." Ph.D. diss., Stanford University, 1937.

Filliozat, Jean. *The Classical Doctrine of Indian Medicine: Its Origins and its Greek Parallels*. Delhi: Manoharlal, 1964.

Fischer, Bonifatius. *Novae concordantiae bibliorum sacrorum iuxta vulgatem versionem critice editam*. Stuttgart: Frommann-Holzboog, 1977.

Fischer, Hermann. *Die heilige Hildegard von Bingen. Die erste deutsche Naturforscherin und Ärztin. Ihr Leben und Werk*. München: Münchner Drucke, 1927.

———. *Mittelalterliche Pflanzenkunde*. München: Münchner Drucke, 1929.

Fischer, Klaus-Dietrich. "Antike Verse in medizinischen Schriften des Mittelalters." *Gesnerus* 39 (1982): 443–450.

———. *Bibliographie des Textes Médicaux Latins*. Saint-Etienne: Publications de l'Université de Saint-Etienne, 2000.

———. "Dr. Monk's Medical Digest." *Social History of Medicine* 13, no. 2 (2000): 239–252.

———. "The Isagoge of Pseudo-Soranus: An Analysis of the Contents of a Medi-eval Introduction to the Art of Medicine." *Medizinhistorisches Journal* 35, no. 1 (2000): 3–30.

———. "Mensch und Heilkunde bei Hildegard von Bingen." *Arzteblatt Rheinland-Pfalz* 51, no. 5 (1998): 165–168.

Flagler, Joel, and Raymond P. Poincelot, eds. *People-Plant Relationships: Setting Research Priorities*. New York: Haworth Press, 1994.

Flanagan, Sabina. "'For God Distinguishes the People of Earth as in Heaven': Hildegard of Bingen's Social Ideas." *Journal of Religious History* 22, no. 1 (1998): 14–34.

———. "Hildegard and the Gendering of Sanctity." In *Hildegard of Bingen and Gendered Theology in Judaeo-Christian Tradition*, edited by Julie Barton and Constant Mews, 81–92. Clayton, Victoria, Australia, 1995.

———. "Hildegard and the Humors: Medieval Theories of Illness and Personality." In *Madness, Melancholy, and the Limits of the Self*, edited by Andrew Weiner and Leonard Kaplan, 14–23. Madison, Wisconsin, 1996.

———. *Hildegard of Bingen, 1098–1179: A Visionary Life*. 2nd ed. London: Routledge, 1998.

———. "Hildegard's Entry into Religion Reconsidered." *Mystics Quarterly* 25, no. 3 (1999): 77–97.

———. "Oblation or Enclosure: Reflections on Hildegard of Bingen's Entry into Religion." In *Wisdom which Encircles Circles, Papers on Hildegard of Bingen*, edited by Audrey Ekdahl Davidson, 1–14. Kalamazoo: Medieval Institute Publications, 1996.

———. *Secrets of God: Writings of Hildegard of Bingen*. Boston: Shambhala, 1996.

———. "Zwischen New Age und wissenschaftlicher Forschung: die Rezeption Hildegards von Bingen in der englisch-sprachigen Welt." In *Prophetin durch die Zeiten: zum 900. Geburtstag*, edited by Edeltraud Forster, 476–484. Freiburg: Herder, 1997.

Fleck, Ludwik. *Genesis and Development of a Scientific Fact*. Translated by Fred Bradley, Thaddeus J. Treun. Edited by Thaddeus J. Treun and Robert K. Merton. Chicago: University of Chicago Press, 1979.

Fleming, Peter. "The Medical Aspects of the Medieval Monastery in England." *Proceedings of the Royal Society of Medicine* 22 (1928): 771–782.

Flemming, Rebecca. *Medicine and the Making of Roman Women: Gender, Nature and Authority from Celsus to Galen*. Oxford: Oxford University Press, 2000.

Flower, Barbara, and Elisabeth Rosenbaum, eds. *The Roman Cookery Book: A Critical Tranlsation of The Art of Cooking by Apicius, for use in the study and the kitchen*. London: P. Nevill, 1958.

Fontaine, Jacques, ed. *Isidore de Seville, Traité de la nature*. Bordeaux: Feret et Fils, 1960.

Fossier, Robert. *Peasant Life in the Medieval West*. Translated by Juliet Vale. Oxford: Basil Blackwell, 1988.

Fox, Matthew. *The Illuminations of Hildegard of Bingen: Text by Hildegard of Bingen with commentary by Matthew Fox, O. P.* Santa Fe: Bear and Company, 1985.

Frankova, Milada. "The Green Knight and the Myth of the Green Man." *Brno Studies in English* 21 (1995): 77–83.

Frazier, James George. *The Dying God*. 3 ed. Vol. 4, *Golden Bough*. London: MacMillan Co., 1911.

French, Roger. "Teaching Aristotle in Medieval English Universities: 'De Plantis' and the Physical 'Glossa Ordinaria.'" *Physis* 34 (1997): 225–296.

———. "The Use of Alfred of Shareshill's Commentary on the 'De Plantis' in University Teachings in the Thirteenth Century." *Viator* 28 (1997): 223–252.

Führkötter, Adelgundis. *Hildegard von Bingen.* Salzburg: Otto Müller Verlag, 1972.

———, ed. *Kosmos und Mensch aus der Sicht Hildegards von Bingen.* Vol. 60. Mainz: Gesellschaft für mittelrheinische Kirchengeschichte, 1987.

Fuhrmann, Horst. *Germany in the High Middle Ages (1050–1200).* Translated by Timothy Reuter. Cambridge: Cambridge University Press, 1986.

Führmann, Joëlle. "Les différentes sources, caractéristiques et fonctions des jardins monastiques au Moyen Age." *Sénéfiance* 28 (1990): 109–124.

Garcia-Ballester, Luis, Roger French, Jon Arrizabalaga, and Andrew Cunningham, eds. *Practical Medicine from Salerno to the Black Death.* Cambridge: Cambridge University Press, 1994.

Gaulin, Jean Louis. "Agronomie antique et élaboration médiévale: De Palladius aux préceptes Cisterciens d'économie rurale." *Médiévales* 26 (1994): 59–83.

———. "Albert le Grand agronome: Notes sur le 'Liber VII De vegetabilibus.'" In *Comprendre et matriser la Nature au Moyen Age. Mélanges d'histoire des sciences offert à Guy Beaujouan,* 155–170. Geneva: Droz, 1994.

Getz, Faye Marie. "Charity, Translation, and the Language of Medical Learning in Medieval England." *Bulletin of the History of Medicine* 64 (1990): 1–17.

———. *Medicine in the English Middle Ages.* Princeton: Princeton University Press, 1998.

———. "Medieval Medicine." *Trends in History* 4, no. 2–3 (1988): 37–54.

Gil-Sotres, Pedro. "Derivation and Revulsion: The Theory and Practice of Medieval Phlebotomy." In *Practical Medicine from Salerno to the Black Death,* edited by Luis Garcia-Ballester, Roger French, Jon Arrizabalaga, and Andrew Cunningham, 110–155. Cambridge: Cambridge University Press, 1974.

Gilles, de Corbeil. "Egidii Corboliensis Viaticus de signis et symptomatibus aegritudium." In *Anecdota Graeca et Graecolatina,* edited by Valentin Rose, 177–201. Berlin: Ferd. Duemmlers Verlagsbuchhandlung, 1864.

Gimpel, Jean. *The Medieval Machine: The Industrial Revolution of the Middle Ages.* New York: Penguin, 1976.

Girault, Pierre-Gilles. *Flores et Jardins: Usages, savoirs et représentations du monde végétal au Moyen Age.* Paris: Le Léopard d'Or, 1997.

Glare, P. G. W. *Oxford Latin Dictionary.* Combined edition first published 1982 ed. Oxford: Clarendon Press, 1984.

Glaze, Florence Eliza. "Medical Writer: 'Behold the Human Creature.'" In *Voice of the Living Light: Hildegard of Bingen and her World,* edited by Barbara Newman, 125–148. Berkeley: University of California Press, 1998.

———. "The Perforated Wall: The Ownership and Circulation of Medical Books in Medieval Europe ca. 800–1200." Ph.D. diss., Duke University, 2000.

Goodey, Robert. "The Medieval Monastic Infirmary in England." Ph.D. diss., University of London, 1987.

Goodich, Michael. *Vita Perfecta: The Ideal of Sainthood in the Thirteenth Century.* Stuttgart: Anton Hiersemann, 1982.

Goody, Jack. *The Culture of Flowers.* Cambridge: Cambridge University Press, 1993.

Gössmann, E., ed. *Hildegard von Bingen: Versuche einer Annährung (Archiv für philosophie- und theologiegeschichtliche Frauenforschung, Sonderband).* München: Iudicum Verlag, 1995.

Gouguenheim, Sylvain. *La Sibylle du Rhin: Hildegarde de Bingen, abbesse et prophétesse rhénane.* Paris: Publications de la Sorbonne, 1996.

Goujarol, Raoul, ed. *Caton de l'agriculture.* Paris: Belles Lettres, 1975.

Gracia Guillén, Diego, and José Luis Vidal. "La Isagoge de Joannitius." *Asclepio* 26–27 (1964–65): 267–382.

Gragson, Ted L., and Ben G. Blount, eds. *Ethnoecology: Knowledge, Resources and Rights.* Athena: University of Georgia Press, 1999.

Grape-Albers, Heidi. *Spätantike Bilder aus der Welt des Arztes.* Wiesbaden: Pressler, 1977.

Grattan, John, and Charles Singer. *Anglo-Saxon Magic and Medicine.* London: Oxford University Press, 1952.

Green, Francis J. "The Archaeology and Documentary Evidence for Plants from the Medieval Period in England." In *Plants and Ancient Man: Studies in paleoethnobotany. Proceedings of the Sixth Symposium of the International Work Group for Paleoethnobotany*, edited by W. van Zeist and W. A. Casparie, 99–114. Rotterdam: A. A. Balkema, 1984.

Green, Monica H. "The Development of the Trotula." *Revue d'Histoire des Textes* 26 (1996): 119–203.

———. "Documenting Medieval Women's Medical Practice." In *Practical Medicine from Salerno to the Black Death*, edited by Luis Garcia-Ballester, Roger French, Jon Arrizabalaga and Andrew Cunningham, 323–352. Cambridge: Cambridge University Press, 1994.

———. "From 'Diseases of Women' to 'Secrets of Women': The Transformation of Gynecological Literature in the Later Middle Ages." *Journal of Medieval and Early Modern Studies* 30, no. 1 (2000): 5–41.

———. "In Search of an 'Authentic' Women's Medicine: The Strange Fates of Trota of Salerno and Hildegard of Bingen." *Dynamis* 19 (1999): 25–54.

———. "Obstetrics and Gynecological Texts in the Middle Ages." *Studies in the Age of Chaucer* 14 (1992): 53–88.

———. "The Transmission of Ancient Theories of Female Physiology and Disease Through the Early Middle Ages." Ph.D. diss., Princeton University, 1985.

———. *The Trotula: A Medieval Compendium of Women's Medicine.* Philadelphia: University of Pennsylvania Press, 2001.

———. *Women's Healthcare in the Medieval West.* Aldershot: Ashgate Variorum, 2000.

————. "Women's Medical Practice and Health Care in Medieval Europe." *Signs* 14 (1989): 434–473.

Greenspan, Kate. "Autohagiography and Medieval Women's Spiritual Autobiography." In *Gender and Text in the Later Middle Ages*, edited by Jane Chance, 216–36. Miami: University Press of Florida, 1996.

Grieco, Allen J., ed. *Le monde végétal (XIIe.-XVIIe. siècles)*. Saint-Denis: Presses Universitaires de Vincennes, 1993.

————. "The Social Politics of pre-Linnaean Botanical Classification." *I Tatti Studies: Essays in the Renaissance* 4 (1991): 131–149.

Grimm, Wilhelm. "Wiesbader Glossen." *Zeitschrift für deutsches Alterthum* 6 (1848): 321–340.

Grmek, Mirko D., ed. *Western Medical Thought from Antiquity to the Middle Ages*, translated by Anthony Shugaar. Cambridge, Mass.: Harvard University Press, 1998.

Gronau, Eduard. *Hildegard von Bingen1098–1179: Prophetische Lehrerin der Kirche an der Schwelle und am Ende der Neuzeit*. Stein-am-Rhein: Christiana, 1985.

Grundmann, Herbert, Steven Rowan, and Robert E. Lerner. *Religious Movements in the Middle Ages: The Historical Links between Heresy, the Mendicant Orders, and the Women's Religious Movement in the Twelfth and Thirteenth Century, with the Historical Foundations of German Mysticism*. Translated by Steven Rowan. Notre Dame: University of Notre Dame Press, 1995.

Guibert, of Gembloux. "Epistola 38." In *Guiberti Gemblacensis Epistolae*, edited by Albert Derolez, 366–379. Turnhout: Brepols, 1989.

Gunther, Robert T. *The Herbal of Apuleius Barbarus from the Early Twelfth-Century Manuscript Formerly in the Abbey of Bury St. Edmunds (MS Bodley 130)*. London: Roxburghe Club Publications, 1925.

Gwei-Djen, Lu, and Joseph Needham. *Celestial Lancets: A History and Rationale of Acupuncture and Moxa*. Cambridge: Cambridge University Press, 1980.

Haeser, Heinrich. *Lehrbuch der Geschichte der Medicin und der epidemischen Krankheiten*. 3 vols. Jena: Hermann Dufft, 1875–1882.

Hagen, Ann. *A Handbook of Anglo-Saxon Food: Processing and Consumption*. Pinner: Anglo-Saxon Books, 1992.

————. *A Second Handbook of Anglo-Saxon Food and Drink: Production and Distribution*. Frithgarth: Anglo-Saxon Books, 1995.

Hagen, Hermannus. *Catalogus codicum Bernensum*. Bern: Haller, 1875.

Hahn, Cynthia. *Portrayed on the Heart: Narrative Effect in Pictorial Lives of Saints from the Tenth through the Thirteenth Century*. Berkeley: University of California Press, 2001.

Hall, Thomas S. "Life, Death, and the Radical Moisture: A Study of Thematic Paths in Medieval Medicine." *Clio Medica* 6 (1971): 3–23.

Halleux, Robert. "Albert le Grand et l'alchimie." *Revue* 66 (1984): 57–80.

Hammond, E. A. "Physicians in Medieval English Religious Houses." *Bulletin of the History of Medicine* 32 (1958): 105–120.

———. "The Westminster Abbey Infirmarer's Rolls as a Source of Medical History." *Bulletin of the History of Medicine* 39 (1965): 261–276.

Hannaway, Caroline. "Environment and Miasmata." In *Companion Encyclopedia to the History of Medicine*, Vol. 1, edited by W. F. Bynum and Roy Porter, 292–308. London: Routledge, 1993.

Harrington, Ann. *Re-enchanted Science: Holism in German Culture from Wilhelm to Hitler*. Princeton: Princeton University Press, 1996.

Hartley, Dorothy. *Lost Country Life*. New York: Pantheon Books, 1979.

Harvey, John H. *Medieval Gardens*. Oregon: Timber Press, 1981.

———. "Westminster Abbey: The Infirmarer's Garden." *Garden History* 20, no. 2 (1992): 97–115.

Haverkamp, Alfred. "Hildegard von Disibodenberg-Bingen: von der Peripherie zum Zentrum." In *Hildegard von Bingen in ihrem historischen Umfeld. Internationaler wissenschaftlicher Kongress zum 900 jährigen Jubiläum, 13–19 September 1998, Bingen am Rhein*, edited by Alfred Haverkamp and Alexander Reverchon, 15–69. Mainz: Philipp von Zabern, 2000.

———. *Medieval Germany 1056–1273*. Translated by Helga Braun and Richard Mortimer. 2 ed. Oxford: Oxford University Press, 1992.

Haverkamp, Alfred, and Alexander Reverchon, eds. *Hildegard von Bingen in ihrem historischen Umfeld. Internationaler wissenschaftlicher Kongress zum 900 jährigen Jubiläum, 13–19 September 1998, Bingen am Rhein*. Mainz: Philipp von Zabern, 2000.

Head, Thomas, ed. *Medieval Hagiography: An Anthology*. New York: Garland Publishing, 2000.

Healy, John, ed. *Natural History: A Selection/Pliny the Elder translated by John Healey*. London: Penguin, 1991.

Heinzelmann, Josef. "Hildegard von Bingen und ihre Verwandtschaft. Genealogische Anmerkungen." *Jahrbuch für westdeutsche Landesgeschichte* 23 (1997): 8–88.

Hennig, John. "Versus de Mensibus." *Traditio* 11, no. 55 (1958): 65–90.

Henschel, L. "Die beschwörung Formeln des Apuleius." *Janus* 1 (1846): 660–669.

Henzen, Walter. "Der Rotulus von Mülinen: Codex 803 der Burgerbibliothek Bern." In *Geschichte, Deutung, Kritik: Literaturwissenschaftliche Beiträge dargebracht zum 65. Geburtstag Werner Kohlschmidts*, edited by Maria Bindschedler and Paul Zinsli, 13–27. Bern: Francke, 1969.

Herrlinger, Robert. *Geschichte der medizinischen Abbildung von der Antike bis um 1600*. Munchen: Heinz Moos Verlag, 1967.

Hertzka, Gottfried, and Wighard Strehlow. *Handbuch der Hildegard-Medizin*. Freiburg im Bresau: Hermann Bauer, 1987.

Heurgon, Jacques, ed. *Varron Economie Rurale*. Vol. 1. Paris: Belles Lettres, 1978.

Hildebrandt, Reiner. "Die deutschsprachige Originalität der Hildegard von Bingen in ihrem Mondphasen-Horoskop." *Orbis Linguarum* 7 (1997): 121–138.

Hildegard of Bingen. *The Book of the Rewards of Life*. Translated by Bruce W. Hozeski. New York: Oxford University Press, 1997.

Hildegard, of Bingen. *Briefwechsel [von] Hildegard von Bingen.* Edited by Adelgundis Führkötter. Salzburg: Otto Müller, 1964.

———. *Das Buch von den Steinen (Physica).* Translated by Peter Riethe. Salzburg: O. Müller, 1979.

———. *Hildegard von Bingen's Physica: The Complete English Translation of her Classic Work on Health and Healing.* Rochester, Vt.: Healing Arts Press, 1998.

———. *Holistic Healing.* Translated by Manfred Pawlik, translator of Latin text, Patrick Madigan, translator of German text, John Kulas, translator of forward. Edited by Mary Palmquist and John Kulas. Collegeville: Liturgical Press, 1996.

———. *The Letters of Hildegard of Bingen.* Translated by Joseph L. Baird, Radd K. Ehrman. Vol. 1. Oxford: Oxford University Press, 1994.

———. *Liber vita meritorum.* Edited by Angela Carlevaris. Vol. 90, *CCCM.* Turnhout: Brepols, 1995.

———. *Scivias.* Edited by Adelgundis Führkötter and Angela Carlevaris. Vols. 43, 43A, *CCCM.* Turnhout: Brepols, 1978.

———. *Scivias.* Translated by Mother Columba Hart and Jane Bishop. New York: Paulist Press, 1990.

———. "Vita Sanctae Rupertis." In *S. Hildegardis Abbatissae Opera Omnia. Patrologia Latina.* Vol. 197, cols. 1081–1094. Paris: Garnier, 1855.

———. "Vita Sancti Disibodi." In *S. Hildegardis Abbatissae Opera Omnia. Patrologia Latina.* Vol. 197, cols. 1095–1116. Paris: Garnier, 1855.

———. 1098–1179. *Hildegard's Healing Plants: from the Medieval Classic 'Physica.'* Translated by Bruce W. Hozeski. Boston: Beacon Press, 2001.

———. *Der Äbtissen Hildegard von Bingen Ursachen und Behandlung der Krankheiten.* Translated by Hugo Schulz. Ulm an der Donau: Karl F. Haugh, 1953.

———. "Liber Simplicis Medicinae [=Physica]." In *S. Hildegardis Abbatissae Opera Omnia, Patrologia Latina.* Vol. 197, cols. 1117–1352. Paris: Garnier, 1855.

———. *Heilkunde. Das Buch von dem Grund und Wesen und der Heilung der Krankheiten (Causae et Curae).* Edited by Heinrich Schipperges. Salzburg: O. Müller, 1957.

Hill, Boyd H. Jr. "The Grain and the Spirit in Medieval Anatomy." *Speculum* 40, no. 1 (1965): 63–73.

Himmelman, P. Kenneth. "Medicinal Body: An Analysis of Medicinal Cannibalism in Europe 1300–1700." *Dialectical Anthropology* 22, no. 2 (1997): 183–203.

Hinkel, Helmut. "Hildegard von Bingen: Nachleben." In *Hildegard von Bingen 1098–1179*, edited by Hans-Jürgen Kotzur, Winfried Wilhelmy, and Ines Koring, 148–154. Mainz: Philipp von Zabern, 1998.

Hoffmann-Krayer, E., and Hans Bächtold-Stäubli. *Handwörterbuch des deutschen Aberglaubens.* Berlin: Walter De Gruyter and Co., 1938/1941.

Höfler, M. "Altgermanische Heilkunde." In *Handbuch der Geschichte der Medizin*, edited by Max Neuburger and Julius Pagel, 453–477. Jena: Fisher, 1902.

Hollister, C. Warren. *Medieval Europe: A Short History*. 8 ed. Boston: McGraw-Hill, 1998.

Holmes, Urban T., Jr., and Frederick R. Weedon. "Peter of Blois as a Physician." *Speculum* 37, no. 2 (1962): 252–256.

Holsinger, Bruce W. *Music, Body, and Desire in Medieval Culture: Hildegard of Bingen to Chaucer*. Stanford: Stanford University Press, 2001.

Holstenius, Lucas, ed. *Codex regularum monasticarum et canonicarum*. 6 vols. Graz: Akademische Druck-U.Verlagsanstalt, 1957(1759).

Holt, Richard. *The Mills of Medieval England*. Oxford: Basil Blackwell, 1988.

Horden, Peregrine. "The Millennium Bug: Health and Medicine around the Year 1000." In *The Year 1000: Medical Practice at the End of the First Millennium*, edited by Peregrine Horden, 201–20. Oxford: Society for the Social History of Medicine, 2000.

———. ed. *The Year 1000: Medical Practice at the End of the First Millennium*. Vol. 13. Oxford: Society for the Social History of Medicine, 2000.

Horn, Walter, and Ernest Born. *The Plan of St. Gall: A Study of the Architecture and Economy of Life in a Paradigmatic Carolingian Monastery*. 3 vols. Berkeley: University of California Press, 1979.

Horowitz, Maryanne Cline. *Seeds of Virtue and Knowledge*. Princeton: Princeton University Press, 1998.

Horst, Eberhard. *Hildegard von Bingen: Die Biographie*. Munich: Verlag GmbH & Co., 2000.

Hotchkin, Julie. "Enclosure and Containment: Jutta and Hildegard at the Abbey of Disibodenberg." *Magistra: A Journal of Women's Spirituality in History* 2, no. 2 (1996): 103–23.

Howald, Ernst, and Henry Sigerist, eds. *Antonii Musae de herba vettonica, liber Pseudo-Apulei herbarius, Anonymi de taxone, liber Sexti Placiti, liber medicinae ex animalibus= Corpus Medicorum Latinorum*. Vol. 4. Leipzig-Berlin: Teubner, 1927.

Huard, Pierre, and Ming Wong. *La Médecine Chinoise*. Paris: Presses Universitaires, 1969.

Hunger, F. W. T., ed. *The Herbal of Pseudo-Apuleius from the Ninth-Century Manuscript in the Abbey of Montecassino: A Facsimile Edition of the Montecassino Manuscript*. Leyden: Brill, 1935.

Hunt, Tony. *Popular Medicine in Thirteenth-Century England: Introduction and Texts*. Cambridge: D. S. Brewer, 1990.

Hutchinson, G. Evelyn. "Attitudes towards Nature in Medieval England: The Alphonso and Bird Psalters." *Isis* 65 (1974): 5–37.

Innes, Matthew. *State and Society in the Early Middle Ages: The Middle-Rhine Valley, 400–1000*. Cambridge, New York: Cambridge University Press, 2000.

Isidore of Seville. "De medicina." In *Etymologiarum*, in *Sancti Isidori opera omnia*, edited by Faustino Arevali. *Patrologia Latina*. Vol. 82, 183–198. Paris: Garnier, 1856.

———. "De ventis." In *De rerum naturae*, in *Sancti Isidori opera omnia*, edited by Faustino Arevali, *Patrologia Latina*. Vol. 82, 479–481. Paris: Garnier, 1856.

Jacquart, Danielle. "Hildegarde et la physiologie de son temps." In *Hildegard of Bingen: the Context of Her Thought and Art*, edited by Charles Burnett and Peter Dronke, 121–134. London: The Warburg Institute, 1998.

———. "L'observation dans les sciences de la nature au Moyen Age: limites et possibilités." *Micrologus* 4 (1996): 55–75.

———. *La Science Médicale Occidentale entre Deux Renaissances*. Brookfield: Variorum, 1997.

———. *Le Milieu Médicale en France du XIIe. au XVe. siècle: En annexe 2e. supplément au Dictionnaire d'Ernest Wickersheimer*. Geneva: Librairie Droz, 1981.

———. "Medieval Scholasticism." In *Western Medical Thought from Antiquity to the Middle Ages*, edited by Mirko Grmek, translated by Anthony Shugaar, 192–240. Cambridge: Cambridge University Press, 1998.

Jacques, Jean-Marie. "La bile noire dans l'antiquité grecque: médecine et littérature." *Revue des Etudes Anciennes* 100, no. 1–2 (1998): 217–234.

Jansen, Ria Sieben. "From Food Therapy to Cookery Book." In *Medieval Dutch Literature in its European Context*, edited by Eric Kooper, 261–279. Cambridge: Cambridge University Press, 1994.

Jenkins, M. "Medicine and Spices with Special Reference to Medieval Monastic Accounts." *Garden History* 4 (1976): 47–49.

Jessen, Carl. "Uber Ausgaben und Handscriften der medicinisch-naturhistorischen Werke der heiligen Hildegard von Bingen." *Sitzungsberichte der Akademie der Wissenschaften in Wien, math.-naturw. Klasse* 45 (1862): 97–116.

Jetter, Dieter. "Klosterhospitäler: St. Gallen, Cluny, Escorial." *Sudhoffs Archiv* 62, no. 4 (1978): 313–338.

Johns, T. *With Bitter Herbs They shall Eat It: Chemical Ecology and the Origins of Human Diet and Medicine*. Tucson: University of Arizona Press, 1990.

Johnson, Mark. *The Body in the Mind: The Bodily Basis of Meaning, Imagination and Reason*. Chicago: University of Chicago Press, 1987.

Jones, Peter Murray. *Medieval Medicine in Illuminated Manuscripts*. Rev. ed. London: The British Library, 1998.

Jones, W. H. S., ed. *Hippocrates' Airs, Waters, and Places*. Vol. 1. Cambridge: Harvard University Press, 1984(1923).

Jordan, D. R. "Two Christian Prayers from Southeastern Sicily." *Greek, Roman, and Byzantine Studies* XXV, no. 3 (1984): 297–302.

Joret, Charles. "Les incantations botaniques des manuscrits f. 277 de la bibliothèque de l'école de médecine de Montpellier et f. 19 de la bibliothèque académique de Breslau." *Romania* 17 (1888): 337–354.

Jouanna, Jacques. "Birth of Western Medical Art." In *Western Medical Thought from Antiquity to the Middle Ages*, edited by Mirko D. Grmek. Translated by Anthony Shugaar. Cambridge: Cambridge University Press, 1998.

———, ed. *Hippocrate Tome II, Airs, Eaux, Lieux*. Paris: Belles Lettres, 1996.

———, ed. *Hippocrate, Des vents, De l'art*. Vol. 5, part 1. Paris: Belles Lettres, 1988.

Jung, Carl G. *Psychology and Alchemy*. Translated by R. F. Hull. Edited by William McGuire. Vol. 12, *Collected Works of C. G. Jung*. Princeton: Princeton University Press, 1968.

Kaibel, G. "Antike Windrosen." *Hermes* 20 (1885): 579–624.

Kaiser, Paul, ed. *Hildegardis Causae et Curae*. Leipzig: Teubner, 1903.

Kaupen-Haas, Heidrun. "Frauenmedizin im deutschen Mittelalter." In *Ethnomedizin und Medizingeschichte*, edited by Joachim Sterly, 169–194. Hamburg: Verlag Mensch und Leben, 1980.

Kay, Sarah, and Miri Rubin, eds. *Framing Medieval Bodies*. Manchester [England]: Manchester University Press, 1994.

Keil, Gundolf. "Arzenibuôch Ipocratis." In *Die deutsche Literatur des Mittelalters Verfasserlexikon*, Vol. 1, edited by Kurt Ruh, Gundolf Keil, Werner Schroder, Burghart Wachinger, and Franz Josef Worstbrock, 505–506. Berlin: Walter de Gruyter, 1978.

———. "Hildegard von Bingen deutsch: Das 'Speyerer Krauterbuch.'" In *Hildegard von Bingen in ihrem historischen Umfeld. Internationaler wissenschaftlicher Kongress zum 900 jährigen Jubiläum, 13.–19. September 1998, Bingen am Rhein*, edited by Alfred Haverkamp and Alexander Reverchon, 441–458. Mainz: Philipp von Zabern, 2000.

———. "Innsbrucker (Prüler) Kräuterbuch." In *Die deutsche Literatur des Mittelalters Verfasserlexikon*, Vol. 4, edited by Kurt Ruh, Gundolf Keil, Werner Schroder, Burghart Wachinger, and Franz Josef Worstbrock, 396–98. Berlin: Walter de Gruyter, 1983.

———. "Innsbrucker Arzneibuch." In *Die deutsche Literatur des Mittelalters Verfasserlexikon*, Vol. 4, edited by Kurt Ruh, Gundolf Keil, Werner Schroder, Burghart Wachinger, and Franz Josef Worstbrock, 395–396. Berlin: Walter de Gruyter, 1983.

———. "Prüler Steinbuch." In *Die deutsche Literatur des Mittelalters Verfasserlexikon*, Vol. 7, edited by Kurt Ruh, Gundolf Keil, Werner Schroder, Burghart Wachinger, and Franz Josef Worstbrock, 875–876. Berlin: Walter de Gruyter, 1977.

Kerby-Fulton, Kathryn. "Prophecy and Suspicion: Closet Radicalism, Reformist Politics, and the Vogue for Hildegardiana in Ricardian England." *Speculum* 75, no. 2 (2000): 318–341.

Kibre, Pearl. *Hippocrates Latinus: Repertorium of Hippocratic Writings in the Latin Middle Ages*. New York: Fordham University Press, 1985.

Kienzle, Beverly Mayne. "Hildegard of Bingen's Gospel Homilies and Her Exegesis of the Parable of the Prodigal Son." In *"Im Angesicht Gottes suche der Mensch sich selbst": Hildegard von Bingen (1098–1179)*, edited by Rainer Berndt, 299–324. Berlin: Academie Verlag GmbH, 2001.

King-Lenzmeir, Anne H. *Hildegard of Bingen: An Integrated Vision*. Collegeville, Minnesota: The Liturgical Press, 2001.

Kitchen, John. *Saints' Lives and the Rhetoric of Gender: Male and Female in Merovingian Hagiography*. New York and Oxford: Oxford University Press, 1998.

Kittredge, George Lynn. *A Study of Gawain and the Green Knight*. Cambridge: Harvard University Press, 1916.

Klaes, Monika, ed. *Vita Sanctae Hildegardis*. Vol. 126, *CCCM*. Turnhout: Brepols, 1993.

———, ed. *Vita Sanctae Hildegardis, Leben der heiligen Hildegard von Bingen; Canonizatio Sanctae Hildegardis, Kanonisation der Heiligen Hildegard*. Freiburg: Herder, 1998.

———. "Zur Schau und Deutung des Kosmos bei Hildegard von Bingen." In *Kosmos und Mensch aus der Sicht Hildegards von Bingen*, edited by Adelgundis Fürhkötter, 37–115. Mainz: Gesellschaft für mittelrheinische Kirchengeschichte, 1987.

Kleinman, Arthur. *Patients and Healers in the Context of Culture*. Berkeley: University of California Press, 1986.

Klibansky, Raymond, Erwin Panofsky, and Fritz Saxl. *Saturn and Melancholy: Studies in the History of Natural Philosophy, Religion, and Art*. New York: Basic Books, 1964.

Klingender, Frances. *Animals in Art and Thought to the End of the Middle Ages*. Edited by Evelyn Antal and John Harthan. Cambridge: MIT Press, 1971.

König, Roderich, ed. *C. Plinius Secundus Naturkunde*. Munich: Artemis and Winkler, 1973.

Koring, Ines. "Hildegard von Bingen 1098–1179, Leben." In *Hildegard von Bingen 1098–11179*, edited by Hans-Jürgen Kotzur, 2–19. Mainz: Philipp von Zabern, 1998.

Kotzur, Hans-Jürgen, Winfried Wilhelmy, and Ines Koring, eds. *Hildegard von Bingen 1098–1179*. Mainz: Philipp von Zabern, 1998.

Kraft, Kent Thomas. "The Eye Sees More than the Heart Knows: The Visionary Cosmology of Hildegard of Bingen." Ph.D. diss., University of Wisconsin, 1977.

Krappe, A. H. "Who was the Green Knight?" *Speculum* 13 (1933): 206–215.

Kristeller, Paul Oskar. "Bartholomaeus, Musandinus, and Maurus of Salerno and Other Early Commentators of the Articella, with a Tentative List of Texts and Manuscripts." *Italia Medioevale e Humanistica* 19 (1976): 57–87.

———. "The School of Salerno. Its Development and Contribution to the History of Learning." *Bulletin of the History of Medicine* 17 (1945): 138–194.

Kroll, Jerome , and Bernard Bachrach. "Sin and the Etiology of Disease in Pre-Crusade Europe." *Journal of the History of Medicine and Allied Sciences* 41 (1986): 395–414.

Kruse, Britta-Julianne. *Verborgene Heilkunste: Geschichte der Frauenmedizin im Spätmittelalter*. Berlin: Walter de Gryter, 1996.

Kühlewein, H. "Die Schrift 'Peri Aeron Udatron Topon' in der lateinischen Ubersetzung des Cod. Paris 7027." *Hermes Zeitschrift für klassische Philologie* 40 (1905): 248–74.

Kühn, C. G., ed. *Klaudiou Galenou Apanta. Claudii Galeni Opera omnia*. 20 vols. Hildesheim: Olms, 1964–1965(1821–1833).

Kupper, Joachim. "(H)er(e)os: Petrarcas Canzoniere und der medizinische Diskurs seiner Zeit." *Romanische Forschungen* 111, no. 2 (1999): 178–224.

Kuriyama, Shigehisa. *The Expressiveness of the Body and the Divergence of Greek and Chinese Medicine*. New York: Zone, 1999.

———. "The Imagination of Winds and the Development of the Chinese Concept of the Body." In *Body, Subject and Power in China*, edited by Angelo Zito, 23–41. Chicago: University of Chicago Press, 1995.

———."Interpreting the History of Bloodletting." *Journal of the History of Medicine and Allied Sciences* 50, no. 1 (1995): 11–46.

———."Varieties of Haptic Experience: A Comparative Study of Greek and Chinese Pulse Diagnosis." Ph.D. diss., Harvard, 1986.

———. "Visual Knowledge in Classical Chinese Medicine." In *Knowledge and the Scholarly Medical Traditions*, edited by Don Bates, 205–234. Cambridge: Cambridge University Press, 1995.

Lactance. *De opificio Dei. L'ouvrage du Dieu créateur*. Translated by Michel Perrin. Edited by Michel Perrin. Paris: Editions du Cerf, 1974.

Lagowski, J. J. "Historical Development of the Concept of Element." In *The New Encyclopedia Brittanica*, edited by Robert McHenry, 933–934. Chicago: Encyclopedia Brittanica, Inc., 1997.

Latour, Bruno. *Laboratory Life*. Princeton: Princeton University Press, 1986.

Lautenschläger, Gabriele. *Hildegard von Bingen. Die theologische Grundlegung ihrer Ethik und Spiritualität*. Bad-Cannstatt: Frommann-Holzboog, 1993.

———. "'Viriditas': Ein Begriff und seine Bedeutung." In *Hildegard von Bingen: Prophetin durch die Zeiten. Zum 900. Geburtstag*, edited by Edeltraud Forster, 224–237. Freiburg: Herder, 1997.

Lauter, Werner. *Das Nachleben der heiligen Hildegard von Bingen*. Bingen: Historische Gesellschaft Bingen, 1981.

———. *Hildegard-Bibliographie. Wegweiser zur Hildegard-Literatur*. 2 vols. Alzey: Rheinischen Druckwerk stätte, 1970–1984.

Lauwaert, Françoise. "Semence de vie, germe d'immortalité." *L'Homme* 129 (1994): 31–57.

Lawrence, Christopher, and George Weisz, eds. *Greater than the Parts: Holism in Biomedicine 1920–1950*. New York: Oxford University Press, 1998.

Lawrence, C. H. *Medieval Monasticism: Forms of Religious Life in Western Europe in the Middle Ages.* 3 ed. New York: Longman, 2001.

Leslie, Charles, ed. *Asian Medical Systems: A Comparative Study.* Berkeley: University of California Press, 1976.

Leslie, Charles, and Allan Young, eds. *Paths to Asian Medical Knowledge.* Berkeley: University of California Press, 1992.

Lewis, Charleton T., and Charles Short. *A Latin Dictionary.* First ed. Oxford: Clarendon Press, 1879.

Leyser, Henrietta. *Hermits and the New Monasticism: A Study of Religious Communities in Western Europe, 1000–1150.* London: Macmillan, 1984.

Liddell, Henry George, and Robert Scott. *A Greek-English Lexicon.* 1 ed. Oxford: Clarendon Press, 1966 (1879).

Lieber, Elinor. "Asaf's 'Book of Medicines': A Hebrew Encyclopedia of Greek and Jewish Medicine, Possibly Compiled in Byzantium on an Indian Model." In *Symposium on Byzantine Medicine,* edited by John Scarborough, 233–249. Washington, D.C.: Dumbarton Oaks Research Library and Collection, 1984.

Liebeschütz, Hans. *Das allegorische Weltbild der heiligen Hildegard von Bingen.* Leipzig: B. G. Teubner, 1930.

Liedtke, Herbert, Gerhard Scharf, and Walter Sperling. *Topographischer Atlas Rheinland-Pfalz.* Neumünster: Karl Wachholtz, 1973.

Lindberg, David. C., ed. *Science in the Middle Ages.* Chicago: University of Chicago Press, 1978.

Lindsay, W. M., ed. *Isidore of Seville, Etymologarium sive originum libri XX.* 2 vols. Oxford: Clarendon Press, 1911, reprint 1985.

Lipinska, Mélanie. *Histoire des femmes médecins depuis l'antiquité jusqu'à nos jours.* Paris: Librairie G. Jacques et Cie, 1900.

Livesey, S. J., and R. H. Rouse. "Nemrot the Astronomer." *Traditio* 37 (1981): 203–266.

Lloyd, George. "Hellenistic Biology and Medicine." In *Greek Science after Aristotle,* 75–90. London: Chatto, 1973.

Logan, Michael H., and Anna R. Dixon. "Agriculture and the Acquisition of Medical Plant Knowledge." In *Eating on the Wild Side: The Pharmacologic, Ecologic, and Social Implications of Using Non-cultigens,* edited by Nina Etkin, 25–45. Tucson: University of Arizona Press, 1994.

Long, R. James. "Alfred of Sareshel's Commentary on the Pseudo-Aristotelian 'De Plantis': A Critical Edition." *Medieval Studies* 47 (1985): 125–167.

———. "A Thirteenth-Century Teaching Aid: An Edition of the Bodleian Abbreviatio of the Pseudo-Aristotelian 'De Plantis.'" In *Aspectus et Affectus,* edited by Gunar Freibergs, 87–103. New York: AMS, 1993.

Lorcin, Marie-Thérèse. "Humeurs, bains et tisanes: l'eau dans la médecine médiévale." *Sénéfiance* 15 (1985): 259–73.

Loux, Françoise. *Pratiques et savoirs populaires: Le corps dans la societé traditionelle.* Paris: Espace des Hommes, 1979.

Loux, Françoise, and Philippe Richard. "Le sang dans les recettes de médecine populaire." In *Affaires du Sang*, edited by Annette Farge. Paris: Imago, 1988.

Lovicz, Simon de, ed. *Macer de herbarum virtutibus*. Warsaw: Wydawnictwa Artystyczne, 1979.

Lulofs, H. J. Drossaart, and E. L. J. Poortman, eds. *Nicolaus Damascenus, De Plantis: Five Translations*. Amsterdam: North Holland Publishing Company, 1989.

Lutz, Alfons. *Initialen aus dem <Liber Antidotarius Magnus>*. Basel: Hageba, 1978.

MacKinney, Loren C. "'Dynamidia' in Medieval Medical Literature." *Isis* 24 (1935): 400–414.

———. *Early Medieval Medicine with Special Reference to France and Chartres*. Baltimore: Johns Hopkins University Press, 1937.

———. "A Half-Century of Medieval Medical History in America." *Medievalia et Humanistica* 11 (1952): 18–42.

———. "Medical Ethics and Etiquette in the Early Middle Ages: The Persistence of Hippocratic Ideals." In *Legacies in Ethics and Medicine*, edited by Chester R. Burns, 173–203. New York: Science History Publications, 1977.

———. *Medical Illustrations in Medieval Manuscripts*. Berkeley: University of California Press, 1965.

Maddocks, Fiona. *Hildegard of Bingen: The Woman of Her Age*. Doubleday: New York, 2001.

Maffi, Luisa. "Domesticated Land, Warm and Cold. Linguistic and Historical Evidence in Tenejapa tzcltal Maya Ethnoecology." In *Ethnoecology: Knowledge, Resources and Rights*, edited by Ted L. Gragson and Ben G. Blount, 41–56. Athens: University of Georgia Press, 1999.

Magnus, Albertus. *De Vegetabilibus, Buch VI, Tractat 2, lateinisch-deutsch*. Edited by Klaus Biewer. Stuttgart: Wissenschaftliche Verlagsgesellschaft mbH, 1992.

Maisonneuve, Roland. "Le symbolisme sacré des couleurs chez deux mystiques mediévales: Hildegarde de Bingen, Julienne de Norwich." *Sénéfiance* 24 (1988): 253–272.

Manders, W. J. A. "Lingua ignota per simplicem hominem Hildegardem prolata." In *Sciencaj Studoj*, 57–60. Copenhagen: Internacia Scienca Asocio Esperantista, 1958.

Manganaro, G. "Nuovi documenti magici della Sicilia orientale." *Rend. Linc.* 8, no. 18 (1963): 52–74.

Manitius, Max. *Geschichte der lateinischer Literatur des Mittelalters*. Vol. 1. Munich: Beck, 1911–1931.

Mannhardt, Wilhelm. *Wald- und Feldkulte*. Vol. 1. Berlin: Verlag von Gebruder Borntraeger, 1904.

Mantuano, Luigi, and Gian Carlo Mancini. "'Semen' nelle opere di Ildegarda di Bingen." *Medicina nei Secoli* 13, no. 2 (2001): 425–440.

Marsolais, Miriam. "'God's Land Is My Land': The Territorial-Political Context of Hildegard of Bingen's Rupertsberg Calling." Ph.D., University of California, Berkeley, 2002.

Martin, Hervé. *Mentalités Médiévales*. Paris: Presses Universitaires de France, 1996.

Martin, René, ed. *Palladius Traité d'Agriculture*. Paris: Belles Lettres, 1976.

———. *Recherches sur les Agronoms Latins et leurs Conceptions Economiques et Sociales*. Paris: Belles Lettres, 1971.

Marzell, Heinrich. "Der Zauber der Heilkräuter in der Antike und Neuzeit." *Sudhoffs Archiv* 29, no. 2 (1936): 3–26.

Massol-Voos, Caroline. "Les jardins: objets d'attentions au Moyen Age." In *Flore et Jardins: usages, savoirs et représentations du monde végétal au Moyen Age*, edited by Pierre-Gilles Girault, 9–35. Paris: Le Léopard d'Or, 1997.

Maurin, Daniel. *Sainte Hildegarde: la santé entre ciel et terre*. Les Tattes: Trois Fontaines, 1991.

Maurmann, Barbara. *Die Himmelsrichtungen im Weltbild des Mittelalters. Hildegard von Bingen, Honorius Augustodunensis und andere Autoren*. Munich: Wilhelm Fink, 1976.

Maurus, Hrbanus. "De Medicina." In *De rerum natura* in *Rabani Mauri Opera omnia*, in the *Patrologia Latina*. Vol. 110, 500–504. Paris: Garnier, 1862.

May, Johannes. *Die heilige Hildegard von Bingen aus dem Orden des heiligen Benedict (1098–1179). Ein Lebensbild*. München: Kösel, 1911.

McCluskey, Stephen C. *Astronomies and Cultures in Early Medieval Europe*. Cambridge: Cambridge University Press, 1998.

McEnerney, John. "Precatio terrae and Precatio omnium herbarum." *Rheinisches Museum für Philologie* 126 (1983): 175–187.

McInerney, Maud Burnett, ed. *Hildegard of Bingen: A Book of Essays*. London: Garland Press, 1998.

McKeon, Richard. "Medicine and Philosophy in the Eleventh and Twelfth Centuries: The Problem of the Elements." *Thomist* 24 (1961): 211–256.

McKitterick, Rosamund, ed. *The New Cambridge Medieval History*. Cambridge: Cambridge University Press, 1995.

McLean, Teresa. *Medieval English Gardens*. London: Barrie and Jenkins, 1989.

McNamara, Jo Ann, and John E. Halborg, eds. *Sainted Women of the Dark Ages*. Durham: Duke University Press, 1992.

McVaugh, Michael R. "Cataracts and Hernias." *Medical History* July (2001): 319–340.

———. "The Development of Medieval Pharmaceutical Theory." In *Arnaldi de Villanova opera medica omnia: Aphorismi de Gradibus*, edited by Michael R. McVaugh, 3–136. Granada-Barcelona: Seminarium Historiae Medicae Granatensis, 1975.

———. "The 'Humidum Radicale' in Thirteenth-Century Medicine." *Traditio* 30 (1974): 259–283.

———. *Medicine before the Plague: Practitioners and Their Patients in the Crown of Aragon, 1285–1345*. Cambridge: Cambridge University Press, 1993.

———. "Quantified Medical Theory and Practice at Fourteenth-Century Montpellier." *Bulletin of the History of Medicine* 43 (1969): 397–413.

Meaney, Audrey. "The Practice of Medicine in England About the Year 1000." In *The Year 1000: Medical Practice at the End of the First Millennium*, edited by Peregrine Horden, 221–38. Oxford: Society for the Social History of Medicine, 2000.

Means, Laurel. "'Ffor as moche as yche man may not have the astralabe': Popular Middle English Variations on the Computus." *Speculum* 67 (1992): 595–623.

———. *Medieval Lunar Astrology: A Collection of Representative Middle English Texts.* Lewiston: E. Mellen, 1993.

Medin, Douglas L., and Scott Atran. *Folk Biology.* Boston: MIT Press, 1999.

Meier, Christel. "Die Bedeutung der Farben im Werk Hildegards von Bingen." *Frühmittelalterliche Studien Jahrbuch des Instituts für Frühmittelalterforschung der Universität Münster* 6 (1972): 245–355.

Meier, Gustav, and Max Burchardt. *Die mittelalterlichen Handschriften der Universität Bibliothek Basel.* Vol. 2. Basel: Verlag Universität Bibliothek, 1966.

Messer, Ellen. "The Hot and Cold in Mesoamerican Indigenous and Hispanicized Thought." *Social Science and Medicine* 25 (1987): 346–399.

———. "Hot-Cold Classification: Theoretical and Practical Implications of a Mexican Study." *Social Science and Medicine* 15B (1981): 133–145.

Mettner, Mattias, and Joachim Müller, eds. *Hildegard von Bingen: 'Renaissance' mit Missverständnissen?* Freiburg: Paulusverlag, 1999.

Mews, Constant. "The Council of Sens (1141): Abelard, Bernard and the Fear of Social Upheaval." *Speculum* 77 (2002): 342–382.

———. "Hildegard and the Schools." In *Hildegard of Bingen: The Context of her Thought and Art*, edited by Charles Burnett and Peter Dronke, 89–110. London: Warburg Institute, 1998.

———. "Hildegard, the Speculum Virginum and Religious Reform in the Twelfth Century." In *Hildegard von Bingen in ihrem historischen Umfeld. Internationaler wissenschaftlicher Kongress zum 900 jährigen Jubiläum, 13.–19. September 1998, Bingen am Rhein*, edited by Alfred Haverkamp and Alexander Reverchon, 237–267. Mainz: Philipp von Zabern, 2000.

———. "Hildegard, Visions and Religious Reform." In *"Im Angesicht Gottes suche der Mensch sich selbst": Hildegard von Bingen (1098–1179)*, edited by Rainer Berndt, 325–342. Berlin: Akademie Verlag, 2001.

Meyer, Ernest, and Carl Jessen, eds. *Alberti Magni . . . De vegetabilibus libri vii, historiae naturalis pars xvii.* Berolini: G. Reimeri, 1867.

Meyer, Heinz. *Die Enzyklopädie des Bartholomaeus Anglicus: Untersuchungen zur Überlieferungs- und Rezeptionsgeschichte von 'de proprietatibus rerum.'* Munich: Fink, 2000.

Meyer-Steineg, Theodore, and Karl Sudhoff. *Geschichte der Medizin.* Jena: Fisher, 1921.

Meyvaert, Paul. "The Medieval Monastic Garden." In *Medieval Gardens*, edited by E. B. MacDougall, 23–53. Washington, D.C.: Dumbarton Oaks, 1986.

Migne, Jacques-Paul, ed. *Patrologiae Cursus Completus [series Latina].* Vols. 221. Paris: Apud Garnier, 1844–1905.

————. ed. *S. Hildegardis abbatissae Opera omnia.* Vol. 197, *Patrologiae cursus completus. Series Latina.* Paris: Garnier, 1855.

Millan-Alvarez, Cristina. "Practice Versus Theory: Tenth-Century Case Histories from the Islamic Middle East." In *The Year 1000: Medical Practice at the End of the First Millennium,* edited by Peregrine Horden, 293–306. Oxford: Society for the Social History of Medicine, 2000.

Millet, Bella. "How Green is the Green Knight?" *Nottingham Medieval Studies* 38 (1994): 138–159.

Mohlberg, Leo Cunibert. *Mittelalterliche Handschriften.* Zurich: Berichthaus, 1952.

Molina, Caroline. "Illness as Privilege: Hildegard von Bingen and the Condition of Mystical Writing." *Women's Studies* 23, no. 1 (1994): 85–91.

Mooney, Catherine M., ed. *Gendered Voices: Medieval Saints and Their Interpreters.* Philadelphia: University of Pennsylvania Press, 1999.

Moore, R. I. *The First European Revolution, c. 970–1215 (The Making of Europe).* Oxford and Malden, Mass.: Blackwell, 1997.

————. *The Formation of a Persecuting Society: Power and Deviance in Western Europe, 950–1250.* Oxford: Basil Blackwell, 1987.

Morton, A. G. *History of Botanical Science.* London: Academic Press, 1981.

Mötsch, Johannes. "Sponheim." In *Die Männer- und Frauenklöster der Benediktiner in Rheinland-Pfalz und Saarland,* edited by Friedhelm Jürgensmeir, 802–803. St. Ottilien: Germania Benedictina IX, 1999.

Moulinier, Laurence. "Abbesse et agronome: Hildegarde et le savoir botanique de son temps." In *Hildegard of Bingen: The Context of Her Thought and Art,* edited by Charles Burnett and Peter Dronke, 135–156. London: The Warburg Institute, 1998.

————. "Deus fragments inédits de Hildegarde de Bingen, copiés par Gerhard von Hohenkirchen." *Sudhoffs Archiv* 83 (1999): 224–238.

————. "Deux jalons de la construction d'un savoir botanique en Allemagne aux XIIe.-XIIIe. siècles: Hildegarde de Bingen et Albert le Grand." In *Le Monde végétal (XIIe. au XIIIe. siècles). Savoirs et usages sociaux,* edited by Allen J. Grieco, O. Redon, and L. Tongiorgi, 89–105. Saint-Denis: Presses Universitaires de Vincennes, 1993.

————. "Ein Präzedenzfall der Kompendien-Literatur. Die Quellen der natur- und heilkundlichen Schriften Hildegards von Bingen." In *Prophetin durch die Zeiten,* edited by Edeltraud Forster, 431–447. Freiburg im Breisgau: Herder, 1997.

————."Hildegarde ou Pseudo-Hildegarde? Réflexions sur l'authenticité du traité 'Cause et cure.'" In *"Im Angesicht Gottes suche der Mensch sich selbst": Hildegard von Bingen (1098–1179),* edited by Rainer Berndt, 115–146. Berlin: Akademie Verlag, 2001.

————, ed. *Hildegardis Bingensis Cause et Cure.* Vol. 1. Berlin: Rarissima mediaevalia, 2003.

————. "La botanique d'Hildegarde de Bingen." *Médiévales* 16–17 (1989): 113–129.

————. *Le manuscrit perdu à Strasbourg: enquête sur l'oeuvre scientifique de Hilde-garde*. Paris: Publications de la Sorbonne, 1995.

————."Magie, médecine et maux de l'ame dans l'oeuvre scientifique de Hilde-garde." In *"Im Angesicht Gottes suche der Mensch sich Selbst,"* edited by Rainer Berndt, 545–599. Berlin: Akademie Verlag, 2001.

Mudry, Philip. "Caelius Aurelianus ou l'anti-Romain: un aspect particulier du traité des maladies aigues et des maladies chroniques." In *Maladie et maladies dans les textes latins antiques et médiévaux*, edited by Carl Deroux, 313–329. Brux-elles: Latomus, 1998.

Müller, Annette. *Krankheitsbilder im Liber de Plantis der Hildegard von Bingen (1098–1179) und im Speyerer Kräuterbuch (1456)*. Hurtenwald: Guido Pressler, 1997.

Müller, Irmgard. "Die Bedeutung der lateinischen Handschrift Ms. Laur. Ashb. 1323 (Florenz, Biblioteca Medicea Laurenziana) für die Rekonstruction der 'Physica' Hildegards von Bingen und ihre Lehre von den naturlichen Wirk-kraften." In *Hildegard von Bingen in ihrem historischen Umfeld. Internationaler wissenschaftlicher Kongress zum 900 jährigen Jubiläum, 13–19 September 1998, Bingen am Rhein*, edited by Alfred Haverkamp and Alexander Reverchon, 421–440. Mainz: Philipp von Zabern, 2000.

————. *Die pflanzlichen Heilmittel bei Hildegard von Bingen*. Freiburg: Herder, 1993(1982).

————. "Die Stellung der Pflanzen in der Heilsordnung bei Hildegard von Bingen." In *Kosmos und Mensch aus der Sicht Hildegards von Bingen*, edited by Adelgun-dis Führkötter, 27–36. Mainz: Gesellschaft für mittelrheinische Kirchenge-schichte, 1987.

————. "Krankheit und Heilmittel im Werk Hildegards von Bingen." In *Hildegard von Bingen 1179–1979: Festschrift zum 800. Todestag der Heiligen*, edited by Anton Ph. Brück, 311–349. Mainz: Gesellschaft für mittelrheinische Kirch-engeschichte, 1979.

————. "Wie 'authentisch' ist die Hildegardmedizin? Zur Rezeption des 'Liber sim-plicis medicinae' Hildegards von Bingen im Codex Bernensis 525." In *Hilde-gard von Bingen. Prophetin durch die Zeiten. Zum 900. Geburtstag*, edited by Edeltraud Forster, 420–430. Freiburg: Herder, 1997.

————. "Zur Verfasserfrage der medizinisch-naturkundlichen Schriften Hildegards von Bingen." In *Tiefe des Gotteswissens: Schönheit der Sprachgestalt bei Hilde-gard von Bingen*, edited by Margot Schmidt, 1–17. Stuttgart-Bad Cannstatt: Frommann-Holzboog, 1995.

Muller-Jahncke, Wolf-Dieter. "Die Pflanzenabbildung im Mittelalter und in der frühen Neuzeit." In *Inter Folia Fructus*, edited by Peter Dilg, 47–64. Eschbon: Govi-Verlag, 1995.

Muller-Rohlfsen, Inge. *Die lateinische ravennatische Ubersetzung der Hip-pokratischen Aphorismen aus dem 5./6. Jahrhundert n. Ch*. Hamburg: Ludke Verlag, 1978.

Murdoch, Brian. "'Peri Hieres Nousou.' An Approach to the Old High German Medical Charms." In *Mit regulu bithungan*, edited by John L. Flood and D. N. Yeandle, 142–160. Gsppingen, 1989.

Murdoch, John E. *Album of Science: Antiquity and the Middle Ages*. New York: Charles Scribner's Sons, 1984.

Mynors, R. A. B., ed. *Cassiodori Senatoris Institutiones*. Oxford: Clarendon Press, 1937.

Nebbiai-Dalla Guarda, Donatella. "Les livres de l'infirmerie." *Revue Mabillon* 5 (1994): 57–81.

Needham, Joseph. "Winds." In *Science and Civilization in China*, edited by Joseph Needham. Vol. 3. 477–479. Cambridge: Cambridge University Press, 1959.

Neuburger, Max. *The Doctrine of the Healing Power of Nature throughout the Course of Time*. Translated by Linn J. Boyd. New York: s.n., 1932.

————. "A Historical Study of the Concept of Nature from a Medical Viewpoint." *Isis* 35 (1944): 16–28.

Neugebauer, Otto. "The Early History of the Astrolabe." *Isis* 40 (1949): 240–256.

Neuser, Kora. *Anemoi: Studien zur Darstellung der Winde und Windgottheiten in der Antike*. Rome: G. Bretschneider, 1982.

Newman, Barbara. "Hildegard and Her Hagiographers: The Remaking of Female Sainthood." In *Gendered Voices: Medieval Saints and Their Interpreters*, edited by Catherine M. Mooney, 16–34. Philadelphia: University of Pennsylvania Press, 1999.

————, ed. *Saint Hildegard of Bingen, Symphonia: A Critical Edition of the Symphonia Armonie Celestium Revelationum*. Ithaca and London: Cornell University Press, 1988.

————. "'Sibyl of the Rhine': Hildegard's Life and Times." In *Voice of the Living Light*, edited by Barbara Newman, 1–29. Berkeley: University of California Press, 1998.

————. *Sister of Wisdom: St. Hildegard's Theology of the Feminine*. Berkeley: University of California Press, 1997(1987).

————. "Three-part Invention: The 'Vita S. Hildegardis' and Mystical Hagiography." In *Hildegard of Bingen: The Context of Her Art and Thought*, edited by Charles Burnett and Peter Dronke, 189–210. London: Warburg Institute, 1998.

————, ed. *Voice of the Living Light: Hildegard of Bingen and Her World*. Berkeley: University of California Press, 1998.

Newmeyer, Stephen T. "Asaph's Book of Remedies: Greek Science and Jewish Apologetics." *Sudhoffs Archiv* 76, no. 1 (1992): 28–36.

Ni, Maoshing. *The Yellow Emperor's Classic of Medicine*. Boston: Shambala, 1995.

Niebyl, Peter H. "Old Age, Fever, and the Lamp Metaphor." *Journal of the History of Medicine* 26 (1972): 351–368.

Nikitsch, Eberhard J. *Kloster Disibodenberg. Religiosität, Kunst und Kultur im mittleren Naheland*. Regensburg: Schnell und Steiner, 1998.

————. "Wo lebte die heilige Hildegard wirklich? Neue Uberlegungen zum ehemaligen Standort der Frauenklause auf dem Disobodenberg." In *Im Angesicht Gottes suche der Mensch sich selbst,*" edited by Rainer Berndt, 147–156. Berlin: Akademie Verlag, 2001.

Noble, Clare, Charles Moreton, and Paul Rutledge. *Farming and Gardening in Late Medieval Norfolk.* Norfolk: Norfolk Record Society, 1997.

North, John D. "The Astrolabe." *Scientific American* 230 (1974): 94–106.

Norwood, Patricia. "Hildegard von Bingen (1098–1179): A Case Study in Methodological Approaches to Medieval Scholarship at the Turn of the Millennium." *Medieval Perspectives* 15, no. 2 (2000): 49–60.

Nutton, Vivian. *Ancient Medicine.* London and New York: Routledge, 2004.

————. "Humoralism." In *Companion Encyclopedia to the History of Medicine,* edited by W. F. Bynum and Roy Porter. Vol. 1, 281–291. London: Routledge, 1993.

————. "Medicine in Late Antiquity and the Early Middle Ages." In *The Western Medical Tradition: 800 B.C.–1800 A.D.,* edited by Lawrence I. Conrad, Michael Neve, Vivian Nutton, Roy Porter, and Andrew Wear, 71–88. Cambridge: Cambridge University Press, 1995.

————. "Medicine in the Greek World, 800–50 B.C." In *The Western Medical Tradition: 800 B.C.–1800 A.D.,* edited by Lawrence I. Conrad, Michael Neve, Vivian Nutton, Roy Porter, and Andrew Wear, 11–38. Cambridge: Cambridge University Press, 1995.

O'Boyle, Cornelius. *The Art of Medicine.* Boston: Brill, 1998.

Obrist, Barbara. "Le diagramme isidorien de l'année et des saisons: son contenu physique et les représentations figuratives." *Mélanges de l'Ecole française de Rome: Moyen Age* 108, no. 1 (1996): 95–164.

————. "Wind Diagrams and Medieval Cosmology." *Speculum* 72 (1997): 33–84.

Olsan, Lea T. "Latin Charms in British Library, MS Royal 12. B. XXV." *Manuscripta* 33 (1989): 119–128.

Olson, Robert. "The Green Man in Hildegard of Bingen." *Studia Mystica* 15, no. 4 (1992): 3–18.

Önnerfors, Alf. "Iatromagisches Beschwörungen in der 'Physica Plinii Sangallensis.'" *Eranos* 83 (1985): 235–252.

————, ed. *Plinii Secundi Iunoris qui furuntur de medicina libri tres.* Vol. 3, *Corpus Medicorum Latinorum.* Berlin: Akadamie Verlag, 1964.

Ots, Thomas. "The Angry Liver, the Anxious Heart and the Melancholy Spleen: The Phenomenology of Perceptions in Chinese Culture." *Culture, Medicine and Psychiatry* 14 (1990): 21–58.

Panofsky, Ernst. "Melancholy in the Physiological Literature of the Ancients." In *Saturn and Melancholy,* edited by R. Klibansky, Fritz Saxl, and Ernst Panofsky, 3–15. London: Nelson, 1964.

Parsons, David. "Byrhtferth and the Runes of Oxford, St. John's College, Manuscript 17." In *Runeninschriften als Quellen interdisziplinärer Forschung,* edited by Kläus Duwel and Sean Nowak, 439–447. Berlin, New York: Walter de Gruyter, 1998.

Pastoureau, Michel. *Couleurs, Images, Symboles: Etudes d'histoire et de l'anthropologie.* Paris: Le Léopard d'Or, 1989.

———. *Figures et Couleurs: Etudes sur la symbolique et la sensibilité médiévale.* Paris: Le Léopard d'Or, 1986.

———, ed. *L'arbre: Histoire naturelle et symbolique de l'arbre, du bois et du fruit au Moyen Age.* Paris: Le Léopard d'Or, 1993.

———. "Le temps mis en couleurs: Des couleurs liturgiques, aux modes vestimentaires XIIe.-XIIIe. s." *Bibliothèque de l'Ecole des Chartes* 157 (1999): 111–135.

———. "Voir les couleurs au XIIIe. siècle." *Micrologus* 6 (1998): 147–165.

Patzelt, Erna. "Moines-Médecins." In *Etudes de Civilisation Médiévale (IXe.–XIIe. s.). Mélanges offerts à Edmond-René Labande*, 577–588. Poitiers: Centre d'Etudes Supérieures de Civilisation Médiévale, 1974.

Pernoud, Regine. *Hildegarde de Bingen: Conscience inspirée du XIIe. siècle.* Monaco: Editions du Rocher, 1994.

———. "La femme et la médecine au Moyen Age." In *Colloque International d'Histoire de la Médecine: Orléans, 4 et 5 mai 1985*, 1038–1043. Orléans: La Societé, 1985.

Phelan, Walter S. *The Christmas Hero and Yuletide Tradition in Sir Gawain and the Green Knight.* Lewiston: Edwin Mueller Press, 1992.

Pilsworth, Clare. "Medicine and Hagiography in Italy (c. 800–c.1000)." In *The Year 1000: Medical Practice at the End of the First Millennium*, edited by Peregrine Horden, 253–64. Oxford: Society for the Social History of Medicine, 2000.

Pinto, Lucille B. "The Folk Practice of Gynecology and Obstetrics in the Middle Ages." *Bulletin of the History of Medicine* 47 (1973): 513–523.

———. "Medical Science and Superstition: A Report on a Unique Medical Scroll of the Eleventh–Twelfth Century." *Manuscripta* 17, no. 1 (1973): 12–21.

Pitra, J. B., ed. *Analecta Sanctae Hildegardis Opera Spicilegio Solesmensi Parata.* Vol. 8(=5), *Analecta Sacra Spicilegio Solesmensi parata*. Montecassino, 1882.

Pliny, the Elder. *Plinii Secondii quae fertur una cum Gargilii Martialis medicina.* Edited by Valentin Rose. Leipzig: Teubner, 1875.

Podehl, Wolfgang, ed. *900 Jahre Hildegard von Bingen. Neuere Untersuchungen und literarische Nachweise.* Wiesbaden: Hessische Landesbibliothek, 1998.

Porkert, Manfred. *The Theoretical Foundations of Chinese Medicine: Systems of Correspondence.* Cambridge: MIT Press, 1985.

Porter, Roy. "History of the Body." In *New Perspectives on Historical Writing*, edited by Peter Burke, 206–232. Cambridge: Polity Press, 1991.

Portmann, Marie-Louise, and Alois Odermatt. *Wörterbuch der unbekannten Sprache (Lingua ignota) in der Reihenfolge der Manuskripte.* Basle: Basler Hildegard-Gesellschaft, 1986.

Pouchelle, Marie-Christine. *The Body and Surgery in the Middle Ages.* Translated by Rosemary Morris. New Brunswick: Rutgers University Press, 1990.

Pouchelle, Marie-Christine. *Corps et Chirurgie à l'apogée du Moyen Age: Savoir et imaginaire du corps chez Henri de Mondeville, chirurgien de Philippe Le Bel.* Paris: Flammarion, 1983.

Poulle, Emmanuel. "Le traité de l'astrolabe d'Adelard de Bath." In *Adelard of Bath: An English Scientist and Arabist of the Early Twelfth Century,* edited by Charles Burnett, 119–132. London: Warburg Institute, 1987.

Power, Eileen. "Some Women Practitioners of Medicine in the Middle Ages." *Proceedings of the Royal Society of Medicine* 15, no. 6 (1922): 20–23.

Pradel, F. *Griechische und suditalienische Beschwörungen, und Rezepte des Mittelalters.* Giessen, 1907.

Priscianus, Theodorus. *Euporiston libri iii cum physicorum fragmento et additamentis pseudo-Theodoreis.* Edited by Valentin Rose. Leipzig: B. G. Teubner, 1894.

Pseudo-Bede. "Humores." In *De Mundi Celestis Terrestrisque Constitutione,* edited by Charles Burnett, 18–19. London: Warburg Institute, 1985.

Quintus, Gargilius Martialis. *Q. Gargilii Martialis de hortis.* Edited by Innocenzo Mazzini. Bologna: Patron, 1978.

Raff, Thomas. "Die Ikonographie der mittelalterlichen Windpersonifikationen." *Aachener Kunstblätter* 48 (1978–79): 71–218.

Raglan, Lady. "The Green Man in Church Architecture." *Folklore* 39 (1939): 45–52.

Rawcliffe, Carole. *Sources for the History of Medicine in Medieval England.* Kalamzoo: Medieval Institute Publications, 1995.

Reeds, Karen. "Albert on the Natural Philosophy of Plant Life." In *Albertus Magnus and the Sciences: Commemorative Essays,* edited by James A. Weisheipl, 341–354. Toronto: Pontifical Institute of Medieval Studies, 1980.

———. *Botany in Medieval and Renaissance Universities.* New York: Garland, 1991.

Reicke, Siegfried. *Das deutsche Spital und sein Recht im Mittelalter.* Amsterdam: P. Schipers, 1961(1932).

Reuter, Timothy. *Germany in the Early Middle Ages, 800–1056.* London: Longman, 1991.

Reynolds, L. D. *Texts and Transmissions: A Survey of Latin Classics.* Oxford: Clarendon Press, 1983.

Reynolds, Philip Lyndon. *Food and the Body: Some Peculiar Questions in High Medieval Theology.* Leiden: Brill, 1999.

Ribémont, Bernard. "Du verger au cosmos: plantes et jardins dans la tradition médiévale." *Sénéfiance* 28 (1990): 313–327.

———, ed. *Le corps et ses énigmes au Moyen Age: Actes du Colloque Orléans 15–16 mai 1992.* Caen: Paradigm, 1993.

———. "Un corps humain animé, un corps humain irrigué: l'encylopédisme et la théorie du corps." In *Le corps et ses énigmes au Moyen Age: Actes du Colloque Orléans 15–16 mai 1992,* edited by Bernard Ribémont, 185–204. Caen: Paradigm, 1993.

Ribémont, Bernard, and Geneviève Sodigné-Costes. "Botanique médiévale: tradition, observation, imaginaire. L'exemple de l'encyclopédisme." In *Le Moyen*

Age et la Science: Approche de quelques disciplines et personalités scientifiques médiévales, edited by Bernard Ribémont, 153–169. Paris: Klinksieck, 1991.

Richter, Will, ed. *Petrus de Crescentiis (Pier de' Crescenzi). Ruralia Commoda. Das Wissen des volkommenen Landwirts um 1300.* 3 vols. Heidelberg: Universitätsverlag C. Winter, 1995.

Riddle, John. "The Introduction and Use of Eastern Drugs in the Early Middle Ages." *Sudhoffs Archiv* (1965): 185–198.

———. "Manuscript Sources for Birth Control." In *Manuscript Sources for Medieval Medicine. A Book of Essays*, edited by Margaret R. Schleissner, 145–158. New York: Garland, 1995.

———. "Medieval Medical Botany." *Journal of the History of Biology* 14, no. 1 (1981): 43–81.

———. "Theory and Practice in Medieval Medicine." *Viator* 5 (1974): 157–184.

Rigotti, Francesca, and Pierangelo Schiera, eds. *Aria, terra, acqua, fuoco: i quattro elementi e le loro metafore.* Bologna: Società editrice il Mulinon, 1996.

Rippere, Vicky. "The Survival of Traditional Medicine in Lay Medical Views: An Empirical Approach to the History of Medicine." *Medical History* 25 (1981): 411–14.

Risse, Guenter B. *Mending Bodies, Saving Souls: A History of Hospitals.* Oxford: Oxford University Press, 1999.

Robertson, Duncan. *The Medieval Saints' Lives: Spiritual Renewal and Old French Literature.* Nicholasville, Kentucky: French Forum, Publishers, 1995.

Robertson, D. J. "Why the Devil Wears Green." *MLN* 69 (1954): 470–472.

Robinson, Henry S. "The Tower of the Winds and the Roman Marketplace." *American Journal of Archaeology* 47 (1943): 291–305.

Rodgers, Robert H. *An Introduction to Palladius.* London: University of London, 1975.

———, ed. *Palladis Rutilis Tauri Ameiliani viri inlustrus opus agriculturae.* Leipzig: Teubner, 1975.

Rosenberg, Charles, and Janet Golden, eds. *Framing Disease: Studies in Cultural History.* New Brunswick: Princeton University Press, 1992.

Rosenberg, Charles E. "The Therapeutic Revolution: Medicine, Man, and Social Change in Nineteenth-Century America." In *Essays in the Social History of American Medicine*, edited by Morris Vogel and Charles E. Rosenberg, 3–23. Philadelphia: University of Pennsylvania Press, 1979.

Roth, F. W. E. "Althochdeutsches aus Trier." *Zeitschrift für deutsches Altertum* 52 (1910): 169–182.

———. "Glossae Hildegardis (Das Wörterverzeichnis der Lingua ignota)." In *Die althochdeutschen Glossen*, edited by Emil Elias Steinmeyer and Eduard Sievers, 390–404. Berlin, 1895.

Royce-Roll, Donald. "The Colors of Romanesque Stained Glass." *Journal of Glass Studies* 33 (1994): 71–80.

Ruh, Kurt, Gundolf Keil, Werner Schroder, Burghart Wachinger, and Franz Josef Worstbrock, eds. *Die deutsche Literatur des Mittelalters. Verfasserlexikon.* Vols. 1–8. Berlin, 1978.

Ruppel, Bernd. "Ein verschollenes Gedicht des 12. Jahrhunderts: Heinrichs von Huntington 'De Herbis.'" *Frühmittelalterliche Studien* 31 (1997): 197–213.

Sabbah, Guy, Pierre-Paul Corsette, and Klaus-Dietrich Fischer. *Bibliographie des textes médicaux latins: Antiquité et haut Moyen Age.* Saint-Etienne: Université de St. Etienne, 1985.

Sacks, Oliver W. *Migraine: The Evolution of a Common Disorder.* Berkeley: University of California Press, 1970.

Sadowski, Piotr. "The Greenness of the Green Knight: A Study of Medieval Color Symbolism." In *The Knight on His Quest: Symbolic Patterns of Transition in Sir Gawain and the Green Knight,* 78–108. Newark: University of Delaware Press, 1999.

Salvat, Michel. "Le traité des couleurs de Barthelemi l'Anglais (XIIIe. s.)." *Sénéfiance* 24 (1988): 359–385.

Santillana, Giorgio de, and Hertha von Dechend. *Hamlet's Mill: An Essay on Myth and the Frame of Time.* Boston: David R. Godine, 1969.

Sarton, George. *Introduction to the History of Science.* Vol. 2. Baltimore: Williams and Wilkins, 1927.

Sassi, Maria Michela. *The Science of Man in Ancient Greece.* Translated by Paul Tucker. Chicago and London: University of Chicago Press, 2001.

Saurma-Jeltsch, Lieselotte E. *Die Miniaturen im 'Liber Scivias' der Hildegard von Bingen: die Wucht der Vision und die Ordnung der Bilder.* Weisbaden: Dr. Ludwig Reichert Verlag, 1998.

Saxl, Fritz. "A Spiritual Encyclopedia of the Later Middle Ages." *Journal of the Warburg and Courtauld Institute* 5 (1942): 82–42.

Schalick, Walton Orvyl, III. "Add One Part Pharmacy to One Part Surgery and One Part Medicine: Jean de Saint-Amand and the Developmentof Medical Pharmacology in Thirteenth-Century Paris." Ph.D. diss., Johns Hopkins University, 1997.

Schibanoff, Susan. "Hildegard of Bingen and Richardis of Stade: The Discourse of Desire." In *Same Sex Love and Desire among Women in the Middle Ages,* edited by Francesca Canade and Pamela Sheingorn, 49–83. New York: Palgrave, 2001.

Schipperges, Heinrich. *Die Benediktiner in der Medizin des frühen Mittelalters.* Leipzig: St. Benno-Verlag, 1964.

———. *Die Welt der Hildegard von Bingen.* Freiburg: Herder, 1997.

———. "Ein unveröffentlichtes Hildegard-Fragment (Codex Berolin. Lat. Qu. 674)." *Sudhoffs Archiv* 40 (1956): 41–77.

———. "Einflüsse arabischen Medizin auf die Microkosmosliteratur des 12. jahrhundert." *Miscellanea Mediaevalia* 1 (1962): 139–142.

———. *Garten der Gesundheit.* München: Artemis, 1985.

————. *Hildegard of Bingen: Healing and the Nature of the Cosmos.* Translated by John Broadwin. Princeton: Princeton University Press, 1997.

————. *Hildegarde de Bingen (1098–1179).* Translated by Pierre Kemner. Paris: Brepow, 1996.

————. "Menschenkunde und Heilkunst bei Hildegard von Bingen." In *Festschrift zum 800. Todestag der Heiligen*, edited by Anton Ph. Brück, 295–310. Mainz: Selbstverlag der Gesellschaft für mittelrheinische Kirchengeschichte, 1979.

Schleissner, Margaret R., ed. *Manuscript Sources of Medieval Medicine: A Book of Essays.* New York/London: Garland Publishing Inc., 1995.

Schmidt, Margot, ed. *Tiefe des Gotteswissens—Schönheit der Sprachgestalt bei Hildegard von Bingen.* Stuttgart-Bad Cannstatt: Fromman-Holzboog, 1995.

Schmitt, Jean-Claude. *The Holy Greyhound: Guinefort, the Healer of Children since the Thirteenth Century.* Translated by Martin Thom. Cambridge: Cambridge University Press, 1983.

Schmitt, Miriam. "Blessed Jutta of Disibodenberg. Hildegard of Bingen's Magistra and Abbess." *American Benedictine Review* 40, no. 2 (1989): 170–189.

————. "Hildegard of Bingen: Viriditas, Web of Greening Life Energy I." *American Benedictine Review* 50, no. 3 (1999): 253–276.

————. "Hildegard of Bingen: Viriditas, Web of Greening Life Energy II." *American Benedictine Review* 50, no. 4 (1999): 353–380.

Schmitz, Rudolf. *Geschichte der Pharmazie.* Eschbon: Govi, 1998.

Schnapp, Jeffrey T. "Virgin Words: Hildegard of Bingen's 'Lingua Ignota' and the Development of Imaginary Languages Ancient to Modern." *Exemplaria, A Journal of Theory in Medieval and Renaissance Studies* 3, no. 2 (1991): 267–298.

Schnell, Bernhard. "Das Prüller Arzneibuch: Zum ersten Herbar in deutscher Sprache." *Zeitschrift für deutsches Altertum und deutsche Literatur* 120 (1991): 184–202.

————. "Vorüberlegungen zu einer 'Geschichte der deutschen Medizinliteratur des Mittelalters' am Beispiel des 12. Jahrhunderts." *Sudhoffs Archiv* 78, no. 1 (1994): 90–97.

Schöner, Erich. "Das Viererschema in der antiken Humoralpathologie." *Sudhoffs Archiv* 4 (1964): 1–111.

Schönfeld, Hildegard, ed. *Hildegard von Bingen. 'Scivias.' Die Miniaturen vom Rupertsberg.* Bingen, 1979.

Schrader, Marianna. *Die Herkunft der Heiligen Hildegard.* Second ed. Mainz: Gesellschaft für mittelrheinische Kirchengeschichte, 1981 (1941).

Schrader, Marianna, and Adelgundis Führkötter. *Die Echtheit des Schrifttums der heiligen Hildegard von Bingen. Quellenkritische Untersuchungen.* Köln: Böhlau, 1956.

Schroeder, Joy. "A Fiery Heat: Images of the Holy Spirit in the Writings of Hildegard of Bingen." *Mystics Quarterly* 30, no. 3/4 (2004): 79–98.

Schulenberg, Jane Tibbetts. *Forgetful of Their Sex: Female Sanctity and Society, ca. 500–1100.* Chicago: University of Chicago Press, 1998.

————. "Strict Active Enclosure and Its Effects on the Female Monastic Experience (500–1100)." In *Medieval Religious Women*, edited by John A. Nichols and Lillian Thomas Shank, 51–86. Indianopolis: Cisterician Publications, 1984.

Scully, Terence. *The Art of Cookery in the Middle Ages*. Woodbridge: The Boydell Press, 1995.

Seaman, Gary. "Winds, Waters, Seeds, and Souls: Folk Concepts of Physiology and Etiology in Chinese Geomancy." In *Paths to Asian Medical Knowledge*, edited by Charles Leslie and Allan Young, 74–97. Berkeley: University of California Press, 1992.

Sears, Elizabeth. *The Ages of Man: Medieval Interpretations of the Life-Cycle*. Princeton: Princeton University Press, 1986.

Secundus, C. Plinius. *Naturkunde: Lateinisch-Deutsch Buch XVII, XVIII, XIX*. Translated by Roderich König. Edited by Roderich König. Darmstadt: Wissenschaftliche Buchgesellschaft, 1987.

Seibrich, Wolfgang. "Geschichte des Klosters Disibodenberg." In *Hildegard von Bingen 1179–1979: Festschrift zum 800. Todestag der Heiligen*, edited by Anton Ph. Brück, 55–75. Mainz: Selbstverlag der Gesellschaft für mittelrheinishen Kirchengeschichte, 1979.

Seide, Jacob. "Medicine and Natural History in the Itinerary of Rabbi Benjamin of Tudela (1100–1177)." *Bulletin of the History of Medicine* 28, no. 5 (1954): 401–407.

Semmler, Joseph. "Die Sorge um der kranken Mitbruder im benediktiner Kloster des frühen und hohen Mittelalter." In *Die Kranke Mensch in Mittelalter und Renaissance*, edited by Peter Wunderli, 45–59. Dusseldorf: Droste Verlag, 1986.

Seyfert, Werner. "Ein Komplexionentext einer Leipziger Inkunabel (angeblich eines Johann von Neuhaus) und seine handschriftliche Herleitung aus der zeit nach 1500." *Arch. Gesch. Med.* 20 (1928): 272–299, 372–389, text at 286–299.

Seymour, M. C., and Colleagues. *Bartholomaeus Anglicus and his Encyclopedia*. Aldershot: Variorum, 1992.

Shatzmiller, Joseph. "Doctors and Medical Practice in Germany around the Year 1200: The Evidence of 'Sefer Asaph.'" *Proceedings of the American Academy for Jewish Research* 50 (1983): 149–164.

————. "Doctors and Medical Practice in Germany around the Year 1200: The Evidence of 'Sefer Hasidim.'" *Journal of Jewish Studies* 33 (1982): 583–593.

————. *Jews, Medicine, and Medieval Society*. Berkeley: University of California Press, 1994.

Sigerist, Henry. "The Latin Medical Literature of the Early Middle Ages." *Journal of the History of Medicine and Allied Sciences* 13 (1958): 127–145.

Silvas, Anna. *Jutta and Hildegard: The Biographical Sources*. University Park: The Pennsylvania State University Press, 1998.

Simek, Rudolf. *Heaven and Earth in the Middle Ages.* Translated by Angela Hall. Great Britain: Boydell Press, 1992.

Simon, Isidore. *Asaph Ha-Jehudi, Médecin et astrologue du Moyen Age.* Paris: Lipschutz, 1937.

Simonini, Roberto. "Herbolarium et materia medici: Codex MS n. 296 della Biblioteca Governmental di Lucca." *Atti e memorie della R. Acc. di sc. lett ed arti di Modena* 5, no. 1 (1936): 183–228.

Singer, Charles. "Allegorical Representations of the Synagogue in a Twelfth-century Illumination of Hildegard of Bingen." *Jewish Quarterly Review* (1914): 267–282.

———. "The Herbal in Antiquity and its Transmission." *Journal of Hellenic Studies* 47 (1927): 1–52.

———. "A Review of the Medical Literature of the Dark Ages, with a New Text of about 1110." *Proceedings of the Royal Society of Medicine: Section of the History of Medicine* 10, part 2 (1916–1917): 107–160.

———. "The Scientific Views and Visions of Saint Hildegard (1098–1179)." In *Studies in the History and Method of Science,* edited by Charles Singer, 1–55. Oxford: Clarendon Press, 1955.

———. "The Visions of Hildegard of Bingen: Essays on the Scientific Twilight." In *From Magic to Science,* edited by Charles Singer, 199–239. London: E. Benn, 1928.

Singer, Charles, and Dorothea Singer. "A Restoration: Byrhtferth of Ramsey's Diagram of the Physical and Physiological Fours." *Bodleian Quarterly Record* 2 (1917): 47–51.

Singer, P. N. *Galen: Selected Works.* Oxford: Oxford University Press, 1997.

Siraisi, Nancy G. *Medieval and Early Renaissance Medicine: An Introduction to Knowledge and Practice.* Chicago: University of Chicago Press, 1990.

Sivin, Nathan. "State, Cosmos, and Body in the Last Three Centuries, B.C." *Harvard Journal of Asiatic Studies* 55, no. 1 (1995): 5–37.

———. *Traditional Medicine in Contemporary China: A Partial Translation of Revised Outline of Chinese Medicine (1972) with an Introductory Study on Changes in Present Day and Early Medicine.* Ann Arbor Center for Chinese Studies: University of Michigan, 1987.

Skeat, W. W. *A Catalogue of the Harleian Manuscripts in the British Museum.* Vol. 2. London: G. Eyre and A. Strahan, 1808–1812.

Skinner, Patricia. *Health and Medicine in Early Medieval Southern Italy.* Leiden: E. J. Brill, 1997.

Skoda, Françoise. *Médecine ancienne et metaphore: Le vocabulaire de l'anatomie et de la pathologie en grec ancien.* Paris: Peeters/Selaf, 1988.

Smith, Susan Warrener. "Bernard of Clairvaux and the Natural Realm: Images Related to the Four Elements." *Cistercian Studies Quarterly* 31, no. 1 (1996): 3–19.

Smith, Wesley D. *The Hippocratic Tradition.* Ithaca: Cornell University Press, 1979.

Smyth, Marina. *Understanding the Universe in Seventh-Century Ireland*. Woodbridge, England: Boyden Press, 1996.

Sodigné-Costes, Geneviève. "Simples et jardins, une histoire." In *Vergers et jardins dans l'univers médiéval*, 329–342. Aix-en-Provence: C. U. E. R. M. A., 1990.

Solmsen, Friedrich. "The Vital Heat, the Inborn Pneuma, and the Aether." *Journal of Hellenic Studies* 77 (1957): 119–123.

Speirs, John. "Sir Gawain and the Green Knight." *Scrutiny* 16, no. 4 (1949): 274–300.

Spink, M. S., and Lewis G. L., eds. *Albucasis, On Surgery and Instruments*. Berkeley: University of California Press, 1973.

Staab, Franz. "Aus Kindheit und Lehrzeit Hildegards. Mit einer Übersetzung der 'Vita' ihrer Lehrerin Jutta von Sponheim." In *Hildegard von Bingen. Prophetin durch die Zeiten. Zum 900. Geburtstag*, edited by Edeltraud Forster, 58–86. Freiburg: Herder, 1997.

———. "Reform und Reformgruppen im Erzbistum Mainz. Vom 'Libellus de Willigis consuetudinibus' zur 'Vita domnae Juttae inclusae.'" In *Reformidee und Reformpolitik im Spätsalisch-Frühstaufischen Reich*, edited by Stefan Weinfurter, 119–187. Mainz: Selbstverlag der Gesellschaft für mittelrheinische Kirchengeschichte, 1992.

Stannard, Jerry. "Albertus Magnus and Medieval Herbalism." In *Albertus Magnus and the Sciences: Commemorative Essays*, edited by James A. Weisheipl, 355–378. Toronto: Pontifical Institute of Medieval Studies, 1980.

———. "Botany." In *Dictionary of the Middle Ages*, edited by Joseph R. Strayer. New York: Scribners, 1984.

———. "The Botany of Albert the Great." In *Albertus Magnus—Doctor Universalis*, edited by Gerbert Meyer OP and Albert Zimmermann, 377–385. Mainz: Mathias Grunewald Verlag, 1980.

———. "Greco-Roman Materia Medica in Medieval Germany." *Bulletin of the History of Medicine* 46 (1972): 455–468.

———. "Medieval Herbals and Their Development." *Clio Medica* 9 (1974): 23–33.

Stannard, Katherine E., and Richard Kay, eds. *Pristina Medicamenta: Ancient and Medieval Medical Botany*. Aldershot: Variorum, 1999.

Stanzl, Gunther. *Die Klosterruine Disibodenberg: Neue baugeschichtliche und archäologische Untersuchungen*. Worms: Wernersche Verlagsgesellschaft, 1992.

Steigen, Christoph von. "Bermerkungen zur Handschrift des Rotulus von Mülinen." In *Geschichte, Deutung, Kritik: Literaturwissenschaftliche Beiträge dargebracht zum 65. Geburtstag Werner Kohlschmidts*, edited by Maria Bindschedler and Paul Zinsli, 28–29. Bern: Francke, 1969.

Steinmeyer, Elias. *Die kleineren althochdeutschen Sprachdenkmäler*. Berlin: Weidmann, 1916.

Stimming, M., ed. *Mainzer Urkundenbuch, Vol. 1. Die Urkunden bis zum Tode Erzbischof Adalberts I (1137)*. Darmstadt: Verlag des historischen Vereins für Hessen, 1932, reprinted 1972.

Stolberg, Michael. "Die Lehre vom 'calor innatus' im lateinischen Canon medicinae des Avicenna." *Sudhoffs Archiv* 77, no. 1 (1993): 33–53.

Story, R. L. *Chronology of the Medieval World 800–1451.* London: Barrie and Jenkins, 1973.

Stow, Kenneth. "The Jewish Family in the Rhineland in the High Middle Ages: Form and Function." *American Historical Review* 92, no. 5 (1987): 1085–1110.

Strabo, Walahfrid. *Des Walahfrid von der Reichenau Hortulus: Gedichte über die kräuter seines Klostergartens vom Jahre 827.* Edited by Karl Sudhoff, H. Marzell and E. Weil. Munich: Verlag des Münchener Druck, 1926.

———. "Hortulus." In *Carolingian Civilization: A Reader*, edited by Paul Edward Dutton, 375–387. Peterborough: Broadview Press, 1993.

———. *Hortulus Walafridi Strabi, das Garchen des Walafridi Strabi ein ehrwürdiges Denkmal des Arznei-Gartenbaus aus dem 9. Jahrhundert.* Edited by Julius Berendes: Pharm. Post, 1908.

Strehlow, Wighard, and Gottfried Hertzka. *Hildegard of Bingen's Medicine.* Translated by Karin Anderson Strehlow. Santa Fe: Bear and Company, 1988.

Strohmaier, Gotthard. "Reception and Tradition: Medicine in the Byzantine and Arab World." In *Western Medical Thought from Antiquity to the Middle Ages*, edited by Mirko D. Grmek, translated by Antony Shugaar, 139–169. Cambridge: Harvard University Press, 1998.

Sudhoff, Karl. "Chirurgie im Mittelalter." *Studien zur Geschichte der Medizin* 10 (1914): 92–133.

———. "Diaeta Theodori." *Archiv für Geschichte der Medizin* 8 (1915): 377–403.

———. "Die medizinischen Schriften, welche Bischof Bruno von Hildesheim 1161 in seiner Bibliothek besass und die Bedeutung des Konstantin von Afrika im 12. Jahrhundert." *Archiv für Geschichte der Medizin* 9 (1916): 348–356.

———. "Eine Verteidigung der Heilkunde aus den Zeiten der Mönchsmedizin." *Archiv für Geschichte der Medizin* 7 (1913): 223–237.

———. "Lateinische Texte über den Rhythmus der Säftebewegung im Gleichklang mit den Jahreszeiten, samt Nachtgleichen- und Sonnwends-Diätetik." *Archiv für Geschichte der Medizin* 11 (1919): 206–211.

Sweet, Victoria. "Body as Plant, Doctor as Gardener: Premodern Medicine in Hildegard of Bingen's *Causes and Cures.*" Ph.D. diss., University of California, San Francisco, 2003.

———. "Hildegard of Bingen and the Greening of Medieval Medicine." *Bulletin of the History of Medicine* 73 (1999): 381–403.

Sylla, Edith, and Michael McVaugh, eds. *Texts and Contexts in Ancient and Medieval Science: Studies on the Occasion of John E. Murdoch's 70th Birthday.* New York: Brill, 1997.

Tabor, Daniel C. "Ripe and Unripe: Concepts of Health and Sickness in Ayurvedic Medicine." *Social Science and Medicine* 15B (1981): 439–455.

Talbot, Charles H. "A Letter from Bartholomew of Salerno to King Louis of France." *Bulletin of the History of Medicine* 30 (1956): 321–328.

Teigen, Philip M. "Taste and Quality in 15th- and 16th-century Galenic Pharmacology." *Pharmacy in History* (1987): 60–68.

Temkin, Owsei. *Galenism: The Rise and Decline of a Medical Philosophy*. Ithaca and London: Cornell University Press, 1973.

Termolen, Rosel. *Hildegard von Bingen: Biographie*. Augsburg: Pattloch, 1990.

Theophilus Presbyter. *On Divers Arts: The Treatise of Theophilus*. Translated by John G. Hawthorne and Cyril Stanley Smith. Chicago: Chicago University Press, 1963.

Theophilus Presbyter. *Theophilus presbyter schedula diversarum artium*. Edited by Albert Ilg. Osnabruck: O. Zeller, 1970(1874).

Theophrastus. *De Ventis*. Translated by Victor Courant and Val L. Eichenlaub. Notre Dame: Notre Dame University Press, 1975.

Thomas, Keith. *Man and the Natural World: A History of the Modern Sensibility*. New York: Pantheon Books, 1982.

Thompson, D'Arcy Wentworth. "The Greek Winds." *Classical Review* 32 (1918): 49–56.

Thomsen, Margaret. *Le jardin symbolique: texte grec tiré du Clarkianus XI*. Paris: Belles Lettres, 1960.

Thorndike, Lynn. *The Herbal of Rufinus*. Chicago: University of Chicago Press, 1946.

———. *A History of Magic and Experimental Science during the First Thirteen Centuries of Our Era*. Vol. 2. New York: Columbia University Press, 1923.

Thorndike, Lynn, and Pearl S. Kibre. *A Catalogue of Incipits of Medieval Scientific Writings in Latin*. 2 ed. Cambridge: Medieval Academy of America, 1963.

Throop, Priscilla. *Hildegard von Bingen's Physica: The Complete English Translation of Her Classic Work on Health and Healing*. Rochester, VT: Healing Arts Press, 1998.

Thundy, Zacharias. "Green and Red: More on Classical Influences in Sir Gawain and the Green Knight." *Classical and Modern Literature: a Quarterly* 12, no. 2 (1992): 169–177.

Tilly, Bertha. *Varro the Farmer. A Selection from the Res rusticae*. London: University Tutorial Press, 1973.

Todd, Malcom. *The Early Germans*. Oxford: Blackwell Publishers, 1992.

Toëlle, Heidi. "Vents, pluies liquides et pluies solides dans le Coran." In *Le temps qu'il fait au Moyen Age: Phénomènes atmospheriques dans la littérature, la pensée scientifique et religieuse*, edited by Claude Thomasset and Joëlle Ducos, 191–208. Paris: Presses de l'Université de Paris-Sorbonne, 1998.

Tolkien, J. R. R., and E. V. Gordon. *Sir Gawain and the Green Knight*. Oxford: Clarendon Press, 1925.

Tombeur, Paul. "The CETEDOC Library of Christian Latin Texts: CDROM." Turnhout: Brepols, 2002.

Touwaide, A. "La thérapeutique médicamenteuse de Dioscoride à Galien." In *Galen on Pharmacology, Philosophy, History and Medicine*, edited by A. Debru, 255–82. Lille, 1995 (1997).

Trithemius, Johannes. *Johannis Trithemii opera historica*. 2 vols. Frankfurt, repr. Frankfurt/Main: Minerva, 1966(1601).

Tromberend, Teresa. *Hildegard von Bingen. Wirkungsstatten*. Regensburg, 1996.

Unschuld, Paul. *Medicine in China: A History of Ideas*. Berkeley: University of California Press, 1985.

———. *Medicine in China: A History of Pharmaceutics*. Berkeley: University of California Press, 1986.

Usuardus, Monachus. "Usuardi Martyrologium." In *Patrologia Latina*. Vol. 123, 453 to end, and Vol. 124, 10–860. Paris: Garnier, 1865.

Valderas, Jose Maria. "Los conceptos de planta, funciòn y diversidad vegetales en las escritos biologicos de San Alberto Magno (c. 1200–1280)." *Asclepio* 40 (1988): 167–186.

Valencius, Conevery Bolton. "The Geography of Health and the Making of the American West: Arkansas and Missouri, 1800–1860." In *Medical Geography in Historical Perspective*, edited by Nicholas Rupke, 121–145. London: The Wellcome Trust Centre for the History of Medicine at UCL, 2000.

Vallentin, Catherine Velay. "Le folklore et l'histoire: Le congrès internationale de folklore de 1939." *Annales* Mars-Avril (1999): 481–506.

Van Acker, Lieven, ed. *Hildegardis Bingensis Epistolarium*. Vol. 91, 91A, *CCCM*. Turnhout: Brepols, 1991, 1993.

Van Acker, Lieven, and M. Klaes-Hachmoller, eds. *Hildegardis Bingensis epistolarium pars tertia CCLI-CCCXC*. Vol. 91B, *CCCM*. Turnhout: Brepols, 2001.

Van Engen, John. "Abbess, Mother, and Teacher." In *Voice of the Living Light: Hildegard of Bingen and Her World*, edited by Barbara Newman, 30–51 and 206–9 (endnotes). Berkeley: University of California Press, 1998.

———. "The Crisis of Cenobitism Reconsidered: Benedictine Monasticism in the Years 1050–1150." *Speculum* 61 (1986): 157–72.

———. "Letters and the Public Persona of Hildegard." In *Hildegard von Bingen in ihrem historischen Umfeld. Internationaler wissenschaftlicher Kongress zum 900 jährigen Jubiläum, 13–19 September 1998, Bingen am Rhein*, edited by Alfred Haverkamp and Alexander Reverchon, 375–418. Mainz: Philipp von Zabern, 2000.

Van Puyvelde, L. *Un Hôpital du Moyen Age et une Abbaye y annexée*. Paris: Champion, 1925.

Venentianer, Ludwig. *Asaf Judaeus, der alteste medizinische Schriftsteller in hebraeischer Sprache*. 3 vols. Budapest: Teubner, 1915–1917.

Vercauteren, Fernand. "Les médecins dans les principautés de la Belgique et du nord de la France, du VIIIe. au XIIIe. siècle." *Le Moyen Age* (1951): 61–91.

Vickery, John B. *The Literary Impact of the Golden Bough*. Princeton: Princeton University Press, 1973.

Vindician. "Epitome altera. Epistula ad Pentadium." In *Theodori Prisciani Euporiston libri III*, edited by Valentin Rose, 485–492. Leipzig: Teubner, 1894.

Vogel, C. "Pratiques superstitieuses au début du XIe. siècle d'après le Corrector sive Medicus de Burchard (965–1025)." In *Etudes de civilisation médiévale, IXe.-XIIe. siècles: Mélanges offerts à Edmond-René Labande à l'occasion de son depart à la retraite et du XXe. anniversaire du C. E. S. C. M. par ses amis, ses collègues, ses élèves*, 731–766. Poitiers: C.E.S.C.M., 1974.

Voigts, Linda E. "Anglo-Saxon Plant Remedies and the Anglo-Saxons." *Isis* 70 (1979): 250–68.

———. "One Anglo-Saxon View of the Classical Gods." *Studies in Iconography, University of N. Kentucky* 3 (1977): 3–16.

Voigts, Linda E., and Michael R. McVaugh. *A Latin Technical Phlebotomy and Its Middle English Translation*. Vol. 74, part 2. Philadelphia: Transactions of the America Philosophical Society, 1984.

Vollmann, Benedikt Konrad. "Auf dem Weg zur authentischen Hildegard. Bemerkungen zu den nur in der Florentiner 'Physica' Handschrift uberlieferten Texten." *Sudhoffs Archiv* 87, no. 2 (2003): 159–72.

Voswinckel, Peter. "Black Bile in the Doctrine of the Four Humours: New Empirical Aspects." In *Actes du XXXIIIe. Congrès International d'histoire de la médecine*, 1275–1285. Anvers: Societas Belgica Historiae Medicinae, 1990.

Waitz, G. W., ed. *Annales Sancti Disibodi 841–1200*. Vol. 17, *Monumenta Germaniae Historica SS*. Hannover, 1861.

Walker-Moskop, Ruth Marie. "Health and Cosmic Continuity in Hildegard of Bingen." Ph.D. diss., University of Texas, 1985.

Wallis, Faith. "The Experience of the Book: Manuscripts, Texts and the Role of Epistemology in Early Medieval Medicine." In *Knowledge and the Scholarly Medical Traditions*, edited by Don Bates, 101–126. Cambridge: Cambridge University Press, 1995.

———. "Signs and Senses: Diagnosis and Prognosis in Early Medieval Pulse and Urine Texts." In *The Year 1000: Medical Practice at the End of the First Millennium*, edited by Peregrine Horden, 265–78. Oxford: Society for the Social History of Medicine, 2000.

Walstra, G. J. J. "Thomas de Cantimpré: 'De naturis rerum.' Etat de la question." *Vivarium* 5, no. 2 (1967): 146–171.

Walstra, G. J. J. "Thomas de Cantimpré: 'De naturis rerum.' Etat de la question." *Vivarium* 6 (1968): 48–61.

Warner, John Harley. "The Nature-Trusting Heresy: American Physicians and the Concept of the Healing Power of Nature in the 1850s and 1860s." *Perspectives in American History* 11 (1977): 291–324.

Watson, Lyell. *Heaven's Breath: A Natural History of the Wind*. Great Britain: Hodden and Stoughton, 1984.

Weinstein, Donald, and Rudolph M. Bell. *Saints and Society: The Two Worlds of Western Christendom, 1000–1700*. Chicago: University of Chicago Press, 1982.

Weinstein, Jessica. "Textualizing and Contextualizing Hildegard's Body in The-odoric's 'Vita.'" *Magistra: A Journal of Women's Spirituality in History* 6, no. 1 (2000): 89–103.

Weiss-Amer, Melitta. "Die 'Physica' Hildegards von Bingen als Quelle für das 'Koch-buch Meister Eberhards.'" *Sudhoffs Archiv* 76, no. 1 (1992): 87–96.

———. "Food and Drink in Medical-Dietetic 'Fachliteratur' from 1050–1400." Ph.D. diss., University of Toronto, 1989.

Wellmann, Max. *Die pneumatische Schule bis auf Archigenes.* Berlin: Weidmannsche Buchandlung, 1895.

White, Lynn, Jr. "Medical Astrologers and Late Medieval Technology." *Viator* 6 (1975): 295–315.

———. *Medieval Religion and Technology: Collected Essays.* Los Angeles: University of California Press, 1978.

Wiebe, S., W. T. Blume, J. P. Girvin, and M. Eliasziw. "A Randomized, Controlled Trial of Surgery for Temporal-Lobe Epilepsy." *The New England Journal of Medicine* 345 (2001): 311–318.

Wilhelm, Friedrich. *Denkmäler deutscher Prosa des 11. und 12. Jahrhunderts.* 2 vols. Munich: G. D. W. Callwey, 1914–1918.

Wilhelmy, Winifried. "Hildegards natur- und heilkundlichen Schriften." In *Hilde-gard von Bingen 1098–179,* edited by Hans-Jürgen Kotzur, 284–308. Mainz: Philipp von Zabern, 1997.

Willerval, Bernard. *Chronologie illustré de l'histoire universelle.* Paris: Eclectis, 1992.

William of Conches. *De philosophia mundi.* In *Patrologia Latina.* Vol. 172, 39–102. Paris: Garnier, 1856.

William of St. Thierry. "De natura corporis et animae." In *Patrologia Latina.* Vol. 180. 695–726. Paris: Garnier, 1867.

Winterfeld, Dethard von. "Kirchen am Lebensweg der Hildegard." In *Hildegard von Bingen in ihrem historischen Umfeld. Internationaler wissenschaftlicher Kongress zum 900 jährigen Jubiläum, 13–19 September 1998, Bingen am Rhein,* edited by Alfred Haverkamp and Alexander Reverchon, 129–159. Mainz: Philipp von Zabern, 2000.

Wise, Thomas Alexander. *Commentary on the Hindu System of Medicine.* Amsterdam: Oriental Press, 1901(1860).

Wlaschky, M. "'Sapientia artis medicinae.' Ein frühmittelalterliches Kompendium der Medizin." *Kyklos* 1, no. 1 (1928): 103–113.

Wogan-Browne, Jocelyn. *Saints' Lives and Women's Literary Culture c. 1150–1300: Virginity and Its Authorization.* New York: Oxford University Press, 2001.

Wolschke-Bulmahn, Joachim, and Gert Goenig. "From Open-mindedness to National-ism: Garden Design and Ideology in Germany during the Early 20th Century." In *People-Plant Relationships: Setting Research Priorities,* edited by Joel Flagler and Raymond P. Poincelot, 133–150. New York: Haworth Press, 1994.

Yanchi, Liu. *The Essential Book of Traditional Chinese Medicine.* Translated by Tin-gyu, Fang, and Chen Laidi. New York: Columbia University Press, 1988.

Zimmermann, Francis. *Généalogie des médecines douces: De l'Inde à l'occident.* Paris: Presses Universitaires de France, 1995.

————. *The Jungle and the Aroma of Meats: An Ecological Theme in Hindu Medicine.* Translated by Janet Lloyd. Berkeley: University of California Press, 1987.

————. *Le discours des remèdes au pays des épices: ênquete sur la médecine hindoue.* Paris: Payot, 1989.

————. "Rtu-Satmya: The Seasonal Cycle and the Principle of Appropriateness." *Social Science and Medicine* 14B (1980): 99–106.

Zirkle, C. "Animals Impregnated by the Wind." *Isis* 25 (1936): 95–130.

Zysk, Kenneth. *Asceticism and Healing in Ancient India: Medicine in the Buddhist Monastery.* New York: Oxford Press, 1991.

Index